PLEASE RETURN TO:
FMC - Ag Chem Info Center
2000 Market Street
Philadelphia, PA 19103

APHIDS ON THE WORLD'S CROPS

APHIDS ON THE WORLD'S CROPS:
An Identification and Information Guide

R.L. Blackman
and
V.F. Eastop
Department of Entomology
British Museum (Natural History)

A Wiley–Interscience Publication

JOHN WILEY & SONS
Chichester · New York · Brisbane · Toronto · Singapore

Copyright © 1984 by British Museum (Natural History)

All rights reserved.

No part of this book may be reproduced by any means, nor transmitted, nor translated into a machine language without the written permission of the publisher.

Library of Congress Cataloging in Publication Data:

Blackman, R.L. (Roger L.), 1941–
  Aphids on the world's crops.

  "A Wiley–Interscience publication."
  Bibliography: p.
  Includes index.
  1. Plant–lice—Host plants. 2. Field crops—
Diseases and pests. I. Eastop, Victor Frank. II. Title.
SB945.P64B5   1984   632'.752   83–25889

ISBN 0 471 90426 0

British Library Cataloguing in Publication Data:

Blackman, R.L.
  Aphids on the world's crops.
  1. Plants, Cultivated   2. Insects, Injurious
  and beneficial
  I. Title   II. Eastop, V.F.
  632'.7   SB931

ISBN 0 471 90426 0

Typeset by MHL Typesetting Ltd., Coventry.
Printed by the Pitman Press, Ltd., Bath, Avon.

Table of Contents

Preface vii

A. Introductory Section 1
    Introduction and aims of the book 1
    Systematics 2
    Life cycles and polymorphism 5
    Host–plant relationships 10
    Geographical distribution 14
    Morphology and key characters 16

B. The Crops and their Aphids 29
    Notes on the use of section 29
    List of crop plants and their English names 31
    Lists and keys of aphids on each crop 35

C. The Aphids 209
    Introduction to the section 209
    Systematic treatment of genera (in alphabetical order) 210

D. Techniques 375
    Collecting 375
    Preservation and mounting 375
    Labelling and storage of slide-mounted preparations 376

E. Sources of information 379
    Regionally classified faunal works 379
    General biology 382
    Morphology, anatomy, and physiology 382
    Genetics and development 383
    Migration and dispersal 383
    Relationships with other insects 383
    Host-plant relationships 384
    Bibliographies 384

F. References 385

G. Photographic Guide 413
    List of photographs of slide-mounted aphids 413

Preface

Ever since Hill (1568) wrote on 'grene fles o ye Gardn', Worlidge (1669) of 'insects and creeping things offending', and Evelyn (1693) of 'green fleas which fatten on the young shoots of peach' and 'little black flies on artichokes in summer', it has been recognized that certain crops were more prone to attack by aphids than others. By the middle of the eighteenth century the concept of plant-specific insects was well established. Linné (1758) lists geographical localities for most of the animals he describes, but for aphids, psyllids, and coccids he lists not localities, but host plants.

These host associations are the basis of the present work.

A. *Introductory Section*

INTRODUCTION AND AIMS OF THE BOOK

This book has two main objectives:

1. To enable workers in agricultural entomology who are not specialist aphid taxonomists to identify most of the aphid species that they are likely to find colonizing commercial crops in agriculture and horticulture throughout the world, without recourse to techniques such as slide-making except where absolutely necessary.
2. To provide concise information on each of the principal aphid pest species, relevant to its identity and economic importance, including geographical distribution, host-plant range, life cycle, and a selective bibliography.

The identification guide makes use of the fact that although over 250 species of the superfamily Aphidoidea feed on agricultural and horticultural crops throughout the world, only a much smaller number of aphid species occur regularly on any one species of crop plant. Since the aphid species feeding on any one crop are seldom very closely related to one another, they can often be distinguished using fairly simple host-plant oriented keys and a binocular microscope. This means that, *with the essential proviso that the aphids to be identified are known to have been colonizing (i.e. feeding and producing young on) a particular, correctly identified crop*, the non-specialist can determine their identity with a satisfactory degree of reliability.

We feel we should stress at the outset, however, that the identification of aphids presents a great many traps for the unwary. Aphids are renowned as a group for the considerable extent to which morphology is influenced by environmental factors. Not only are particular morphs with discrete morphological differences triggered by environmental cues, but the range of continuous morphological variation is wider than in many other insect groups. Increases or decreases in size due to nutritional effects, for example, can accumulate over several parthenogenetic generations, since the size of the mother can affect the size of her offspring. The structural distinction between the apterous and alate morphs is not always as clear as might be expected; apterous individuals with some tendency towards alatiform characters may occur, and vice versa. Likewise, intermediates between other morphs can sometimes occur; for example, between viviparous parthenogenetic females and oviparous sexual females. A 'fundatrix effect', whereby the generation after the fundatrix has a tendency towards 'fundatriciform' characters which is progressively lost in subsequent generations, is widespread. In aphids there is an almost general tendency for certain morphological features—for example, degree of pigmentation, development of wax glands—to change with the progression of the seasons.

An additional problem is provided by the occurrence of 'abnormal' morphological features, which may be due to mutation but are often no more than character states near the extremes of a range of continuous variation, and are part of the natural variation of all living organisms. In most organisms with any degree of outbreeding, any particular abnormal character state is likely to occur in only the occasional, rare individual within a population, and thus be instantly recognizable as an abnormal or extreme condition. In aphids, morphological abnormalities need not necessarily be restricted to single individuals. Because aphid colonies and populations may be clonal, it is possible to find a whole colony of individuals all with the same anomalous characters. This could lead to a very misleading idea of the range of variation, and the 'normal' state of the character, in the species.

Most of these difficulties are avoided by the choice of characters used in the keys, and provided that only the identification of apterous viviparous females is attempted. However, it is clearly advisable for anyone working with, or seeking to identify, aphids, to have some basic knowledge of their biology. The nature of aphid polymorphism, and the evolution of life cycles and host-plant relationships, are the features of aphid biology which seem most relevant to their taxonomy and identification, and are summarized in the following pages. Many other important and much-studied aspects of aphid biology, particularly those related to their developmental physiology and ecology, are not considered at all in this volume. However, an annotated bibliography to principal sources of information about aphid biology in general is provided on pages 379–384.

SYSTEMATICS

Aphids belong to the series Sternorrhyncha within the order Hemiptera, suborder Homoptera, along with scale insects, psyllids, and whitefly. The name Sternorrhyncha refers to the position of the rostrum, which has its base between the fore-coxae, and in its resting position is deflected back along the ventral surface of the body. Other characters shared by all four Sternorrhynchous groups of Homoptera are relatively well-developed antennae and 1- or 2-segmented tarsi.

The scale insects (superfamily Coccoidea) are generally regarded as the sister group of the aphids (superfamily Aphidoidea). Within the Aphidoidea, it is quite clear that two groups—the adelgids and phylloxerids— are distinct from other aphids and retain a number of more primitive features, of which the most obvious are the absence of viviparity and the absence of siphunculi. Some aphid taxonomists prefer to place the Adelgidae and Phylloxeridae in a superfamily of their own (Phylloxeroidea). The other major groups of aphids are then classified as families (Lachnidae, Pemphigidae, Aphididae, etc.) within the

Introductory section

superfamily Aphidoidea, as opposed to the system used here, where they are regarded as subfamilies (Lachninae, Pemphiginae, Aphidinae, etc.) of the family Aphididae. This disagreement about levels of classification can be a source of confusion to non-taxonomists and taxonomists alike.

Although there is still some disagreement over phylogenetic relationships (for example, whether the Lachnidae are a primitive or an advanced group, and whether Chaitophorinae should be included within Drepanosiphinae or regarded as a sister group), the main groupings within the Aphidoidea are fairly clear and their characteristic features can be summarized as follows:

Adelgidae (Conifer Woolly Aphids; e.g. *Adelges, Pineus*). All female morphs are oviparous and have an ovipositor. Antennae of alata 5-segmented, with usually 3 primary sensoria. Cubital and anal veins with separate origin in forewing. Wings held roof-like in repose. Siphunculi absent, and no evident cauda. Sexuales with mouthparts. Host-alternating species have a two-year cycle, with *Picea* species as primary hosts and other conifers as secondary hosts.

Phylloxeridae (e.g. *Phylloxera, Moritziella, Viteus*). All female morphs are oviparous. Antennae of alata 3-segmented with 2 primary sensoria. Cubital and anal veins in forewing with a common base. Wings held horizontally in repose. Siphunculi absent, and no evident cauda. Sexuales without mouthparts. Life cycle variable, without distinct host alternation; always on dicotyledons.

Aphididae (e.g. *Lachnus, Callaphis, Aphis*). Parthenogenetic forms viviparous. Antennae of alatae usually 5- or 6-segmented with two primary sensoria, the last segment with a distinct terminal process. Siphunculi present or secondarily absent. Cauda often developed. Life cycle variable, but usually annual, with host alternation in some groups. On various plants.

The Aphididae include the following major groups:

Lachninae (e.g. *Pterochloroides, Trama*—Plates 1–6). Antennae 6-segmented, with a very short terminal process. Secondary sensoria on antennae of alatae rounded or transverse oval. Siphunculi if present either very broad hairy cones or chitinized rings. Cauda broadly rounded. Life cycle typically annual, without host alternation, on various woody plants (Cinarini, Lachnini) or on roots (Tramini). Often ant-attended. Mainly holarctic.

Chaitophorinae (e.g. *Atheroides, Sipha*—Plates 7–9). Antennae 5- or 6-segmented, with terminal process longer than base of last segment. Secondary sensoria on antennae of alatae rounded. Siphunculi truncate, often with polygonal reticulation. Cauda either broadly rounded or knobbed (but anal plate never bilobed). Life cycle annual, without host alternation, on certain trees (*Salix, Populus, Acer*—the Chaitophorini) or on grasses (the Siphini); often ant-attended. Mainly holarctic.

Drepanosiphinae (e.g. *Myzocallis, Therioaphis*—Plates 10–13). Antennae 6-segmented, usually long, with terminal process of variable length. Secondary sensoria on antennae of alatae usually transverse oval or elongate, sometimes rounded. Siphunculi often truncate, but sometimes ring-like or tubular. Cauda also variable but often knobbed, and then the anal plate is usually bilobed. A biologically very variable group, but the tree-dwelling species in summer often occur mainly or entirely as an active alate morph. Life cycle annual, without host alternation, mainly on various dicotyledonous trees but also on Cyperaceae, Juncaceae, Gramineae, and Papilionaceae. World wide.

Aphidinae (e.g. *Aphis, Macrosiphum, Myzus*—Plates 14–128). Antennae long and usually 6-segmented, with terminal process of variable length but rarely very short and in the majority of species at least twice as long as the base of the last segment. Secondary sensoria on antennae of alatae usually rounded. Siphunculi usually tubular. Cauda usually tongue-shaped, sometimes short and broad, but never knobbed. Life cycle annual. Many species have host alternation between woody and herbaceous angiosperms, in which case the males are winged and migrate independently to the primary host; but often the primary woody host is lost and the entire life cycle is passed on the usually herbaceous secondary host. A few species live on Coniferae and ferns, the majority on higher angiosperms. Widespread, particularly holarctic and oriental.

Greenideinae (e.g. *Cervaphis, Greenidea*—Plates 130 and 131). Antennae 5- or 6-segmented, terminal process often shorter than and rarely more than twice as long as base of last segment. Siphunculi either conical or long and tubular, bearing many hairs. Cauda usually either broadly rounded or triangular, never tongue-shaped or knobbed. On dicotyledonous trees and shrubs, without host alternation. Mainly oriental.

Anoeciinae (e.g. *Anoecia, Aiceona*—Plate 129). Antennae of medium length, usually 6-segmented, with a short terminal process. Secondary sensoria on antennae of alatae rounded. Siphunculi are small, low cones. Cauda rounded. Oriental on Lauraceae (Aiceonini), or holarctic with an annual alternation between *Cornus* and the roots of grasses (Anoeciini). In the latter case the autumn migrants are winged female sexuparae which produce small (but rostrate) apterous males and sexual females on the primary host.

Pemphiginae (e.g. *Pemphigus, Eriosoma, Tetraneura, Forda*—Plates 136–150). Antennae rather short, 5- or 6-segmented with a very short terminal process. Secondary sensoria on antennae of alata usually transverse elongate, often ringing the segment. Wax gland pores often grouped in large plates. Siphunculi merely pores, or absent. Cauda usually broadly rounded. Life cycle annual or (in Fordini) biennial, usually with host alternation, often inducing gall formation on the woody primary host, and often colonizing the roots of the secondary host. Autumn migrants are winged female sexuparae which produce very small, arostrate apterous males and sexual females on the primary host.

(Sexuparae of Pemphiginae can be recognized in microscope slide preparations because the embryos seen through the body wall have no stylets.) Holarctic.

Hormaphidinae (e.g. *Cerataphis, Astegopteryx*—Plates 132–135). Antennae short, 3- to 5-segmented. Secondary sensoria on antennae of alata ringing the antenna. Frons often with a pair of horn-like processes. Siphunculi merely pores, or absent. Cauda knobbed and anal plate bilobed. Apterae on secondary hosts are often coccid- or aleyrodid-like, with very short legs concealed beneath body. Wax gland pore-plates often present. Life cycle annual or biennial, usually with host alternation involving species of Hamamelidaceae or Styracaceae as primary hosts and various secondary hosts including palms and bamboos. Males and sexual females apterous with well-developed mouthparts. Mainly oriental.

LIFE CYCLES AND POLYMORPHISM

Cyclical parthenogenesis or amphigony, the alternation of a phase of sexual reproduction and a phase of parthenogenetic reproduction in the life cycle of one species, is a primitive feature of the Aphidoidea, although the various types of regular annual or biennial holocycle which are found in present-day species seem to have become elaborated independently in the different major groups within the superfamily.

No species are known which have secondarily lost the parthenogenetic phase of the life cycle, although it is sometimes abbreviated to two or three generations. Loss of the sexual phase is, however, a widespread tendency, although very often it is a within-species tendency found only in certain populations of a species or in certain genotypes within a population. The number of species which appear to be wholly or permanently parthenogenetic is relatively small.

Another significant feature in the evolution of aphid life cycles is host alternation (heteroecy). That is, a regular seasonal migration between two, often distantly related, host plants, one of which (termed the primary host) is used for sexual reproduction while the other (the secondary host) is colonized only by parthenogenetic morphs. Host alternation is not primitive to the group as a whole; it has apparently arisen several times independently in different families and subfamilies of Aphidoidea.

Both cyclical parthenogenesis and heteroecy are major evolutionary developments which have enabled aphids to exploit their food plants, particularly short-lived herbaceous plants such as most agricultural crops, perhaps to a greater extent than any other group of insects.

Evolutionary consequences of cyclical parthenogenesis in aphids

Separation of sexual and multiplicative functions

Cyclical parthenogenesis separates the two basic functions of any organism which are essential to its full exploitation of the environment: (a) sexual reproduction involving gene recombination and segregation to produce an array of new genotypes; and (b) increase of biomass to fully realize the potential of each genotype in space and time and thus maximize its chances of contributing genes to the next sexual phase. The separation of these functions means that their development can proceed with a degree of independence, so that the morphology of the parthenogenetic and sexual generations can diverge. This divergence is seen to its greatest extent in the Pemphiginae, where the first parthenogenetic generation (the fundatrix) is a relatively large, viviparous, highly fecund, gall-forming plant parasite, and the female of the sexual generation is a small, non-feeding individual which produces just one egg about as large as itself (Fig. 1). Other aphid groups all show this divergence of form and function between the sexual and parthenogenetic females to a variable but lesser extent.

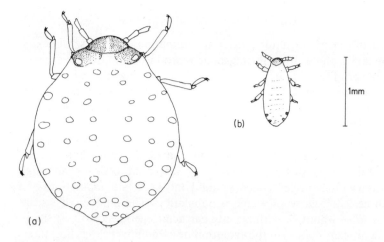

Fig. 1 (a) fundatrix, (b) sexual female, of *Pemphigus* sp.

In Adelgidae and Phylloxeridae both sexual and parthenogenetic females lay eggs. In all Aphididae, however, parthenogenesis is combined with viviparity. With little doubt, viviparity was a very significant development in the divergence of the sexual and multiplicative functions of aphids. It enabled generations to be telescoped into one another, so that embryos could start to develop even before birth of the mother, and be ready for independent existence outside the mother as soon as she became adult. The shortening of the generation time and overlapping of generations resulting from the development of viviparity enabled a very dramatic increase in the multiplicative potential of aphids during their parthenogenetic phase.

Separation of multiplicative and dispersive functions

An aphid's maximum potential will only be fully realized if the increase in biomass has a spatial dimension; in other words, if there is an effective means of dispersal whereby it can colonize fresh host plants in new localities. The parthenogenetic phase of the aphid life cycle enables very efficient dispersal mechanisms to develop, whereby some species can fully exploit even the very transient food sources provided by short-lived annual and ephemeral plants. The dispersal mechanisms of aphids usually involve a further separation of function between two distinct parthenogenetic female morphs: one apterous, more sedentary, highly fecund, and the other alate, more active, often less fecund.

Thus, a typical young, growing colony of aphids on a herbaceous plant consists mainly or entirely of apterous females, committed to the business of extracting nutrients from plant sap and converting them to aphid biomass as quickly and efficiently as the system permits. As the colony ages, alate females start to appear, with behaviour oriented first to migration and dispersal, and secondly to colonizing new host plants. The great reproductive potential of the apterous morph provides sufficiently large numbers of alate females to compensate for the high mortality associated with migration and dispersal.

Host alternation (heteroecy)

The division of function between apterous and alate parthenogenetic females also paves the way for host alternation. In host-alternating aphids the highly fecund, apterous fundatrices found colonies which rapidly attain large size, on the spring growth flush of the primary host, providing the necessary 'springboard' for the hazardous dispersal of alate spring migrants to found colonies on secondary hosts. A successful return migration, similarly hazardous in terms of high mortality of individual alatae, depends on the mid-season productivity of one or more generations of apterous females actively growing on secondary host plants.

In host-alternating members of the subfamily Aphidinae, the return migrants comprise alate males and alate parthenogenetic females which produce egg-laying sexual females when they reach the primary host, and are therefore termed gynoparae. In other host-alternating aphids only one morph is involved on the return migration: an alate parthenogenetic female which is termed a sexupara because it produces apterous sexuales of both sexes when it reaches the primary host (Fig. 2). The alate parthenogenetic females of the return migration, whether sexuparae or gynoparae, often show small structural differences from the alate females of the spring migration.

Loss of host alternation (secondary monoecy)

Host alternation seems to have provided the stepping-stone to the present-day widespread exploitation by aphids of herbaceous plants. Most of the aphid species which today go through their entire life cycle on herbaceous plants have probably evolved through a host-alternating (heteroecious) phase. In many cases an aphid species with a complete life cycle on a herbaceous plant

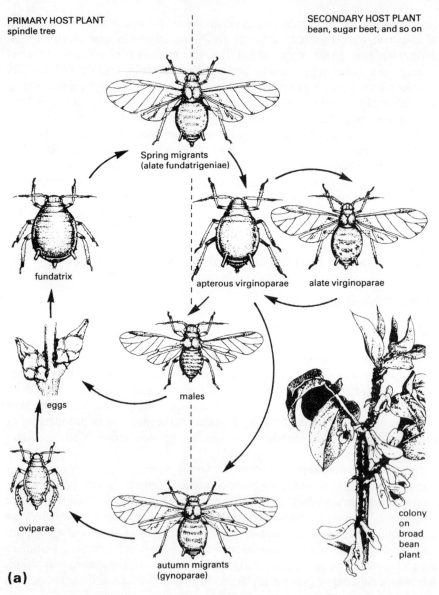

Fig. 2 Life cycles of (a) heteroecious Aphidine (*Aphis fabae*) and (b) heteroecious Pemphigine (*Pemphigus bursarius*) ((a) from Blackman, 1975; reproduced by permission of Pergamon Press Ltd)

(a monoecious species) is closely related to a heteroecious species which uses a related herbaceous plant as its secondary host. This seems a sure indication that loss of host alternation is a frequent and persistent phenomenon in the recent as well as the more distant evolutionary history of aphids.

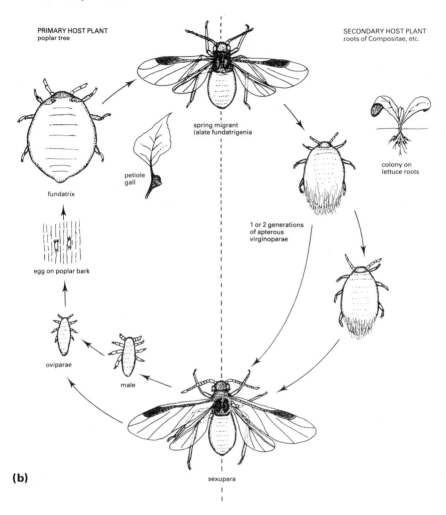

(b)

In the Aphidinae, the males of heteroecious species have to be alate, since they are produced on the secondary host and must migrate back to the primary host for breeding. When host alternation is lost, wings cease to be a necessity and many Aphidinae with a complete life cycle on herbaceous plants produce apterous males. Wing dimorphism in males of Aphidinae, rather than being environmentally triggered as in females, seems often to be under a simple form of genetic control; in several species both apterous and alate males are produced.

Loss of the sexual phase (anholocycly)

One other frequent and persistent evolutionary event in the phylogeny of aphid life cycles is the loss of the sexual phase. The widespread occurrence of anholocycly in aphids can be seen as another consequence of cyclical parthenogenesis. Aphids have acquired their very successful and elaborate system for exploiting the advantages of parthenogenetic reproduction through a long period of evolutionary time. During all of this time the sexual process has been retained as part of the life cycle of the vast majority of aphids. Sexual reproduction is apparently essential to the long-term evolutionary success, and indeed survival, of all organisms, to judge from its universal occurrence in all extant major groups of animals and plants. There seems little doubt that the development and refinement of the parthenogenetic system in aphids has only been possible through this long association with a regular sexual phase.

However, the onset of the annual sexual phase of most aphids is triggered by a seasonal change in the environment (in temperate climates this is usually decreasing photoperiod or temperature). If the environment does not provide the right stimulus (in the tropics, or in an artificially controlled environment in a greenhouse, for example), then parthenogenesis will continue. Alternatively, a relatively small genetic change in the aphid itself, the lowering of a response threshold or the 'switching off' of a photoperiodic timing mechanism, for example, could have the same effect. Thus, several species in all the major aphid groups have apparently become completely anholocyclic. A still greater number have become partially anholocyclic, with anholocyclic clones or races existing either sympatrically with the holocyclic form of the same species, or in another part of its distribution area. Partial anholocycly can occur in heteroecious aphids as well as monoecious ones, the anholocyclic forms staying all the year round on the secondary host plant.

To summarize this subsection briefly, aphids have an evolutionarily ancient and well-established system by which parthenogenetic and sexual reproduction are combined in the life cycle. The advantages of this system have been exploited by allocating different basic functions (e.g. sexual reproduction, multiplication, dispersal) to specialized morphs, and, in the Aphididae, by combining parthenogenesis with viviparity. It has also led several times independently to the phenomenon of host alternation, and from there to the all-year-round exploitation by aphids of herbaceous plants. Loss of the sexual phase (anholocycly) is also a frequent occurrence in all modern aphid groups, although apparently of little long-term evolutionary significance.

HOST PLANT RELATIONSHIPS

More detailed consideration of the host plant relationships of Aphidoidea can be found in various review articles (see p. 384). The topic will only be treated in outline here, with emphasis on some of the more immediately relevant points.

The origins of the Aphidoidea almost certainly antedate those of the angiosperms, which they now exploit so effectively. Possibly they arose on some now extinct group of gymnosperms, and then later at various times some of the ancestors of modern aphids moved to Coniferae (the Adelgidae, and perhaps the ancestors of *Mindarus* and *Neophyllaphis*), and others to early families of woody angiosperms (the Phylloxeridae, and most Aphididae). Most of the Aphididae which today feed on Coniferae, and the few species found on ferns and mosses, have almost certainly acquired these host plants secondarily in relatively recent times. The Cinarini in particular, a Lachnine tribe of about 270 species feeding exclusively on conifers, are thought to be a rather recent adaptive radiation.

Most of the older groups within the Aphididae have clear and probably primitive associations with particular families of woody angiosperms. This applies to aphid groups which have apparently never developed host alternation, such as the Greenideini on Fagaceae, the Drepanosiphini on Cupuliferae (Fagaceae + Betulaceae), the Chaitophorini on Salicaceae and Aceraceae, and the Pterocommatinae on Salicaceae, as well as host-alternating groups, where the oldest association is usually with the primary host plant; for example, Hormaphidini with Hamamelidaceae, Eriosomatini with Ulmaceae, Pemphigini with Salicaceae, and Fordini with Anacardiaceae. The galls formed by species of Hormaphidinae and Pemphiginae as adaptations to their life on their primary hosts are further evidence that these are long-standing aphid/plant associations. In general, the secondary host plants utilized by Pemphiginae in particular are in various more advanced plant families, suggesting that they are more recently acquired.

However, one notable exception is an evolutionary line within the Pemphigini, comprising *Prociphilus* and closely-related genera which all use the roots of conifers as secondary hosts. Some genera within this line, e.g. *Pachypappa* and *Gootiella*, still have *Populus* as primary hosts (like other Pemphigini, comprising *Prociphilus* and closely-related genera which use the phase relatively recently to more advanced 'primary' hosts such as *Crataegus*, *Lonicera*, and *Fraxinus*.

In the Lachninae and Aphidinae, the ancestral host plant relationships are often more obscure, as there have clearly been a great many host plant changes often involving captures of new hosts quite unrelated to the old. No extant Lachninae have host alternation, but the Tramini are anholocyclic on the roots of Compositae, and may have acquired this habit after undergoing a period of alternation to an unknown, perhaps extinct, woody primary host at some time in their evolutionary past. The Lachnini are mostly on deciduous trees, predominantly in the older angiosperm families, but at some time there was a 'capture' of a conifer by a Lachnine, which led to the explosive evolution of the Cinarini already mentioned.

The Aphidinae seemed to have expanded relatively recently, since the Oligocene, to judge from the fossil record, and Hille Ris Lambers (1939, 1950a) pointed out how much of the evolution of this group can be associated with that of the Rosaceae. Many genera of Aphidinae are clearly primitively associated with woody Rosaceae, either as their primary or their only host plants. This is very evident in the subtribe Rhopalosiphina, with the genera *Rhopalosiphum*, *Schizaphis*, *Hysteroneura*, *Melanaphis*, and *Hyalopterus* all showing alternation between primary hosts in the Rosaceae and monocotyledonous secondary hosts. In some of these genera, especially *Schizaphis*, certain species have lost the primary host. *Rhopalosiphum nymphaeae* is unusual in having a wide range of secondary hosts in many plant families, all of them water plants.

In the Macrosiphini, an evolutionary line associated with Rosaceae as primary hosts can be detected from the more primitive genera such as *Anuraphis*, with short siphunculi and rounded cauda, to the more advanced genera with long siphunculi and a long, finger-like cauda such as *Macrosiphum*. The more primitive genera within this series form leaf pseudo-galls on their Rosaceous primary hosts, an indication that the relationship is probably an ancient one, whereas the more advanced genera produce no reaction at all from their host plants. Important genera such as *Dysaphis*,

Many of the aphid genera of the 'Rosaceous series' contain monoecious species which have lost their Rosaceous primary host. One may conjecture that similar events in the past have given rise to a large number of genera and groups of genera which are clearly related to those of the Rosaceous series, and part of the same phylogeny, but have lost all connection with the Rosaceae as host plants. Some of these genera, such as *Uroleucon*, are composed entirely of monoecious species on herbaceous plants, whereas others, such as *Capitophorus*, are still heteroecious but have undergone a change of primary host.

Most genera within the Aphidinae, whether or not they relate clearly to the Rosaceous series, have evident associations with particular groups of host plants usually at or below the family level. In many cases the association is virtually complete, e.g. most *Uroleucon* and *Macrosiphoniella* species feed on the related families Compositae and Campanulaceae, all *Lipaphis* and *Brevicoryne* on Cruciferae, all *Cavariella* migrate from *Salix* to Umbelliferae or the related family Araliaceae, and so on. It is difficult to decide which of such associations are the result of co-evolution, i.e. speciation of the host plants and contemporaneous speciation of the aphids, and which have arisen because aphids undergoing a change of host plant in the past have been more likely to establish themselves successfully—and subsequently speciate—if the new hosts were in the same family as the old.

However, there are other aphidine genera in which the host plant range is markedly discontinuous, showing that the 'capture' of a new host quite unrelated to the old is certainly not a particularly rare event in the evolution of aphid-host plant relationships. In *Amphorophora*, for example, although most species live on *Rubus*, there are some species on Geraniaceae and Primulaceae, and some on ferns. One may surmise that the *Amphorophora* species on ferns in future may become rather rapidly differentiated from the *Rubus*-feeding species, and that present morphological similarity may be because the acquisition of ferns as host plants occurred relatively recently. Similar marked changes of host plant in the more distant evolutionary past may have led to many groups of species now recognized as separate genera—e.g. *Corylobium* on *Corylus*, *Elatobium* on conifers, *Micromyzus* on ferns, and *Myzodium* on mosses. For

Host plant specificity

As the genera of aphids tend to be associated with particular plant families, so each species within an aphid genus tends to restrict its feeding to a certain genus or species of host plant, or at least to certain plant species within a clearly defined group of genera. But for this high degree of host-plant specificity it would not, of course, be feasible to produce a host-plant-orientated identification guide. Although the primary and secondary host plants of host-alternating aphids are botanically distinct, this does not make such aphids any less specific in their host-plant relationships. For example, *Hyperomyzus lactucae* migrates from *Ribes* (usually *nigrum*) to *Sonchus*, *Cryptomyzus ribis* from *Ribes rubrum* to *Stachys*, and so on. Usually specificity to the primary host is even greater than to the secondary host. The generally high degree of aphid/host-plant specificity is emphasized here because workers only encountering economically important aphids may easily gain a false impression of the overall picture. Pest aphids, particularly those on agricultural crops, tend to have a wider host range than related economically unimportant species, and they also include a number of species which feature prominently in the keys in this book because they are polyphagous. The term 'polyphagous' is used to mean that they can feed on plants in many different families, rather than that they are totally non-specific, since even the most polyphagous aphid (*Myzus persicae*) has not been recorded even from as much as 1% of flowering plant species.

Polyphagy in aphids may perhaps be regarded as an extension of the 'reserve host phenomenon' discussed by Stroyan (1957). Even quite specific aphids may occasionally be found colonizing certain plants with which they are not normally associated. Often such reserve hosts have no aphids specific to them, but seem to provide vacant ecological niches for a number of aphids whose true host associations are elsewhere. Non-indigenous plants, particularly cultivated varieties of crop plants, are most likely to provide vacant ecological niches and thus be used as reserve hosts.

The majority of pest aphids restrict their feeding to species within one plant family (e.g. summer generations of *Cavariella aegopodii* on Umbelliferae, *Sitobion avenae* on Gramineae) but perhaps what makes them pests is that they are able to colonize a wider range of reserve hosts, including economically important plants, than congeneric species. The few truly polyphagous aphids are characterized by their ability to exploit reserve hosts in many different plant families, so that even the family with which they were originally associated is not always clear. Eastop (1981) reviews the wild plants which are most often used as alternative hosts by the main aphid pest species.

GEOGRAPHICAL DISTRIBUTION

The major zoogeographical regions of the world each have their own characteristic aphid fauna on native plants. However, most major agricultural crops are not indigenous to the countries in which they are grown, and the aphids which infest such crops are usually equally exotic. Information on the present distribution of each of the main aphid pests is given in Section C of the book. This subsection outlines the natural geographical distribution of the Aphidoidea; this is followed by a general discussion of the causes and consequences of human activities.

Zoogeography

The Aphidoidea is predominantly a northern temperature group, richest in species in North America, Europe, and Central and East Asia. The general features of the aphid life cycle which are common to all families and subfamilies of Aphidoidea—parthenogenetic generations exploiting actively growing plants and a sexual generation resulting in an overwintering diapause egg—strongly suggest that the evolution of the group as a whole has been primarily in relation to a temperate climate with cold winters, or at least with marked seasonal changes of temperature and day length.

South-east Asia is the only area where all the major groups within the Aphidoidea are well represented. The tropical forest aphid fauna in South-east Asia, although small, is larger than anywhere else. The present-day distributions of both Greenideinae and Hormaphidinae seem to be centred there, although in the past these groups may have been more widely distributed. Other South-east Asian groups which may be relicts of wider distributions in the past are the Aiceonini (living on Lauraceae) and the Melaphidina (migrating from sumac to mosses). South-east Asia also has many indigenous genera of Aphidinae which are not, or scarcely, represented elsewhere in the world.

Some of the largest aphid genera (*Pemphigus, Chaitophorus, Cinara, Aphis, Uroleucon, Macrosiphum*) are well represented in all parts of the Northern Hemisphere. Europe and North America, in particular, have many genera in common and there are numerous parallels between the different species on the native vegetation of each continent. There is, however, a characteristic Mediterranean aphid fauna, including the Fordina associated with *Pistacia* and *Therioaphis* on legumes, which is not represented in North America except by introduced species.

Compared with the Northern Hemisphere, the poverty of the aphid faunas of the southern continents is remarkable. The relatively few native genera seem for the most part to be relicts of an old southern fauna, differentiated from northern aphids at about the tribal level, with affinities between the faunas of each southern continent demonstrating the old land links. For example, there are *Neophyllaphis* on *Podocarpus* in South America, Africa, and Australasia, and *Neuquenaphis* on Andean *Nothofagus* species is related to *Sensoriaphis* on *Nothofagus* in New Zealand and Tasmania. The few species of Aphidinae

Introductory section 15

native to the southern continents are congeneric, or nearly so with Northern Hemisphere species; for example, *Picturaphis* on South American Leguminosae is closely related to *Microparsus* on legumes in North America, and there seems to have been rather recent speciation of *Sitobion* on both grasses and dicotyledons in Africa.

The Pacific islands seem to have no indigenous aphid species.

Introduced species

Few aphids encountered by applied entomologists are native. Almost all the major aphid pests of North American agriculture are introduced Old World species, *Macrosiphum euphorbiae* being a notable exception. Many of these species have now also spread to South America, southern Africa, and Australasia, and indeed it is a fairly safe prediction that certain aphid species will eventually find (or be transported to) their crop host plants wherever in the world they are grown. (In 1983, *Therioaphis trifolii maculata* has just reached New Zealand, five years after its first appearance in Australia.) A few other aphid pests have spread around the tropics, for example *Cerataphis palmae* (= *variabilis*) on palms and *Pentalonia nigronervosa* on bananas.

The Old World also has some important introduced Nearctic aphids, such as *Eriosoma lanigerum* (American blight), and more recently *Nearctaphis bakeri* on clover; and of course the best known of all is the vine phylloxerid, *Viteus vitifolii*.

The distribution of some heteroecious aphid pest species is limited by that of their primary hosts, even though the crops which are their secondary host plants are grown more widely. However, most of the cosmopolitan pest species are anholocyclic, that is able to live all year round parthenogenetically on crop hosts, and some have spread in this way to areas where their primary hosts do not occur. There may be three factors contributing to the predominance of anholocycly in cosmopolitan aphid pest species:

(1) the absence of primary host plants in the case of heteroecious aphids;
(2) the absence of the necessary conditions to trigger production of sexual forms, as most introductions occur into warmer climates than those in the country of origin; and
(3) the more likely establishment of a species in new territory if it can reproduce without sexual reproduction, thus overcoming the problem of mate-finding when populations of the introduced species are initially low.

Aphids introduced into new territory can show rather obvious founder effects; that is, the mean and range of various characters may be different from that of the same species in its original distribution area. The introduced population may even stem from a single parthenogenetic female and thus be initially clonal. Even when introduced aphids are anholocyclic, however, new 'biotypes' can arise by mutation and be selected out rapidly by the new environment. The separate introductions of *Therioaphis trifolii* into North America, probably from two different parts of Europe, provide classic examples of founder effects and their consequences (p. 363).

MORPHOLOGY AND KEY CHARACTERS

This account of aphid morphology is rather superficial and will concentrate on the apterous parthenogenetic female, because this is the morph referred to almost exclusively in the keys. Particular attention is given to key characters, since it is essential that they are interpreted correctly. Brief descriptions are then given of the other aphid morphs and their differences from the apterous viviparae.

Body size and shape

Body length is measured from the centre of the frons in dorsal view to the end of the abdomen, not including the cauda. Most aphids are 1.5–3.5 mm long. Aphid species in which most individuals are less than 2 mm long are considered to be 'small'; those of body length usually above 3 mm are large. *Body width* is estimated across the widest part of the body.

Apterous morphs of Phylloxeridae have a pear-shaped (pyriform) body, broadest anteriorly. In the apterous viviparae of Aphididae the abdomen is usually the widest part of the body, being expanded to accommodate a large number of developing embryos. Body form of apterous Aphididae may be described as globose (e.g. certain Pemphiginae on roots), ovate (most Aphidini), spindle-shaped (many Macrosiphini), or elongate (e.g. *Atheroides*). The greatest specialization of form is found in certain Hormaphidinae, e.g. *Cerataphis*, with apterae superficially resembling whitefly pupae, with an almost circular flattened, wax-fringed body completely hiding the reduced legs and antennae beneath.

In aphids where the abdomen of the prelarviposition female becomes greatly distended with embryos and there is only light sclerotization of the abdominal cuticle, the posterior segments in older females may become somewhat telescoped into one another after the young are born. Measurement of body length of such older females will therefore be misleading.

Segmentation

The segmentation of the body is usually quite distinct. The head and prothorax are usually separate, but are fused in some groups (e.g. *Thelaxes*). The abdomen has eight visible segments, terminating in a cauda and anal plate which may represent the modified tergite and sternite of the ninth or the tenth segment (Fig. 3). The anal plate and the sternite anterior to the genital aperture (the subgenital plate) are usually pigmented. The numbers and positions of hairs on the subgenital plate, and also on the eighth tergite, are of taxonomic value, but are not used as characters in this book.

Introductory section 17

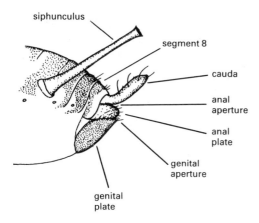

Fig. 3 Lateral view of end of abdomen of a member of the Aphidinae

Antennae

The antennae of most aphids are 5- or 6-segmented and the segments are numbered from I to V or I to VI, I and II being the scape and pedicel, the remainder comprising the flagellum. The last antennal segment comprises a wider basal part and a narrower terminal process (= unguis). Where the terminal process arises there is usually a rather large primary sensorium, with sometimes an adjacent group of smaller accessory sensoria. The length of the terminal process in comparison with the base of the last segment is very frequently used as a key character. The basal part of the last antennal segment is measured to the apex of the primary sensorium (Fig. 4). Other characters of the antennae used in the keys include their total length relative to the body; the relative proportions of particular segments; the length of hairs, especially those on segment III

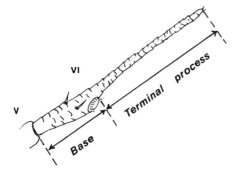

Fig. 4 Measurement of base and terminal process of last antennal segment

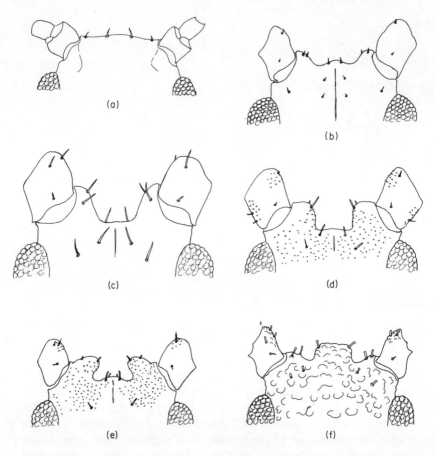

Fig. 5 Dorsal views of heads of various Macrosiphini, to show the degree of development of the antennal tubercles: (a) undeveloped—*Pseudaphis abyssinica*; (b) diverging, with median frontal prominence—*Metopolophium dirhodum*; (c) diverging, well-developed—*Macrosiphum euphorbiae*; (d) parallel—*Aulacorthum solani*; (e) convergent—*Myzus persicae*; (f) with well-developed median frontal projection—*Myzaphis rosarum*

which are usually compared to the diameter of this segment at its basal articulation; the presence or absence, numbers, and distribution of secondary sensoria, which are usually present at least on segment III in alatae but absent or in reduced numbers in apterae of the same species.

Head

The front of the head is bent downward quite sharply so that the clypeus is anterio-ventral and the base of the rostrum is on the posterior part of the underside of the head, near the bases of the fore coxae. In the Aphidinae, and particularly in Macrosiphini, there is often marked development of the latero-

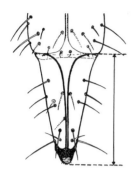

Fig. 6 Measurement of length of last segment of rostrum

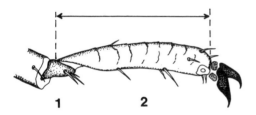

Fig. 7 Measurement of length of second segment of hind tarsus

frontal part of the head so that the antennal bases are elevated to a varying extent as 'tubercles' (Fig. 5). The degree of development and shape of the antennal tubercles is often difficult to observe in slide-mounted specimens, where the front of the head is often flattened and distorted, but it is a very useful character for specimens in alcohol. The aphid should be viewed in dorsal aspect with the head tilted downwards slightly to compensate for the upward slant of the tubercles when they are well developed. The degree of development of the antennal tubercles is frequently compared with that of the medial part of the frons, which projects forwards in some aphids to give the front of the head between the antennae a sinuate or W-shaped outline in dorsal view (Fig. 5b). In a few species (e.g. *Myzaphis rosarum*, Fig. 5f) the medial frontal projection is more pronounced than the antennal tubercles.

Other characters of the head used in the keys include the length of hairs on the frons or vertex, often compared with the length of antennal segment I; whether the eyes are compound, multi-faceted, or of only three facets (a triommatidium); whether the triommatidium is present (the usual case) or absent as a posterior process (ocular tubercle) of the compound eye; and the length of the last segment of the rostrum (Fig. 6), usually in comparison with the length of the second segment of the hind tarsus (Fig. 7). (It is often easy to see whether the last rostral segment is longer or shorter than the second hind tarsal segment, but in some cases accurate measurement will be necessary and this is more easily accomplished with slide-mounted specimens.) The chaetotaxy of the last rostral segment, especially the number of accessory hairs on the basal part, is an important character in aphid taxonomy, but it is not used in this book since it can only be observed in prepared specimens.

Legs

The general form of the aphid leg does not vary greatly. The hind legs are usually longest, perhaps because they have to extend far enough back to balance a large and heavy abdomen. The forelegs are used to push against the plant when the stylets are withdrawn, and sometimes to enable the aphid to spring away from the plant in an escape reaction. The fore coxae are greatly enlarged in some genera (e.g. *Therioaphis*) to increase the speed or power of this process. The tarsi are usually 2-segmented, the second segment much longer than the first; in Tramini the second segment of the hind tarsus is greatly elongated. One-segmented tarsi are found in apterae of certain Pemphiginae, and a few genera of Aphidinae (none of economic importance) have completely lost their tarsi.

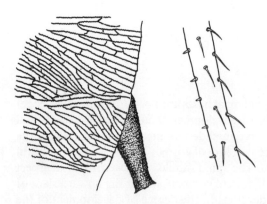

Fig. 8   Stridulatory apparatus of *Toxoptera aurantii*

Apart from the length of the hind tarsal segments, few characters of the legs are used in the keys. The level of pigmentation is only used where there are obvious differences, as this can vary greatly within species. The presence or absence of long, finely-pointed hairs on the hind femur is used to separate some *Aphis* species. In *Toxoptera* species there is a row of evenly spaced peg-like hairs on the hind tibiae which are the comb of the stridulatory apparatus (Fig. 8). Tarsal chaetotaxy is important in aphid taxonomy, but has to be observed in slide-mounted specimens.

Introductory section

Dorsal abdomen

Sclerotization of the abdominal tergite is usually associated with considerable deposition of melanin, but not always; for example, some species of *Sitobion* have the whole tergum lightly sclerotized but only intersegmental muscle sclerites darkened. Patterns of sclerotization and associated pigmentation are used wherever possible in the keys as they are easily observed, but there is considerable variation within species in the form and extent of dorsal abdominal pigmentation, and this should be kept in mind when comparing specimens with the photographs in this book. Immature aphids rarely have any pigmentation of the dorsal cuticle.

In *Toxoptera* species a pattern of cuticular ridges ventrolaterally on sternites 5 and 6 is particularly well developed and, together with the comb of peg-like hairs on the hind tibiae, functions as a stridulatory apparatus (Fig. 8). Lateral abdominal tubercles are usually readily seen in aphids in alcohol, and their presence or absence from certain segments is used in some keys.

Siphunculi

These structures, characteristic of aphids, are nevertheless secondarily lost, especially in the apterae of certain groups. If present, they are usually situated dorso-laterally on the posterior margin of abdominal segment 5. They adopt a wide variety of forms in different species, varying from simple pores through broad conical structures to long tubes (Fig. 9). The central part may be swollen, or more usually the swelling is on the distal half, making the siphunculi club-shaped (clavate). The extreme apex is often widened by a flange. The siphuncular cuticle often has small ridges (imbrications) but may also have rows of minute spinules or a pattern of polygonal reticulation. The siphunculi often point upwards, so that they may need to be deflected downwards with a mounted needle in order to observe their true length relative to other structures in dorsal view.

Cauda and anal plate

The shape of the cauda in dorsal view (Fig. 10) and the number of hairs it carries are frequently-used characters. In unmounted specimens the cauda often has an upward tilt, so it may be necessary to raise the front of the aphid to observe the length and shape of the cauda in dorsal view without foreshortening. If the cauda is longer than its basal width in dorsal view, it is usually conveniently described as tongue-shaped. A shorter cauda may be broadly rounded, rounded, or pentagonal (in the latter case a convenient description for its appearance in dorsal view is 'helmet-shaped'). In some groups the cauda has a constriction, separating its basal part from a knob-like apex. In Drepanosiphinae a knobbed cauda often occurs in conjunction with a bi-lobed anal plate (Fig. 10d).

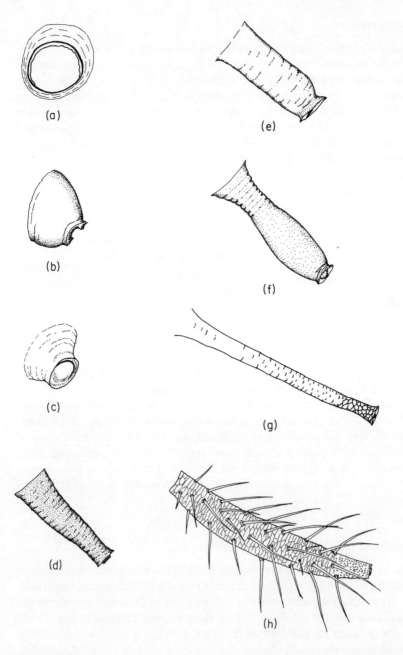

Fig. 9 Various aphid siphunculi: (a) pore-like—*Ceratovacuna* sp.; (b) mammariform—*Diuraphis noxia*; (c) truncate—*Sipha flava*; (d) tapering—*Aphis fabae*; (e) swollen proximal to flange—*Rhopalosiphum insertum*; (f) clavate—*Hyadaphis foeniculi*; (g) with subapical zone of polygonal reticulation—*Macrosiphum euphorbiae*; (h) Greenideine—Greenidea ficicola

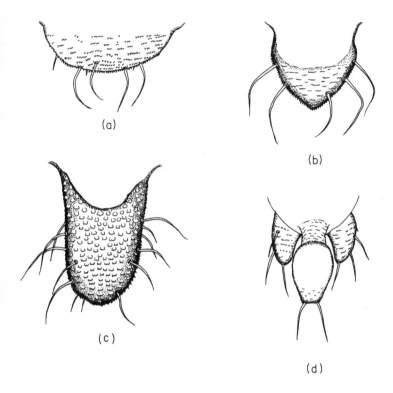

Fig. 10 Cauda of various aphids: (a) broadly rounded—*Brachycaudus amygdalinus*; (b) helmet-shaped—*Dysaphis crataegi*; (c) tongue-like—*Aphis fabae*; (d) knobbed, with bilobed anal plate—*Therioaphis trifolii*

The presence of a distinct cauda is usually the most convenient way of picking out the adults in a sample containing both immature and adult apterous aphids. When the cauda is short and rounded the distinction may not be so clear, but in most crop-inhabiting species of the main subfamily Aphidinae with a very short cauda the adult aptera is recognizable by its development of dorsal abdominal pigmentation.

In certain root-dwelling aphids (*Baizongia, Geoica, Smynthurodes*) the anus is displaced to a dorsal position, and the large anal plate is surrounded on three sides by a horseshoe-shaped sclerite which is the eighth abdominal tergite (Fig. 11). The hairs on this 'trophobiotic organ' hold droplets of honeydew until they are imbibed by ants.

Hairs

The chaetotaxy of aphids is more readily observed in slide-mounted specimens. However, the presence of long capitate hairs, often with tuberculate bases, is a

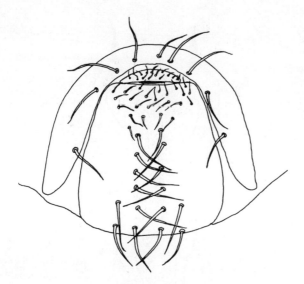

Fig. 11 Posterior view of end of abdomen of *Geoica lucifuga*

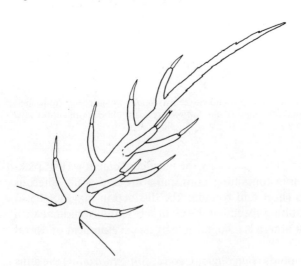

Fig. 12 Hair-bearing abdominal process of *Cervaphis* sp.

characteristic feature of some aphids and such hairs are easily seen under the stereomicroscope, especially those projecting anteriorly from the front of the head and the proximal antennal segments. In some aphids the hairs on body and antennae are so small that they are difficult to see except in slide-mounted preparations. In other species the hairs may be long and fine, thick and spine-like, spatulate or fan-shaped, and in certain Greenideinae such as *Cervaphis* the hairs are borne on remarkably large abdominal processes (Fig. 12).

Tubercles

The term 'tubercle' in aphid morphology covers a diversity of structures. Tubercles in Aphidinae are generally wart-like, thin-walled, and rather small, even as small as the hair bases (e.g. in *Rhopalosiphum*). In Aphidini (*Aphis, Rhopalosiphum*, etc.) there are lateral tubercles as a general rule on abdominal segments 1 and 7, and variably on other segments, whereas in Macrosiphini (*Macrosiphum, Myzus*, etc.) such tubercles, if present at all, are usually only found on abdominal segments 2–5. In other aphid subfamilies, single or paired conical or finger-like tubercles may be characteristic of particular genera or species, and are most frequently spinal in location (that is, near the dorsal midline).

Wax glands

Specialised wax glands which produce solid wax particles or filaments are common in aphids. In Aphidinae the wax gland pores are scattered over certain areas of the cuticle and usually produce a fine particulate wax which may give the aphid a dull waxy bloom or, if produced in large quantities, a mealy appearance. In Pemphiginae and some Drepanosiphinae the glands are often concentrated on well-defined wax pore-plates which may be grouped in characteristic arrangements, and usually secrete a white flocculent wax. In some Hormaphidinae wax pore-plates are situated in a row all around the edge of the body, producing a fringe of wax.

Spiracles

Aphids usually have two pairs of thoracic spiracles and one pair on each of abdominal segments 1–7. Spiracles are usually placed beside small lateral sclerites and their apertures may be circular, oval, or kidney-shaped (reniform). Sometimes they are partially or almost completely covered by a cowl-like operculum. In Aphidinae, the relative positions of the spiracles on abdominal segments 1 and 2 is of taxonomic significance; in the Aphidini the first and second abdominal spircales on each side are quite widely spaced (with a lateral tubercle placed more or less between them), whereas in most Macrosiphini they are much closer together, less than half as far apart as is the second abdominal spiracle from the third.

Alate vivipara

Alatae may differ from apterous viviparae of the same species in numerous characters, including chaetotaxy and relative lengths of antennal segments, siphunculi, and cauda, as well as more obvious differences such as sclerotization of the thorax and increased sensoriation of the antennae. In most taxonomic works apterae and alatae have to be keyed separately. The present keys are generally based on apterae, except where exclusively alate populations are likely to be collected; for example, many Drepanosiphinae and gall-formers such as species of Melaphidina on Sumac. Characters of the alatae are also used in a few cases where they show obvious differences between species with very similar apterae.

Fundatrix

The morph which develops from the overwintering egg, to form the first parthenogenetic generation of the year, differs from the individuals of later parthenogenetic generations in a whole range of morphological characters. It is apterous, except in some Drepanosiphinae, and generally has a broader, more rounded body and shorter antennae, legs, cauda, and siphunculi (the latter may be appreciably different in shape; for instance, in *Myzus persicae* fundatrices on peach the siphunculi are not even slightly clavate). The antennal tubercles are also greatly reduced in comparison with the apterae of later generations, and there may be many other differences. It follows that keys to apterous viviparae will not work for fundatrices. However, fundatrices will not be encountered on most crops since these are usually the secondary host plants of the aphids. Where the crop is the primary host, there is only a short period in early spring when the only adult aphids in small, young colonies are fundatrices, and large colonies will usually have predominantly second or even third generation adult apterae.

Fundatrices of most Pemphiginae and some Aphidinae (e.g. *Anuraphis*) form galls or 'pseudogalls' on their woody host plants, and their progeny are then frequently all alate, spring migrants. In such cases the fundatrix or its alate daughters are the only morphs available for identification, and the keys take account of this.

Oviparae

The sexual, egg-laying females of Aphididae are almost invariably apterous and may differ considerably in size, shape, colour, and other features from the apterous vivipara of the same species. However, oviparae only occur at a certain time of year and, in aphids with host alternation, where the differences between oviparae and viviparae are usually greatest, oviparae are restricted to the primary host. Oviparae and viviparae are most alike in Lachninae and Aphidinae; in some species the two morphs can only be readily distinguished by the more or less swollen hind tibiae of the oviparae, bearing scent glands which secrete a sex pheromone.

Males

The males of Pemphiginae, Anoeciinae, Hormaphidinae, and some Lachninae are apterous and dwarfish. In the Aphidinae the males of host-alternating

Introductory section

species are invariably alate as they have to migrate back to the primary host. In monoecious species of Aphidinae the males can be either apterous or alate, or both. The external male genitalia are heavily pigmented and quite large, easily visible to the naked eye, but not so far used in taxonomy because males are relatively uncommon and only found during a limited period of the year. Males of Aphidinae, whether apterous or alate, generally have a pattern of body markings similar to that of alate viviparous females of the same species, but males are usually somewhat smaller and darker, and have many more secondary sensoria on the antennae.

B. *The Crops and their Aphids*

NOTES ON THE USE OF THIS SECTION

Arrangement

This section contains:

(1) **An alphabetical list of latin names of crops, with the English names for them used in this book.**
(2) **A list of crops in alphabetical order of their English names.** Under each crop, the aphid species most likely to be found colonizing the crop are listed, in most cases with a code letter in front of each species name (A, B, etc.). If several polyphagous aphids may be found on one crop, then they are listed with a single code letter and the user is referred to the separate keys for these species at the end of the section. A key then follows wherever necessary. In the key each species (or group of species) is identified by its code letter.
At the end of each list of aphids there may be some additional species in square brackets, listed alphabetically. These are aphids which have been recorded only rarely from the crop in question, and therefore are not included in the key. Some are well-known polyphagous or oligophagous species, which use the crop rather infrequently as a 'reserve host'. Others are little-known species which were originally described from the crop, but in most cases have not been seen by the present authors. Such species may not be referred to elsewhere in the book, in which case the author(s) and date of the original reference are given.
(3) **Separate keys to the most economically important and widespread polyphagous aphids.** Keys are provided to the 10 main cosmopolitan Aphidinae (5 *Aphis* spp. and 5 species of Macrosiphini), and to 7 species of aphid which occur rather widely in the tropics on a range of trees and shrubs.

Limitations

Reliable identifications will only be possible if certain important limitations are continually borne in mind when using the keys in this section:

(1) **The keys refer only to characters of adult apterous viviparous females**

(apt.), except where it is specifically stated that they refer to alate viviparous females (al.).
(2) The keys should only be applied to aphids which have been found actually colonizing the crop (i.e. occurring in numbers, feeding and producing young). Aphids caught by trapping or sweeping in the vicinity of the crop, or even within it, may have originated on other plants and cannot be reliably identified using this book.
(3) The keys are not meant to stand on their own, but to be used in conjunction with the photographs at the end of the book, and the textual accounts of the species in the next section. These will often include information that will help to confirm identifications.
(4) The keys are designed for use in general with specimens preserved in alcohol and viewed at 10–35 times magnification under a good quality stereoscopic microscope. As well as transmitted light, a high intensity source of reflected light is sometimes useful, and a calibrated micrometer eyepiece will be needed for some characters.
(5) In a few instances the separation of certain species within a genus may require specimens to be macerated, cleared, and mounted on microscope slides for viewing with a compound microscope. The techniques involved are described on pages 375–6.
(6) In temperate climates it must be kept in mind that holocyclic aphid populations on their overwintering host plants may contain fundatrices in early spring, and sexual morphs in late summer and autumn. The keys will not usually work satisfactorily for these morphs. The life cycle and polymorphism of aphids (pp. 5–10) should always be borne in mind when looking at their morphology.
(7) Although the main polyphagous aphid species are included in the lists and keys under all crops which they colonize at all frequently, the possibility of their occurrence on other crops needs to be borne in mind.

Anyone working regularly with aphids will soon learn to recognize these species, but initially the keys to polyphagous aphids provided at the end of this section should help with this problem.

Summary of procedure for using this section

(1) The sample should be composed of aphids collected from a colony on a (correctly identified) crop.
(2) A single adult aptera is selected from the sample to take through the appropriate key. Adults are distinguishable by their distinct cauda, or by the development of dorsal pigmentation. There may be several species in one sample, so it is important to work with single aphids until familiar with the distinguishing characters.
(3) If the key provides a name for the aphid, the brief account of the species in the next section is consulted to see if information on appearance in life, etc. supports the identification.

List of crop plants

(4) The photograph of a slide-mounted specimen, if provided, can be used to check whether the specimen conforms in general appearance, body proportions, etc. to the species indicated. (Of course, unless specimens are macerated, cleared, and mounted on slides, the comparison can only be a superficial one.)

(5) If there are no conflicting points, the identification is probably correct. However, when accurate identification is of great importance (i.e. in biological control work, or intensive research programmes on particular aphid species) it is obviously advisable to have the identification checked by a specialist taxonomist.

(6) If the key does not provide a name for the aphid, or if there is any reason to doubt whether the species indicated by the key is correct after studying the information provided in the text and photographs, then the aphid may be of a species which does not generally occur on that crop. If there are species names in square brackets at the end of the list of keyed species, the possibility that it is one of these can be investigated. The main polyphagous aphids, whether or not they are listed under that crop, can also be checked out to see if it is one of them. If doubt remains, specimens should be sent to a specialist aphid taxonomist for examination.

(7) Whether or not it is necessary to prepare and mount specimens on microscope slides for identification, anyone working regularly with aphids is strongly advised to build up a reference collection of such permanent preparations, not only to assist in future identifications but also to provide a permanent record of the specimens concerned, so that any future doubts about their identity can be resolved. This is particularly important with pest species, where in some cases the application of a particular name could change if, for example, future work demonstrated either that more than one species was involved under the same name, or that meaningful subspecific categories could be introduced based on differences in life cycle, host-plant relationships, etc. Techniques for making permanent slide collections of aphids are summarized on pages 375–6.

LIST OF CROP PLANTS AND THEIR ENGLISH NAMES

Actinidia	Kiwi Fruit
Aleurites spp.	Tung Oil
Allium ascalonicum	Shallot
Allium cepa	Onion
Allium porum	Leek
Allium sativum	Garlic
Allium schoenoprasum	Chive
Alocasia indica	Taro, Giant
Aloe spp.	Aloe
Alpinia	Galangal
Amaranthus spp.	Amaranth
Anacardium occidentale	Cashew Nut
Ananas comosus (= *sativus*)	Pineapple
Anethum graveolens	Dill

Angelica archangelica	Angelica
Annona spp.	Custard Apple
Apium graveolens	Celery
Arachis hypogea	Ground Nut
Armeniaca vulgaris	Apricot
Armoracia lapathifolia (= *rusticana*)	Horseradish
Artocarpus altilis (= *incisa*)	Breadfruit
Arundinaria	Bamboo
Asparagus officinalis	Asparagus
Avena sativa	Oats
Azalea	Rhododendron
Bambusa	Bamboo
Bertholletia excelsa	Brazil Nut
Beta vulgaris	Beet, Common, etc.
Brassica chinensis	Cabbage, Chinese
Brassica napo-rapa	Swede
Brassica napus	Rape
Brassica nigra	Mustard
Brassica oleracea	Cabbage
Brassica oleracea var. *botrytis*	Cauliflower
Brassica pekinensis	Cabbage, Pak-choi
Brassica rapa	Turnip
Cajanus cajan	Pigeon Pea
Camellia sinensis	Tea
Canavalia ensiformis	Bean; Jack, Sabre, Sword
Cannabis sativa	Hemp
Capparis spinosa	Capers
Capsicum frutescens (= *annuum; baccatum*)	Chili
Carica papaya	Pawpaw
Carthamus tinctoria	Safflower
Carum carvi	Caraway
Carya pecan (= *illinoiensis*)	Pecan
Castanea sativa (= *vesca*)	Chestnut
Cattleya spp.	Orchids
Ceiba pentandra	Kapok
Chloris guyana	Grass, Rhodes
Chrysanthemum spp.	Chrysanthemum
Chrysanthemum cinerariaefolium	Pyrethrum
Cicer arietinum	Chick Pea
Cichorium intybus	Chicory
Cinchona spp.	Quinine
Cinnamomum camphora	Camphor
Cinnamomum cassia	Cassia Bark
Cinnamomum zeylandicum	Cinnamon
Citrullus vulgaris (= *lanatus*)	Water Melon
Citrus spp.	Citrus
Citrus paradisi	Grapefruit
Cochlearia armoracia	Horseradish
Cocos nucifera	Coconut
Coffea spp.	Coffee
Cola spp.	Cola
Colocasia antiquorum	Coco
Colocasia esculenta	Taro
Convallaria majalis	Lily of the Valley
Coriandrum sativum	Coriander
Corylus spp.	Hazelnut
Crocus sativus	Saffron
Cydonia oblonga (= *vulgaris*)	Quince
Cynara scolymus	Artichoke, Globe
Cucumis melo	Melon
Cucumis sativus	Cucumber

List of crop plants

Cucurbita pepo	Pumpkin
Cynodon dactylon	Grass, Bermuda
Cyphomandra betacea	Tree Tomato
Cypripedium spp.	Orchids
Dahlia spp.	Dahlia
Daucus carota	Carrot
Dendrobium spp.	Orchids
Dendrocalamus spp.	Bamboo
Derris spp.	Derris
Dianthus barbata	Sweet William
Dianthus caryophyllus	Carnation
Dianthus plumarius	Pinks
Dioscorea spp.	Yam
Dolichos lablab	Bean, Lablab
Durio zebethinus	Durian
Elaeis guineensis	Oil Palm
Elettaria cardamomum	Cardamon
Eleusine coracana	Ragi
Eriobotrya japonica	Loquat
Eugenia caryophyllata	Cloves
Festuca spp.	Grasses of Temperate Pastures
Ficus carica	Fig
Foeniculum vulgare	Fennel
Fortunella spp.	Kumquat
Fragaria chiloensis	Strawberry
Gaylussacia baccata	Huckleberry
Gladiolus spp.	Gladiolus
Glycine max	Soybean
Glycyrrhiza glabra	Liquorice
Gossypium spp.	Cotton
Grossularia spp.	Gooseberry
Helianthus tuberosus	Artichoke, Jerusalem
Hemerocallis spp.	Lily, Day
Hevea braziliensis	Rubber
Hibiscus esculentus	Okra
Hordeum vulgare (= *sativum*)	Barley
Humulus lupulus	Hop
Imperata arundinacea (= *cylindrica*)	Lalang
Indigofera spp.	Indigo
Ipomoea batatas	Potato, Sweet
Iris spp.	Iris
Juglans regia	Walnut
Lablab vulgaris	Bean, Lablab
Lactuca sativa	Lettuce
Laurus nobilis	Bay laurel
Lavendula officinalis	Lavender
Lens esculentum	Lentil
Levisticum officinale	Lovage
Lilium spp.	Lily
Linum usitatissimum	Flax
Litchi chinensis	Litchi
Lolium spp.	Rye-Grass
Luffa cylindrica	Loofah
Lupinus spp.	Lupin
Lycopersicon esculentum	Tomato
Macadamia spp.	Macadamia
Majorana hortensis	Marjoram, Sweet
Malus pumila	Apple
Mangifera indica	Mango
Manihot esculenta (= *utilissima*)	Cassava
Manihot glaziovii	Ceara Rubber

Matthiola spp.	Stock
Medicago sativa	Lucerne (Alfalfa)
Mentha spp.	Mint
Mespilus germanica	Medlar
Metroxylon spp.	Sago Palms
Morus nigra	Mulberry
Mucuna deeringiana	Bean, Velvet
Musa × *sapientum*	Banana
Musa textilis	Manila Hemp
Myristica fragrans	Nutmeg
Myrrhis odorata	Ciceley, Sweet
Napus rapifera	Swede
Narcissus spp.	Daffodil
Nasturtium officinale	Watercress
Nephelium litchi	Litchi
Nicotiana tabacum	Tobacco
Ocimum basilicum	Basil
Olea europea	Olive
Origanum majorana	Marjoram, Sweet
Oryza sativa	Rice
Palmae spp.	Palms
Panicum maximum	Grass, Guinea
Panicum miliaceum	Millett, Common
Papaver somniferum	Poppy, Opium
Papaya carica	Pawpaw
Pastinaca sativa	Parsnip
Paullinia cupana	Guarana
Pennisetum clandestinum	Grass, Kikuyu
Pennisetum purpureum	Grass, Elephant
Pennisetum spicatum	Millett, African
Pennisetum typhoides	Millett, Pearl
Persea americana (= *gratissima*)	Avocado Pear
Petroselinum spp.	Parsley
Phaseolus spp.	Bean, Common, etc.
Phaseolus aureus	Gram, Golden
Phaseolus mungo	Gram, Black
Phoenix dactylifera	Date
Phyllostachys spp.	Bamboo
Pimpinella anisum	Anise
Piper nigrum	Pepper, Black
Pistacia vera	Pistachio
Pisum sativum	Pea, Garden
Poa spp.	Grasses of Temperate Pastures
Polygonum fagopyron	Buckwheat
Prunus amygdalus	Almond
Prunus armeniaca	Apricot
Prunus avium	Cherry, Sweet
Prunus cerasus	Cherry, Morello
Prunus domestica	Plum
Prunus domestica × *insititia*	Greengage
Prunus insititia	Damson
Prunus mume	Apricot, Japanese
Prunus persica	Peach
Prunus persica var. *nectarina*	Nectarine
Prunus spinosus	Sloe
Psidium guajava	Guava
Punica granatum	Pomegranate
Pyrus communis	Pear
Pyrus malus	Apple
Raffia spp.	Raffia
Raphanus sativus	Radish

Abaca

Rheum spp.	Rhubarb
Rhododendron spp.	Rhododendron
Rhus spp.	Sumac
Ribes grossularia	Gooseberry
Ribes nigrum	Currant, Black
Ribes rubrum	Currant, Red
Ricinus communis	Castor Oil
Rubus fruticosus	Blackberry, European
Rubus idaeus	Raspberry, European
Rubus idaeus strigosus	Raspberry, American Red
Rubus laciniatus	Blackberry, Cut-leaved
Rubus × *loganobaccus*	Loganberry
Rubus occidentalis	Raspberry, Black
Rubus procerus	Blackberry, Himalayan
Rubus vitifolius	Blackberry, Cory or Pacific
Rudbeckia spp.	Rudbeckia
Saccharum officinale	Sugarcane
Salvia officinalis	Sage
Scorzoneura hispanica	Salsify, Black
Secale cereale	Rye
Sechium edule	Chayote
Sesamum indicum (= *orientale*)	Sesame
Solanum melongena	Egg Fruit
Solanum nigrum	Huckleberry, Garden
Solanum tuberosum	Potato
Sorghum spp.	Guinea Corn
Sorghum halepense	Grass, Johnson
Spinacia oleracea	Spinach
Symphytum spp.	Comfrey
Theobroma cacao	Cocoa
Thymus vulgaris	Thyme
Tilia spp.	Lime
Tragopogon porrifolius	Salsify
Trifolium spp.	Clover
Triticum spp.	Wheat
Tulipa spp.	Tulip
Vaccinium spp.	Blueberry *and* Vaccinium
Vaccinium macrocarpon	Cranberry
Vanilla fragrans	Vanilla
Vicia faba	Bean, Broad, etc.
Vicia sativa	Vetch
Vigna sesquipedalis	Bean, Asparagus
Vigna sinensis	Cow-pea
Vitis vinifera	Grape Vine
Zea mays	Maize
Zingiber officinale	Ginger

LISTS AND KEYS OF APHIDS ON EACH CROP

Abaca	see **Manila Hemp**

Agave	no aphids recorded

Alfalfa	see **Lucerne**

Allspice	no aphids recorded

Almond	A	*Pterochloroides persicae*	E	*Brachycaudus (Thuleaphis)*
	B	*Hyalopterus amygdali*		*amygdalinus*
	C	*Aphis gossypii*		[*Aphis fabae; Brachycaudus*
	D	*Brachycaudus helichrysi*		*persicae; Myzus persicae;*
				Rhopalosiphum nymphaeae]

1. Terminal process much shorter than base of last antennal segment. Siphunculi in form of dark, hairy cones ... **A**

   Terminal process longer than base of last antennal segment. Siphunculi tubular ... 2

2. Cauda clearly longer than its basal width in dorsal view ... 3
   Cauda short and broad, not longer than its basal width in dorsal view ... 4

3. Siphunculi very small, shorter and much thinner than cauda, and without a flange ... **B**

   Siphunculi longer than cauda, tapering, with a small flange ... **C**

4. Cauda helmet-shaped in dorsal view, almost as long as its basal width, which is less than the length of the last rostral segment. Abdomen without any dark dorsal markings ... **D**

   Cauda very broadly rounded, much broader than long, its basal width in dorsal view greater than the length of the last rostral segment. Posterior abdominal tergites often with dark markings ... **E**

Aloe	*Aloephagus myersi* *Myzus persicae*	

Amaranth	*Aphis citricola* *Aphis gossypii*	try key to polyphagous aphids

Angelica	A	*Cavariella aegopodii*	D	*Aphis fabae*
	B	*Cavariella archangelicae*		[*Cavariella konoi; Dysaphis angelicae; Dysaphis ossiannilssoni; Macrosiphum euphorbiae*]
	C	*Hyadaphis foeniculi*		

Anise 37

1. Eighth abdominal tergite with posteriorly projecting supracaudal process ... 2

   No supracaudal process ... 3

2. Terminal process about as long as base of antennal segment VI ... **A**

   Terminal process 1.5–2.0 times longer than base of antennal segment VI ... **B**

3. Siphunculi clavate ... **C**

   Siphunculi tapering from base to flange ... **D**

Anise A *Cavariella aegopodii*   C *Hyadaphis coriandri*
         B *Hyadaphis foeniculi*   D *Aphis fabae*
                                      E *Macrosiphum euphorbiae*

---

1. Eighth abdominal tergite with supracaudal process ... **A**

   No supracaudal process ... 2

2. Siphunculi clavate, similar in length to cauda ... **B**

   Siphuncili not clavate, shorter or longer than cauda ... 3

3. Siphunculi a little shorter than cauda, slightly swollen before flange ... **C**

   Siphunculi longer than cauda, rarely of similar length, tapering or cylindrical ... 4

4. Body ovate. Siphunculi and cauda black ... **D**

   Body spindle-shaped. Siphunculi and cauda mainly pale, siphunculi with a subapical zone of polygonal reticulation ... **E**

Apple

	A	*Watabura nishiyae*	M	*Myzus persicae*
	B	*Eriosoma lanigerum*	N	*Ovatus crataegarius*
	C	*Nearctaphis bakeri*	O	*Ovatus malisuctus*
	D	*Dysaphis* spp.—see separate key		[*Aphis fabae; Eriosoma herioti; Longistigma xizangensis* Zhang 1981; *Macrosiphum rosae; Ovatus insitus; Prociphilus sasakii* Monzen 1927; *Pterochloroides persicae; Pyrolachnus pyri; Radisectaphis gyirongensis* Zhang 1981]
	E	*Allocotaphis quaestionis*		
	F	*Rhopalosiphum insertum*		
	G	*Aphis craccivora*		
	H	*Toxoptera aurantii*		
	I	*Aphis gossypii*		
	J	*Aphis pomi*		
	K	*Aphis citricola*		
	L	*Macrosiphum euphorbiae*		

1. Siphunculi absent. Antennae 4-segmented (illustr.) Tarsi with only one claw well developed .... **A**

   Siphunculi present, even if only pores Antennae 5- or 6-segmented. Tarsi with 2 equally developed claws 2

2. Terminal process shorter than base of last antennal segment. Siphunculi merely pores .... **B**

   Terminal process longer than base of last antennal segment. Siphunculi tubular 3

3. Antennal tubercles weakly developed, not exceeding height of medial part of frons 4
   Antennal tubercles well developed 11

4. Cauda rounded, helmet-shaped or triangular in dorsal view, no longer than its basal width 5
   Cauda tongue-shaped, clearly longer than its basal width in dorsal view 7

5. Siphunculi pale, a little shorter than cauda .... **C**

   Siphunculi dark, longer than cauda 6

6. Abdomen without marginal tubercles .... **D**
   Abdomen with marginal tubercles on most segments .... **E**

Apple

7.	Siphunculi slightly swollen subapically and constricted before flange	F
	Siphunculi tapering from base to flange	8
8.	Dorsal abdomen with extensive black patch	G
	Dorsal abdomen unpigmented	9
9.	Terminal process more than 3.5 times longer than base of last antennal segment. Stridulatory apparatus present	H
	Terminal process less than 3.5 times longer than base of last antennal segment. No stridulatory apparatus	10
10.	Cauda paler than siphunculi, with usually 4–6 hairs	I
	Cauda and siphunculi both dark; cauda with usually more than 6 hairs	11
11.	Lateral tubercles present on abdominal segments 2–4. Cauda rarely with less than 14 hairs. Length of last rostral segment more than 130 μm	J
	Lateral tubercles absent from abdominal segments 2–4. Cauda usually with less than 12 hairs. Length of ultimate rostral segment less than 120 μm	K
12.	Inner faces of antennal tubercles divergent. Siphunculi with a subapical zone of polygonal reticulation	L
	Inner faces of antennal tubercles parallel or convergent. Siphunculi without polygonal reticulation	13
13.	Siphunculi slightly swollen on distal half	M
	Siphunculi tapering from base to flange	14

14.	Siphunculi pale	N
	Siphunculi dark	O

Key to Dysaphis spp. forming spring colonies on apple

1.	Galled leaves yellow to greenish yellow, irregularly or transversely coiled. Antennae in apt. about as long as the distance from frons to bases of siphunculi, in al. about as long as body. Apt. without pigmentation of abdominal tergites anterior to siphuncili	*Dysaphis (Pomaphis) plantaginea*	
	Galled leaves either wholly or partly reddish, longitudinally rolled. Antennae relatively shorter. Apt. with dorsal abdominal pigment spots or patches anterior to siphunculi		2

2.	Colonies containing many alatiform apterae with sclerotized thorax; normal alatae much less common	*Dysaphis devecta*	
	Colonies mainly consisting of alatoid nymphs and normal adult alatae		3

3.	Al. with more than 70 sensoria on antennal segment III	*Dysaphis brancoi*	
	Al. with less than 70 sensoria on antennal segment III		4

4.	Al. with longest hair on antennal segment III of individual specimens ranging in length from 40 to 71 µm, 1.5–2.7 times longer than diameter of segment at base	*Dysaphis chaerophylli*
	Al. with longest hair on antennal segment III in range 20–48 µm, 1.0–1.7 times longer than diameter of segment at base	*Dysaphis anthrisci*
	Al. with longest hair on antennal segment III in range 9–29 µm, 0.5–1.3 times longer than diameter of segment at base	*Dysaphis radicola*

Apricot

	A	*Pterochloroides persicae*	E	*Hyalopterus pruni*
	B	*Asiphonaphis pruni*	F	*Hyalopterus amygdali*
	C	*Myzus persicae*		[*Brachycaudus cardui; Myzus mumecola*]
	D	*Rhopalosiphum nymphaeae*		

1.	Terminal process shorter than base of last antennal segment	A
	Terminal process longer than base of last antennal segment	2

2.	Siphunculi absent, lateral abdominal tubercles large and conspicuous	B
	Siphunculi present, lateral abdominal tubercles absent or inconspicuous	3

Apricot, Japanese

3.	Siphunculi much longer than cauda, slightly clavate	4
	Siphunculi small, shorter than cauda	5

4.	Antennal tubercles well developed, with inner faces convergent in dorsal view. Siphunculi pale	C
	Antennal tubercles weakly developed. Siphunculi dark	D

5.	Siphunculi narrow, 3–4 times longer than their maximum diameter	E
	Siphunculi broader, about twice as long as maximum diameter	F

Apricot, Japanese

- A *Brachycaudus helichrysi*
- B *Myzus cerasi umefoliae*
- C *Phorodon humuli japonensis*
- D *Myzus persicae*
- E *Myzus mumecola* [*Myzus yamatonis*]

1.	Cauda helmet-shaped, about as long as its basal width in dorsal view	A
	Cauda triangular or tongue-shaped, longer than its basal width in dorsal view	2

2.	Abdominal dorsum dark, sclerotic	B
	Abdominal dorsum pale	3

3.	Antennal tubercles with long, forwardly-directed processes	C
	Antennal tubercles without processes, their inner faces parallel or convergent	4

4. Siphunculi cylindrical or slightly swollen on distal half, diameter at mid-length about the same as diameter at flange. Hairs on body and antennae usually short and blunt

D

Siphunculi tapering from base to flange, diameter at mid-length greater than diameter at flange. Hairs on body and antennae long and pointed, the longest exceeding the basal diameter of antennal segment III

E

Artichoke, Globe

A	*Trama troglodytes*	F	*Aphis fabae/solanella*
B	*Protrama radicis*	G	*Myzus persicae*
C	*Brachycaudus cardui*	H	*Capitophorus elaeagni*
D	*Dysaphis cynarae*	I	*Capitophorus horni*
E	*Aphis craccivora*		[*Aphis maidi-radicis;* *Capitophorus carduinus*]

---

1. Terminal process shorter than base of last antennal segment. Hind tarsus elongate, at least half as long as hind tibia ... 2

Terminal process longer than base of last antennal segment. Hind tarsus of normal length ... 3

2. Siphunculi absent ... A
   Siphunculi present as shallow cones ... B

3. Antennal tubercles little developed ... 4
   Antennal tubercles well developed, their inner faces divergent or convergent ... 7

4. Cauda short, helmet-shaped and not longer than width at base in dorsal view ... 5
   Cauda triangular or tongue-shaped, longer than width at base ... 6

5. Dorsal abdomen with an extensive solid black patch anterior to siphunculi ... C
   Dorsal abdomen without a solid black patch, only scattered dark markings anterior to siphunculi ... D

Artichoke, Jerusalem

6. Dorsal abdomen with solid black patch.
    Terminal process less than 3 times longer
    than base of last antennal segment ... **E**
    Dorsal abdomen mainly unsclerotized.
    Terminal process usually more than 3 times
    longer than base of last antennal segment ... **F**

7. Antennal tubercles converging. Dorsal body
    hairs inconspicuous, mainly short and blunt ... **G**

    Antennal tubercles diverging. Many of dorsal
    body hairs rather long and capitate, arising
    from tuberculate bases ... **8**

8. Siphunculi with pigmented apices. Dorsal
    abdomen anterior to siphunculi with one pair
    of spinal hairs per segment, forming a single
    longitudinal row on each side of mid-line ... **H**
    Siphunculi with pale apices. Dorsal abdomen
    anterior to siphunculi with most spinal hairs
    duplicated, not arranged in two clear
    longitudinal rows ... **I**

Artichoke, Jerusalem

A *Uroleucon (Uromelan) helianthicola*
B *Trama troglodytes*
C *Protrama penecaeca*

1. Terminal process of antenna long, siphunculi
    long and dark with a subapical zone of
    polygonal reticulation. Hind tarsus normal ... **A**

    Terminal process of antenna very short,
    siphunculi absent or inconspicuous. Hind
    tarsus elongate, at least half as long as
    hind tibia ... **2**

2. Siphunculi absent ... **B**
    Siphunculi present as small, pigmented cones ... **C**

Asparagus

	A	*Brachycorynella asparagi*	
	B	*Myzus persicae*	
	C	*Aphis craccivora*	
	D	*Aphis gossypii* [*Aulacorthum (Neomyzus) circumflexum; Sitobion chanikiwiti*]	

1. Siphunculi very small, truncated cones, much shorter than cauda .... **A**

   Siphunculi much longer than cauda **2**

2. Antennal tubercles well developed, with convergent inner faces. Siphunculi slightly clavate .... **B**

   Antennal tubercles not developed. Siphunculi tapering from base to flange **3**

3. Solid black dorsal abdominal patch centred on tergites 4–5 .... **C**
   No dorsal abdominal pigmentation .... **D**

Avocado Pear

	A	*Sinomegoura citricola*	C *Toxoptera aurantii*
	B	*Myzus persicae*	D *Aphis gossypii*
			E *Aphis citricola*

1. Siphunculi pale at base, dark on distal part, a little shorter than the very long, dark cauda .... **A**

   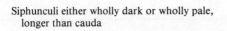

   Siphunculi either wholly dark or wholly pale, longer than cauda **2**

Azalea

2. Antennal tubercles well developed, with convergent inner faces. Siphunculi pale, slightly clavate .... B

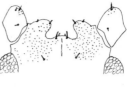

Antennal tubercles weakly developed. Siphunculi dark, tapering from base to flange 3

3. Terminal process more than 3.5 times longer than base of last antennal segment. Stridulatory apparatus present C

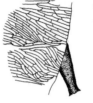

Terminal process less than 3.5 times longer than base of last antennal segment. No stridulatory apparatus 4

3. Cauda paler than siphunculi, with usually 4–6 hairs C
Cauda dark, usually with more than 6 hairs .... D

Azalea   a subgenus of *Rhododendron*, q.v.

Bamboos

The world fauna of bamboo aphids comprises about 50 species; only the more widely distributed species are included here. Liao (1976) gives an account of 21 species in Taiwan, and A.K. Ghosh (1976) gives a key to bamboo-feeding Hormaphidinae in India.

A	*Takecallis taiwanus*	G	*Pseudoregma alexanderi*
B	*Takecallis arundinariae*	H	*Pseudoregma bambusicola*
C	*Takecallis arundicolens*	I	*Astegopteryx* spp.
D	*Glyphinaphis bambusae*	J	*Ceratovacuna* spp.
E	*Ceratoglyphina bambusae*	K	*Melanaphis bambusae*
F	*Chaitoregma tattakana*	L	*Hysteroneura setariae*

1. All adult viviparous females alate. Antenna
long and thin and conspicuously banded 2
Colonies consisting mainly of apterous
viviparous females. Antennae if long not
conspicuously banded 4

2. (Al.) Antennae shorter than body, with only the
distal part of segment III dark .... A
(Al.) Antennae longer than body, with dark
pigmentation of segment III not confined to
distal part 3

3. (Al.) Thorax and abdomen with a pair of
elongate dark patches on each tergite.
Pigmentation of sensoriated part of antennal
segment III extending almost to base. Cauda
pale or dusky .... B
(Al.) Thorax and abdomen without dark patches.
Antennal segment III with pale section at
base, proximal to dark, sensoriated part.
Cauda black C

4. Eyes with 3 facets. Antennae shorter than
forelegs 5
Eyes with many facets. Antennae longer than
forelegs 11

5. Head with a pair of forwardly-directed frontal
horns. Body hairs short 6
Head without frontal horns. Body hairs long
and spine-like .... D

6. Cauda and anal plate rounded. Frontal horns
longer than antennal segments I and II
combined (on *Bambusa*) .... E

Cauda knob-like, anal plate bi-lobed. Frontal
horns not longer than antennal segments I
and II combined (on various bamboos) 7

7. Whole body strongly sclerotized. Frontal
horns rounded. (On *Arundinaria*) .... F

At least abdominal tergites mainly membranous. Frontal horns conical 8

Banana 47

8.	Fused head and pronotum both sclerotized. Large sclerotic patches on meso- and metanotum and laterally on abdomen	9
	Pronotum not sclerotized. No sclerotic patches on rest of thorax and abdomen	10

9.	Body rather elongate, often more than 3.2 mm long. Wax plates on abdominal segments 6–8. (On *Dendrocalamus* and *Miscanthus*)	G
	Body oval, usually less than 3.2 mm long. Wax plates absent or only on abdominal segment 8. (On *Bambusa*)	H

10.	Siphunculi in form of shallow cones surrounded by fine hairs	I
	Siphunculi in form of raised pores with rims	J

11.	Siphunculi short, truncated cones, only a little longer than width at base. Cauda dark	K
	Siphunculi over 4 times longer than width at base. Cauda very pale	L

Banana *Pentalonia nigronervosa*

Barley as for **Wheat**, except that (E) *Diuraphis (Holcaphis) tritici* can probably be neglected.

Bay Laurel no aphids recorded

Bean (Common, Dwarf, French, Kidney, Haricot, Lima, Pole, Navy, Scarlet Runner, String)

A	*Smynthurodes betae*	F	*Picturaphis* species
B	*Aphis craccivora*	G	*Acyrthosiphon gossypii*
C	*Aphis fabae*	H	*Macrosiphum euphorbiae*
D	*Aphis gossypii*		[*Aphis lhasaensis* Zhang 1981]
E	*Myzus persicae*		

1.	Terminal process much shorter than base of last antennal segment. Siphunculi absent. (On roots, dusted with white wax)	A
	Terminal process much longer than base of last antennal segment. Siphunculi present. (On aerial parts)	2
2.	Antennal tubercles poorly developed	3
	Antennal tubercles well developed	5
3.	Solid black dorsal patch centred on abdominal segments 4–5	B
	Abdominal tergites 4–5 mainly or wholly unsclerotized	4
4.	Cauda dark, with more than 10 hairs	C
	Cauda pale, with 4–7 hairs	D
5.	Inner faces of antennal tubercles convergent	E
	Inner faces of antennal tubercles divergent	6
6.	Antennal segments I and II black. Siphunculi with proximal half pale, distal half pigmented and slightly clavate. (Al. with wing veins conspicuously black-bordered)	F
	Antennal segments I and II pale. Siphunculi wholly pale or with pigmentation increasing gradually towards apex	7
7.	Siphunculi very long and thin, about 3 times longer than cauda and pale except for slight darkening at extreme apex	G
	Siphunculi about twice as long as cauda, with a subapical zone of polygonal reticulation	H

Bean (Broad, Field, Horse)

A	*Smynthurodes betae*	**E**	*Myzus ornatus*
B	*Brachycaudus helichrysi*	**F**	*Acyrthosiphon pisum*
C	*Aphis craccivora*	**G**	*Megoura viciae*
D	*Aphis fabae*	**H**	*Megoura crassicauda* [*Aphis craccae*]

1. Terminal process much shorter than base of last antennal segment. Siphunculi absent — **A**

   Terminal process much longer than base of last antennal segment. Siphunculi present — 2

2. Cauda helmet-shaped, as short as broad in dorsal view — **B**

   Cauda tongue-shaped, longer than its basal width — 3

3. Antennal tubercles little developed — 4
   Antennal tubercles well developed — 5

4. Solid black dorsal abdominal patch — **C**
   No solid black dorsal patch — **D**

5. Small aphid (body less than 2 mm long). Inner faces of antennal tubercles convergent, dorsal abdomen with a pattern of dark ornamentation — **E**
   Large aphid (body length generally much more than 2 mm). Inner faces of antennal tubercles divergent — 6

6. Siphunculi pale, very attenuate distally. Cauda pale — **F**

   Siphunculi black, swollen in middle. Cauda black — 7

7. Antennal segment III with 10–18 small secondary sensoria, not in a row, on basal three-quarters of segment only — **G**
   Antennal segment III with 15–50 strongly protruding secondary sensoria, in a row extending over three-quarters to nine-tenths of segment — **H**

Bean, Asparagus   *Aphis craccivora*

Bean (Jack, Sabre, Sword)
[*Aphis robiniae canavaliae* Zhang 1981]

Bean, Lab-Lab
Aphis craccivora
Picturaphis brasiliensis
Acyrthosiphon gossypii
[*Aphis robiniae canavaliae* Zhang 1981]

Use relevant couplets in first bean key.

Bean, Soya   see Soybean

Beets (Common, Mangold, Red, Sugar)
A *Smynthurodes betae*
B *Pemphigus* species
C *Aphis gossypii*
D *Aphis fabae*
E *Rhopalosiphoninus staphyleae tulipaellus*
F { *Myzus persicae*
     Macrosiphum euphorbiae   } see key to
     Aulacorthum solani   polyphagous Macrosiphini
   [*Brachyunguis plotnikovi* (Nevsky 1928); *Prociphilus erigeronensis*]

1. Terminal process much shorter than base of last antennal segment. Siphunculi absent 2

   Terminal process much longer than base of last antennal segment. Siphunculi present 3

2. Body and appendages with numerous hairs. Abdomen without wax pore plates .... A

   Body and appendages with very sparse, inconspicuous hairs. Abdomen with wax pore plates on posterior segments .... B

Bermuda Grass

3.	Antennal tubercles poorly developed	4
	Antennal tubercles well developed	5
4.	Cauda paler than siphunculi, with 4–7 hairs	C
	Cauda as dark as siphunculi, with more than 10 hairs	D
5.	Dorsal abdomen with a large dark patch. Siphunculi markedly clavate	E
	Dorsal abdomen unpigmented. Siphunculi slightly clavate, tapering or cylindrical	F

Bermuda Grass   see **Grass, Bermuda**

Blackberry (European)

	A	*Aphis ruborum*	D	*Macrosiphum funestum*
	B	*Amphorophora rubi*		[*Acyrthosiphon rubi*
	C	*Sitobion fragariae*		Narzikolov 1957;
				Macrosiphum rosae;
				Maculolachnus rubi A.K.
				Ghosh & Raychaudhuri 1972]

1.	Small aphid (under 2 mm long). Antennal tubercles not developed	A
	Larger aphids (usually over 3 mm long). Antennal tubercles well developed, with inner faces divergent	2
2.	Siphunculi clavate, without a subapical zone of polygonal reticulation	B
	Siphunculi tapering from base to flange, with a subapical zone of polygonal reticulation	3
3.	Antennae shorter than body. Femora pale	C
	Antennae longer than body. Femora with dark apices	D

Blackberry (Cory, Pacific)

Amphorophora rubitoxica

Blackberry (Cut-leaved)
- A *Aphis ruborum*
- B *Amphorophora rubitoxica*
- C *Sitobion fragariae*

Use key under **European Blackberry**, neglecting (D) and noting the species substituted at (B).

Blueberry
- A *Aphis vaccinii*
- B *Aphis gossypii*
- C *Fimbriaphis scammelli/ pernettyae*
- D *Illinoia pepperi*
- E *Illinoia azaleae*
- F *Illinoia borealis*

1.	Antennal tubercles weakly developed	2
	Antennal tubercles well developed	3
2.	Cauda black with more than 10 hairs	A
	Cauda pale with 4–7 hairs	B
3.	Siphunculi tapering from base to flange, without any polygonal reticulation	C
	Siphunculi swollen on distal half, but with a cylindrical subapical region with polygonal reticulation	4
4.	Last segment of rostrum shorter than second segment of hind tarsus	D
	Last segment of rostrum longer than second segment of hind tarsus	5
5.	First segments of tarsi with 5 hairs	E
	First segments of tarsi with 3 hairs	F

Brazil Nut
no aphids recorded

Breadfruit
Greenidea artocarpi

Brussels Sprout
as for **Cabbage**

Buckwheat

Buckwheat
A { *Aphis fabae* group
Aphis nasturtii
Aphis gossypii } see key to polyphagous *Aphis* spp.

B { *Macrosiphum euphorbiae*
Aulacorthum solani } see key to polyphagous Macrosiphini

[*Acyrthosiphon rubi* Narzikulov 1957]

1. Antennal tubercles weakly developed .... A

   Antennal tubercles well developed .... B

Cabbage

A	*Myzus persicae*	E	*Smynthurodes betae*
B	*Myzus ascalonicus*	F	*Pemphigus populitransversus*
C	*Brevicoryne brassicae*		[*Aphis gossypii*]
D	*Lipaphis erysimi*		

1. Terminal process longer than base of last antennal segment. Siphunculi present. (On aerial parts) 2
   Terminal process much shorter than base of last antennal segment. Siphunculi absent. (On roots) 5

2. Siphunculi pale, more than 1.5 times longer than cauda 3
   Siphunculi dusky or dark, less than 1.5 times longer than cauda 4

3. Siphunculi a little longer than antennal segment III, with minimum diameter of proximal part greater than middle diameter of hind tibia .... A

   Siphunculi shorter than antennal segment III, and minimum diameter of proximal part a little less than middle diameter of hind tibia .... B

4. Cauda broadly triangular in dorsal view. Dorsal abdomen with segmental pattern of pigmentation   C

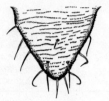

Cauda tongue-shaped. Dorsal abdomen without pigmentation anterior to siphunculi   D

5. Body and appendages with numerous hairs. Wax pore plates absent   E

Body and appendages with sparse, inconspicuous hairs. Wax pore plates on posterior abdominal segments   F

Cabbage, Chinese as for **Cabbage**

Cabbage, Pak-Choi as for **Cabbage**

Cacao see **Cocoa**

Camphor A *Aiceona* sp(p).
B *Sinomegoura citricola*
C *Toxoptera citricidus*

1. Terminal process shorter than base of last antennal segment. Cauda broadly rounded. Siphunculi consisting of large pores situated on shallow, pigmented hairy cones   A

Terminal process longer than base of last antennal segment. Cauda tongue-shaped, black. Siphunculi tubular, tapering from base to flange   2

Cantaloupe

2. Frons with well-developed antennal tubercles, inner faces divergent in dorsal view. Cauda pointed at apex, a little longer than siphunculi .... **B**

   Frons broad with poorly developed antennal tubercles. Cauda rounded at apex, a little shorter than siphunculi .... **C**

Capers [*Anuraphis capparidis* Nevsky 1929]

Cantaloupe   see **Melon**

Caraway

A	*Pemphigus passeki*	**C**	*Hyadaphis foeniculi*
B	*Cavariella aegopodii*	**D**	*Myzus ornatus*

1. Terminal process shorter than base of last segment. Siphunculi absent. (On roots) .... **A**

   Terminal process longer than base of last antennal segment. Siphunculi present. (On aerial parts) **2**

2. Eighth abdominal tergite with posteriorly projecting supracaudal process .... **B**

   No supracaudal process **3**

3. Siphunculi clavate, about as long as cauda. No dorsal abdominal pigmentation .... **C**

   Siphunculi much longer than cauda, tapering from base to flange. Dorsal abdomen with dark intersegmental ornamentation .... **D**

Cardamon *Pentalonia nigronervosa*
  [*Aphis craccivora;*
  Micromyzus kalimpongensis]

Carnation (and Pinks)

A	*Aphidura pujoli*	F	*Myzus persicae*
B	*Aphis sambuci*	G	*Myzus dianthicola*
C	*Aphis fabae*	H	*Aulacorthum solani*
D	*Myzus ascalonicus*	I	*Macrosiphum stellariae*
E	*Myzus certus*		[*Aphidura picta;*
			Smynthurodes betae]

1. Antennal tubercles weakly developed, not
    exceeding height of medial part of frons ... 2
    Antennal tubercles well developed ... 4

2. Cauda and siphunculi pale. Siphunculi slightly
    clavate ... A
    Cauda and siphunculi dark. Siphunculi
    tapering from base to flange ... 3

3. Siphunculi more than twice as long as cauda.
    Abdominal segments 1–5 and 7 with large
    flat lateral tubercles. (On roots) ... B
    Siphunculi less than twice as long as cauda.
    Lateral abdominal tubercles papillate and
    usually only on segments 1 and 7. (On aerial
    parts) ... C

4. Siphunculi slightly to moderately clavate.
    Antennal tubercles usually convergent in
    dorsal view, sometimes almost parallel. (Al.
    with solid black dorsal abdominal patch) ... 5
    Siphunculi not clavate. Inner faces of
    antennal tubercles parallel or divergent in
    dorsal view. (Al. with dorsal abdominal
    pigmentation either segmentally divided or
    absent) ... 8

5. Siphunculi rather small, much shorter than
    antennal segment III, with minimum diameter
    of basal part less than diameter of middle
    part of hind tibia ... D

    Siphunculi usually a little longer than antennal
    segment III, with minimum diameter of basal
    part greater than diameter of middle part
    of hind tibia ... 6

Carrot

6. Terminal process usually less than 3.2 times longer than base of last antennal segment. (Reddish brown in life.) (Al. with 0–3 sensoria on antennal segment IV) — **E**

   Terminal process usually more than 3.2 times longer than base of last antennal segment. (Variable in colour but usually some shade of green.) (Al. without sensoria on antennal segment IV) — **7**

7. Siphunculi usually as pale as body and only slightly clavate. (Mid-green, pale green, pale yellow or pink in life, not causing leaf chlorosis) — **F**

   Siphunculi a little darker than body colour, particularly at apices, and more obviously clavate. (Deep yellow-green in life, causing leaf chlorosis) — **G**

8. Inner faces of antennal tubercles parallel. Siphunculi without polygonal reticulation — **H**

   Inner faces of antennal tubercles divergent. Siphunculi with a subapical zone of polygonal reticulation — **I**

Carrot

A	*Pemphigus phenax*	
B	*Cavariella aegopodii*	
C	*Semiaphis dauci*	
D	*Semiaphis heraclei*	
E	*Myzus persicae*	
F	*Myzus ornatus*	
G	*Dysaphis crataegi*	
H	*Dysaphis foeniculus*	
I	*Hyadaphis coriandri*	
J	*Hyadaphis foeniculi*	
K	*Aphis lambersi*	
L	*Aphis armoraciae*	
M	*Aphis helianthi*	
N	*Aphis fabae* [*Aphis citricola; Aphis gossypii; Rhopalosiphoninus latysiphon*]	

Aphids on the world's crops

1.	Terminal process shorter than base of last antennal segment. Siphunculi absent	A
	Terminal process longer than base of last antennal segment. Siphunculi present	2
2.	Eighth abdominal tergite with posteriorly projecting supracaudal process	B
	Supracaudal process absent	3
3.	Siphunculi very small and flangeless, only about half the length of the cauda or less, with aperture slanted towards the mid-line	4
	Siphunculi at least two-thirds of the length of the cauda	5
4.	Siphunculi more than 1.5 times longer than their basal width. Hairs very short; longest hairs on hind femur rarely exceeding 20 μm	C
	Siphunculi less than 1.5 times longer than their basal width. Longest hairs on hind femur up to 60 μm, more than half as long as the basal diameter of the femur	D
5.	Antennal tubercles well developed, their inner faces convergent in dorsal view	6
	Antennal tubercles weakly developed	7
6.	Siphunculi slightly clavate. No dorsal abdominal pigmentation	E
	Siphunculi tapering from base to flange. Dorsal abdomen with intersegmental pattern of dark ornamentation	F

Cashew Nut

7. Cauda helmet-shaped in dorsal view, a little shorter than its basal width, with angular apex 8

    Cauda tongue-shaped, at least as long as its basal width in dorsal view, with rounded apex 9

8. Siphunculi 2.5–3.0 times longer than their basal width. Hairs on frons very short and blunt .... **G**

    Siphunculi short and broad-based, rarely more than twice as long as their basal width. Hairs on frons long and fine, longest almost as long as antennal segment I .... **H**

9. Siphunculi shorter than or of similar length to cauda; if a little longer then clavate 10

    Siphunculi longer than cauda, tapering 11

10. Siphunculi a little shorter than cauda, slightly swollen before flange .... **I**

    Siphunculi clavate, about equal in length to cauda or a little longer .... **J**

11. Large lateral tubercles present on abdominal segments 2–5 as well as 1 and 7 .... **K**

    Lateral tubercles only usually on abdominal segments 1 and 7 12

12. Antennal segment III without sensoria. Hairs on hind femur long and fine, the longest exceeding the basal width of the hind femur 13

    Antennal segment III with 5–12 sensoria. Hairs on hind femur all shorter than its basal width .... **L**

13. Terminal process shorter (0.6–0.9 times) than antennal segment III .... **M**

    Terminal process similar in length to antennal segment III or longer (1.0–1.6 times) .... **N**

Cashew Nut
   A *Toxoptera aurantii*
   B *Toxoptera odinae*
      [*Aphis gossypii*]

1. Siphunculi as long as or longer than cauda.
   Terminal process more than 3.5 times longer
   than base of last antennal segment .... **A**
   Siphuncili shorter than cauda. Terminal
   process less than 3 times longer than base
   of last antennal segment .... **B**

Cassava   *Aphis craccivora*

Cassia Bark   no aphids recorded

Catjang   *see* **Pigeon Pea**

Cauliflower   as for **Cabbage**

Cayenne Pepper see **Chili**

Ceara Rubber   no aphids recorded

Celery
- A *Cavariella aegopodii*
- B *Cavariella konoi*
- C *Aphis sambuci*
- D *Aphis armoraciae*
- E *Dysaphis apiifolia*
- F *Dysaphis foeniculus*
- G *Aphis helianthi*
- H *Semiaphis heraclei*
- I *Hyadaphis coriandri*
- J *Hyadaphis foeniculi*
- K { *Myzus ornatus*
      Myzus persicae
      Aulacorthum solani
      Macrosiphum euphorbiae } see key to polyphagous Macrosiphini
     [*Aphis apigraveolens* Essig
       1938; *Aphis citricola*;
       Aphis fabae; *Aphis gossypii*;
       Rhopalosiphoninus latysiphon]

---

1. Eighth abdominal tergite with posteriorly
   projecting supracaudal process 2

   No supracaudal process 3

Celery

2.	Terminal process about equal in length to base of last antennal segment. Last rostral segment 0.08–0.13 mm long (and usually without any accessory hairs)	**A**
	Terminal process 1.5 or more times longer than base of last antennal segment. Last rostral segment 0.12–0.18 mm long (with 2 accessory hairs)	**B**

3.	Cauda dark, short and broad, not clearly longer than its basal width in dorsal view	4
	Cauda pale or dark, clearly longer than its basal width	7

4.	Cauda broadly tongue-shaped in dorsal view, rounded at apex	5
	Cauda helmet-shaped in dorsal view, with angular apex	6

5.	Siphunculi more than 3.5 times longer than their basal width. Large flat lateral tubercles present on abdominal segments 2–5 as well as 1 and 7	**C**
	Siphunculi less than 3.5 times longer than their basal width. Lateral tubercles usually only on abdominal segments 1 and 7	**D**

6.	Siphunculi 2.5–3.5 times longer than their basal width. Hairs on frons very short and blunt	**E**
	Siphunculi short and broad-based, rarely more than twice as long as their basal width. Hairs on frons long and fine	**F**

7.	Siphunculi black, tapering from base to flange. Cauda dark	**G**
	Siphunculi pale or dusky, clavate or tapering. Cauda pale or a little dusky	8

8.	Siphunculi very small and flangeless, only about half the length of the cauda or less, with aperture slanted towards the mid-line	H
	Siphunculi more than two-thirds of length of cauda	9
9.	Siphunculi a little shorter than cauda	I
	Siphunculi longer than cauda or, if about equal in length to cauda, then obviously clavate	10
10.	Siphunculi about equal in length to cauda or a little longer, and clavate	J
	Siphunculi much longer than cauda; tapering, cylindrical or slightly swollen on distal half	K

Chayote Aphis gossypii
Aulacorthum magnoliae

Cherry, sweet Myzus cerasi
[Brachycaudus cardui;
Brachycaudus helichrysi;
Myzus persicae; Myzus yamatonis; Rhopalosiphum nymphaeae; Tinocalloides montanus Basu 1970]

Cherry, Morello Myzus cerasi
[Brachycaudus cerasicola (Mordvilko ex Nevsky 1929); Rhopalosiphum insertum]

Chestnut
A Myzocallis castanicola
B Patchia virginiana
C Thelaxes suberi
D Lachnus roboris
E Lachnus tropicalis
F Moritziella castaneivora
G Phylloxera ?castaneae

1.	Antennae 5- or 6-segmented	2
	Antennae 3-segmented. (Small insect, body pyriform, broadest anteriorly)	6

Chick Pea

2. Terminal process either about equal to, or longer than, base of last antennal segment ... 3
   Terminal process much shorter than base of last antennal segment ... 4

3. All adults alate. Body elongate, head and thorax with longitudinal dorsal stripe, abdomen with paired pigment spots and patches. Antennal segment III thin with very short, sparse hairs ... **A**

   Al. with oval body, wing veins dusky-bordered. Abdomen with broad transverse bands or solid patch of pigment. Antennal segment III thicker than fore tibia with numerous long hairs ... **B**

4. Small aphid with 5-segmented antennae and a knobbed cauda. Eyes 3-faceted. Abdomen sclerotized bearing long spine-like hairs ... **C**
   Very large aphid with 6-segmented antennae and rounded cauda. Al. with pigmented wings ... 5

5. Terminal process about one-third as long as base of antennal segment VI. Al. with clear area at base of forewing ... **D**
   Terminal process at least one half as long as base of antennal segment VI. Al. with forewings pigmented to base ... **E**

6. Many dorsal tubercles on head, thorax, and abdominal segments 1–6, decreasing in size and number posteriorly ... **F**
   Only four dorsal tubercles present, confined to head ... **G**

Chick Pea	*Aphis craccivora*
	Acyrthosiphon pisum

Chicory	A	*Pemphigus bursarius*	I	*Uroleucon ambrosiae/ pseudambrosiae*
	B	*Trama troglodytes*		
	C	*Protrama flavescens*	J	*Nasonovia ribisnigri*
	D	*Neotrama caudata*	K	*Macrosiphum euphorbiae*
	E	*Brachycaudus helichrysi*	L	*Aulacorthum solani*
	F	*Aphis intybi*		[*Aphis albella* Nevsky 1951; *Aphis armoraciae*; *Neotrama pamirica* (Narzikulov 1962); *Prociphilus erigeronensis*]
	G	*Myzus persicae*		
	H	*Uroleucon cichorii*		

1. Terminal process shorter than base of last
    antennal segment (on roots) 2

   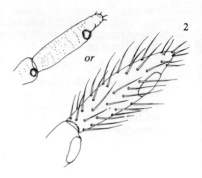

   Terminal process longer than base of last
   antennal segment (on aerial parts) 5

2. Second segment of hind tarsus of normal
    length. Body hairs very short and sparse .... A

   Second segment of hind tarsus elongate, much
   longer than other tarsi. Body hairs numerous,
   long and fine 3

3. Siphunculi present as shallow cones. Eyes
    usually with more than 12 facets 4
    Siphunculi absent. Eyes with only 3 facets .... B

4. Antennal segment VI similar in length to V
    or a little longer, with terminal process at
    least one-third as long as base. Many apt.
    alatiform with dark transverse bands on
    dorsal abdomen C

   Antennal segment VI a little shorter than V,
   with terminal process very short. Dorsal
   abdominal pigmentation never present .... D

5.	Cauda short, helmet-shaped, no longer than its basal width in dorsal view 	E
	Cauda tongue-shaped, longer than its basal width in dorsal view	6
6.	Antennal tubercles weakly developed	F
	Antennal tubercles well developed	7

7.	Inner faces of antennal tubercles convergent. Siphunculi slightly clavate 	G
	Inner faces of antennal tubercles parallel or divergent. Siphunculi not clavate	8
8.	Siphunculi black, even at their bases, and 1.0–1.5 times longer than cauda, with zone of polygonal reticulation extending over distal one-quarter to one-third of length 	9
	Siphunculi often darker than body but paler at base, darker towards apex, more than 1.5 times longer than cauda, and if with sub-apical polygonal reticulation then this extends over no more than one-fifth of length	10

9. All dorsal abdominal hairs arising from small dark sclerites, and also with a larger sclerite of irregular shape anterior to the base of each siphunculus H

Only some of dorsal abdominal hairs arising from sclerites, and without a large ante-siphuncular sclerite I

10. Dorsal abdomen with paired dark intersegmental sclerites. Terminal process more than 6 times longer than base of last antennal segment J
Dorsal abdomen without dark pigmentation. Terminal process less than 6 times longer than base of last antennal segment 11

11. Inner faces of antennal tubercles divergent. Siphunculus with a subapical zone of polygonal reticulation. Cauda long, one-seventh to one-fifth of length of body K

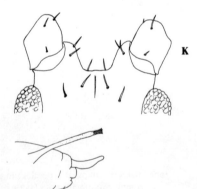

Inner faces of antennal tubercles parallel. Siphunculus without polygonal reticulation. Cauda shorter, one-tenth to one-eighth of length of body L

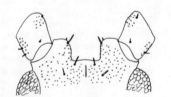

Chili

Chili
 A *Aphis gossypii*
 B *Myzus persicae*
 C *Macrosiphum euphorbiae*

1.	Antennal tubercles weakly developed. Siphunculi dark	A
	Antennal tubercles well developed. Siphunculi mainly pale	2
2.	Inner faces of antennal tubercles convergent. Siphunculi slightly clavate, without polygonal recticulation	B
	Inner faces of antennal tubercles divergent. Siphunculi not clavate, with subapical zone of polygonal reticulation	C

Chives
 A *Myzus ascalonicus* C *Neotoxoptera oliveri*
 B *Neotoxoptera formosana* D *Myzus (Sciamyzus) cymbalariae*

1.	Inner faces of antennal tubercles approximately parallel in dorsal view, or only slightly convergent apically	2
	Inner faces of antennal tubercles clearly convergent in dorsal view	3
2.	Antennal segments I and II and distal parts of femora pale. 'Stem' of siphunculus narrow (illustration), minimum diameter slightly less than middle diameter of hind tibia. (Al. with wing veins not black-bordered.)	A
	Antennal segments I and II and distal parts of femora black. Minimum diameter of 'stem' of siphunculus greater than middle diameter of hind tibia. (Al. with black-bordered wing veins.)	B

3. Siphunculus with swollen part only lightly
    imbricated, and with apical flange well
    developed. (Al. with dark-bordered wing veins,
    and with a black central dorsal abdominal
    patch.)

C

Siphunculus with swollen part scabrous (coarsely
imbricated), and with apical flange only weakly
developed. (Al. with wing veins not black-
bordered, and with separate transverse dark
bars on abdominal tergites.)

D

Chrysanthemum

- **A** *Brachycaudus cardui*
- **B** *Brachycaudus helichrysi*
- **C** *Coloradoa rufomaculata*
- **D** *Aphis gossypii*
- **E** *Pleotrichophorus chrysanthemi*
- **F** *Macrosiphoniella oblonga*
- **G** *Macrosiphoniella sanborni*
- **H** *Macrosiphoniella tanacetaria*
- **I** *Macrosiphoniella yomogifoliae*
- **J** *Aulacorthum (Neomyzus) circumflexum*
- **K** *Myzus ascalonicus*
  Myzus persicae
  Myzus ornatus
  Aulacorthum solani
  Macrosiphum euphorbiae   } see key to polyphagous Macrosiphini
  [*Macrosiphum centranthi;*
  Protrama flavescens;
  Uroleucon compositae]

---

1. Cauda helmet-shaped, no longer than its basal
    width in dorsal view 2

    Cauda tongue-shaped, longer than its basal
    width in dorsal view 3

2. Dorsal abdomen with extensive black
    pigmentation .... A
    Dorsal abdomen without pigmentation .... B

3. Antennal tubercles absent or weakly developed 4
    Antennal tubercles well developed 5

Chrysanthemum

4.	Rostrum sharply pointed. Siphunculi slightly clavate. Terminal process less than twice as long as base of last antennal segment		C
	Rostrum with blunt apex. Siphunculi tapering from base to flange. Terminal process more than twice as long as last antennal segment		D
5.	Body hairs with conspicuously knobbed or funnel-shaped apices		E
	Body hairs with blunt or pointed apices		6
6.	Siphunculi less than 1.5 times longer than cauda, with polygonal reticulation extending over distal two-fifths to two-thirds		7
	Siphunculi more than 1.5 times longer than cauda, with polygonal reticulation, if present, restricted to distal one-fifth		9
7.	Body large, elongate. Siphunculi pale at base, darker towards apex, a little longer than cauda, which is pale		F
	Body medium-sized, broadly spindle-shaped. Siphunculi and cauda wholly dark. Siphunculi a little shorter than cauda		8
8.	Length of siphunculus less than 2.5 times its basal width. Antennal segment III, proximal half of femur and middle section of tibia pale		G
	Length of siphunculus more than 3 times its basal width. Antennae and legs mainly black		9
9.	Antennal segment III wholly black, bearing 12–26 sensoria on basal half		H
	Antennal segment III paler at base, bearing 3–13 sensoria on basal half		I
10.	Dorsal abdomen with a large dark patch, usually roughly horseshoe-shaped, and dark transverse bands on thoracic tergites		J
	Body without extensive dorsal pigmentation		K

Cicely, Sweet (Garden Myrrh)

A	*Cavariella aegopodii*	D	*Aphis fabae*
B	*Cavariella konoi*	E	*Myzus ornatus*
C	*Dysaphis crataegi*	F	*Macrosiphum gei*
			[*Cavariella archangelicae*]

1. Eighth abdominal tergite with supracaudal process. Siphunculi clavate — 2

   No supracaudal process. Siphunculi not clavate — 3

2. Terminal process about equal in length to base of last antennal segment, or not more than 1.25 times longer — **A**

   Terminal process clearly rather longer than base of last antennal segment, by 1.5 times or more — **B**

3. Antennal tubercles weakly developed — 4
   Antennal tubercles well developed — 5

4. Cauda helmet-shaped, shorter than its basal width in dorsal view — **C**

   Cauda tongue-shaped, longer than its basal width in dorsal view — **D**

5. Small aphid with inner faces of antennal tubercles convergent and a dark intersegmental pattern on dorsal abdomen — **E**

   Rather large aphid with inner faces of antennal tubercles divergent, and no dorsal pigmentation. Siphunculi with a subapical zone of polygonal reticulation — **F**

Cinnamon   *Sinomegoura citricola*

Citrus

Citrus

	A	*Brachyunguis harmalae*	J	*Myzus persicae*
	B	*Brachycaudus helichrysi*	K	*Sinomegoura citricola*
	C	*Aphis craccivora*	L	*Aulacorthum magnoliae*
	D	*Toxoptera odinae*	M	*Aulacorthum solani*
	E	*Toxoptera citricidus*	N	*Macrosiphum euphorbiae*
	F	*Toxoptera aurantii*		[*Aphis arbuti*; *Aphis*
	G	*Aphis nerii*		*fabae*; *Brachycaudus*
	H	*Aphis gossypii*		*cardui*; *Pterochloroides*
	I	*Aphis citricola*		*persicae*; *Rhopalosiphum*
				maidis]

1.	Antennal tubercles weakly developed	2
	or	
	Antennal tubercles well developed	10

2.	Terminal process a little shorter than base of last antennal segment. Siphunculi much shorter than cauda	A
	Terminal process much longer than base of last antennal segment. Siphunculi shorter or longer than cauda	3

3.	Cauda helmet-shaped in dorsal view, not longer than its width at base	B
	Cauda tongue-shaped or triangular in dorsal view, longer than its basal width	4

4.	Dorsal abdomen with an extensive solid black patch	C
	Dorsal abdomen without a black central patch	5

5.	Siphunculi much shorter than cauda	D
	Siphunculi longer than or at least as long as cauda	6

6.	Terminal process more than 3.5 times longer than base of last antennal segment. Cauda with not less than 10 hairs	7
	Terminal process less than 3.5 times longer than base of last antennal segment. Cauda usually with less than 10 hairs	9

7.	Cauda with usually more than 20 hairs. Hairs on antennal segment III longer than diameter of this segment at base (illustr.). Thoracic tergites often partly sclerotized	E
	Cauda with usually less than 20 hairs. Hairs on antennal segment III often shorter than diameter of this segment at its base. Thoracic tergites usually unsclerotized	8
8.	Siphunculi less than 1.5 times longer than cauda. Stridulatory apparatus present	F
	Siphunculi more than 1.5 times longer than cauda. Stridulatory apparatus absent	G
9.	Cauda paler than siphunculi, with 4–7 hairs. Femoral hairs all rather short, less than width of femur at base	H
	Cauda dark, with 6–12 hairs. Some femoral hairs long and fine, exceeding width of femur at base	I
10.	Inner faces of antennal tubercles convergent	J
	Inner faces of antennal tubercles parallel or divergent	11
11.	Siphunculi a little shorter than the long dark cauda	K
	Siphunculi much longer than cauda	12
12.	Head, legs, and antennae mainly dark; femora basally pale but with distal one-half to three-quarters black. Siphunculi slightly swollen over distal two-thirds. Cauda with a constriction	L
	Head, legs, and antennae mainly pale. Siphunculi tapering or parallel over most of length. Cauda without a constriction	13

Clover

13. Inner faces of antennal tubercles parallel. Siphunculi without polygonal reticulation. Cauda only one-tenth to one-eighth of length of body ... **M**

Inner faces of antennal tubercles divergent. Siphunculi with a subapical zone of polygonal reticulation. Cauda longer, one-seventh to one-fifth of length of body ... **N**

Clover

A	*Therioaphis subalba*	L	*Aulacorthum solani*
B	*Therioaphis luteola*	M	*Acyrthosiphon pisum*
C	*Therioaphis trifolii*	N	*Subacyrthosiphon cryptobium*
D	*Brachycaudus helichrysi* forma *warei*		[*Aphis fabae*]
E	*Nearctaphis bakeri*		[*Aphis gossypii*]
F	*Nearctaphis crataegifoliae*		[*Aphis nasturtii*]
G	*Nearctaphis crataegifoliae occidentalis*		[*Acyrthosiphon malvae* group]
			[*Macrosiphum euphorbiae*]
H	*Aphis coronillae*		[*Myzus persicae*]
I	*Aphis craccivora*		[*Myzus (Sciamyzus) cymbalariae*]
J	*Myzus ornatus*		[*Prociphilus erigeronensis*]
K	*Sitobion akebiae*		[*Smynthurodes betae*]

1. Cauda with a constriction and a knob-like apex in dorsal view, anal plate bilobed. Body hairs long and capitate, arising from tubercles ... 2

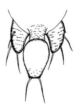

Cauda helmet-shaped, triangular, or tongue-shaped in dorsal view, anal plate entire. Body hairs mainly short, not capitate ... 4

2. Abdominal tergites 1–5 each usually with only 4 long capitate hairs arising from large tubercles (1 pair of marginal hairs and 1 pair of spinal hairs per segment) ... **A**

Abdominal tergites 1–5 each with usually at least 6 long capitate hairs arising from large tubercles, all of similar size ... 3

3. Capitate hairs on abdominal tergites 1–5 rather regularly 6 per segment and arranged so that they form 6 clear longitudinal rows (2 marginal rows and 4 spinal). Tuberculate bases of hairs only faintly pigmented, dusky .... B

Capitate hairs on abdominal tergites 1–5 numbering 7–10, mostly 8, per segment and less regularly arranged so that longitudinal rows are not evident. Tuberculate bases of hairs usually deeply pigmented C

4. Siphunculi short, truncated cones. Cauda short and broad 5
   Siphunculi tubular, much longer than the cauda which is tongue-shaped, much longer than its basal width in dorsal view 8

5. Siphunculi quite smooth. Abdominal spiracles open, large, and rounded. Tibiae pale. Dorsal abdomen without any dark sclerites .... D

   Siphunculi strongly imbricated with rows of spinules. Abdominal spiracles small, kidney-shaped and partially occluded. Tibiae pigmented at least distally. Dorsal abdomen often with small dark hair-bearing sclerites 6

6. Last segment of rostrum similar in length to second segment of hind tarsus, or a little shorter. Head and siphunculi pale. Abdominal tergites 1–5 with scattered mainly small hair-bearing sclerites .... E
   Last segment of rostrum clearly longer than second segment of hind tarsus. Either head and siphunculi dark, *or* abdominal tergites 1–5 are without hair-bearing sclerites 7

7. Head and siphunculi pale. Abdominal tergites 1–5 without sclerites, and little sclerotization of tergites 6 and 7 .... F
   Head and siphunculi dark. Abdominal tergites 1–5 with some small hair-bearing sclerites, especially marginally, and broad transverse sclerotic bars on 6 and 7 .... G

8. Antennal tubercles weakly developed. Dorsal abdomen with extensive black patch 9

   Antennal tubercles well developed. Dorsal abdomen unsclerotized 10

Cloves 75

9.	Each abdominal segment with a pair of large, flat transparent marginal tubercles. Cauda usually with more than 7 hairs	**H**
	Usually only abdominal segments 1 and 7 with small marginal tubercles. Cauda with usually less than 7 hairs	**I**

10.	Inner faces of antennal tubercles convergent. Dorsal abdomen with an intersegmental pattern of dark ornamentation	**J**
	Inner faces of antennal tubercles parallel or divergent. No dark intersegmental ornamentation	11

11.	Siphunculi wholly black, with a subapical zone of polygonal reticulation extending over distal one-fifth	**K**
	Siphunculi pale or dusky, never very dark except at extreme apices, and without any polygonal reticulation	12

12.	Inner faces of antennal tubercles parallel. Siphunculi more than twice as long as cauda	**L**
	Inner faces of antennal tubercles divergent. Siphunculi less than twice as long as cauda	13

13.	Antennal tubercles smooth. Siphunculi and cauda very long; 7–10 and 4.5–6 times longer respectively than the last rostral segment. Dorsal abdomen completely without pigmentation	**M**
	Antennal tubercles scabrous. Siphunculi and cauda about 4 and 2 times longer respectively, than the last rostral segment. Small dark sclerites present anterior and posterior to bases of siphunculi, and some sclerotization of abdominal tergites 7 and 8	**N**

Cloves	no aphids recorded

Cobnut	see **Hazel Nut**

Coco, Coco-yam	*Aphis gossypii* *Pentalonia nigronervosa* [*Patchiella reaumuri*]

Cocoa

| | A | Cervaphis rappardi/ schouteniae | C | Toxoptera aurantii |
| | B | Macrosiphum martorelli | D | Aphis gossypii [Aphis citricola] |

1. Body with very long, branched, hair-bearing lateral processes ... **A**

   Body without long hair-bearing lateral processes ... 2

2. Body rather elongate, spindle-shaped. Antennal tubercles well developed. Siphunculi long and dark except at bases, with a subapical zone of polygonal reticulation on about the distal one-fifth ... **B**

   Body broadly oval. Antennal tubercles weakly developed. Siphunculi dark including bases, without polygonal reticulation ... 3

3. Cauda dark with more than 10 hairs. Terminal process more than 3.5 times longer than base of last antennal segment. Stridulatory apparatus present ... **C**

   Cauda paler than siphunculi, with 4–7 hairs. Terminal process less than 3.5 times base of last antennal segment. Stridulatory apparatus absent ... **D**

Coconut

see key to aphids on **Palms**)

Coffee

Toxoptera aurantii

Cola

no aphids recorded

Comfrey

| | A | Macrosiphum euphorbiae | C | Aphis fabae |
| | B | Aphis symphyti | | [Brachycaudus cardui] |

Coriander

1. Body elongate oval. Antennal tubercles well developed. Siphunculi pale basically, often darkening towards apex where there is a subapical zone of polygonal reticulation ... **A**

   Body broadly oval. Antennal tubercles weakly developed. Siphunculi wholly dark ... 2

2. Cauda pale ... **B**
   Cauda black ... **C**

Coriander

A *Hyadaphis coriandri*
B *Semiaphis heraclei*

1. Siphunculi 0.7–1.0 times as long as cauda, pale except at apex, swollen in middle, with a distinct flange ... **A**

   Siphunculi about 0.5 times as long as cauda or less, darker than body, tapering from base to apex; flangeless, with aperture slanted towards mid-line ... **B**

Cotton

A *Smynthurodes betae*
B *Aphis maidi-radicis*
C *Rhopalosiphum rufiabdominalis*
D { *Aphis craccivora* / *Aphis fabae* / *Aphis gossypii* } see key to polyphagous *Aphis* spp.
E *Myzus persicae*
F *Acyrthosiphon gossypii*
G *Macrosiphum euphorbiae*
[*Aphis alvata* Zhang 1981]
[*Brachyunguis plotnikovi* Nevsky 1928]

1. Terminal process much shorter than base of last antennal segment. Siphunculi absent ... **A**

   Terminal process longer than base of last antennal segment. Siphunculi present ... 2

78   *Aphids on the world's crops*

2. Cauda bluntly conical, about as long as its
   basal width in dorsal view   B

   Cauda tongue-shaped, longer than its basal
   width in dorsal view   3

3. Antennae 5-segmented, bearing long, fine hairs,
   many of them over twice as long as diameter
   of segment III   C

   Antennae usually 6-segmented, with much
   shorter hairs   4

4. Antennal tubercles weakly developed   D

   Antennal tubercles well developed   5

5. Inner faces of antennal tubercles convergent
   in dorsal view. Siphunculi slightly clavate   E

   Inner faces of antennal tubercles divergent.
   Siphunculi tapering or cylindrical   6

6. Siphunculi very long and thin, about 3 times
   longer than cauda, without a subapical zone
   of polygonal reticulation   F

   Siphunculi about twice as long as cauda, with
   a subapical zone of polygonal reticulation   G

Cow-pea   A   *Aphis craccivora*   C   *Myzus ornatus*
                B   *Aphis gossypii*   [*Aulacorthum solani*]

1. Dorsal abdomen with an extensive solid black
   patch centred on tergites 4–5   A
   Dorsal abdomen without a solid black patch   2

Cranberry

2. Antennal tubercles weakly developed. Abdomen without dorsal abdominal pigmentation .... B

Antennal tubercles well developed, with inner faces convergent. Dorsal abdomen with an intersegmental pattern of dark ornamentation C

Cranberry see also **Vaccinium**
Fimbriaphis scammelli   Smynthurodes betae
Illinoia species

Cucumber

A { *Aphis fabae*, *Aphis citricola*, *Aphis nasturtii*, *Aphis gossypii* } see key to polyphagous *Aphis* spp.

B { *Myzus persicae*, *Aulacorthum solani*, *Macrosiphum euphorbiae* } see key to polyphagous Macrosiphini

[*Aphis craccivora*]
[*Brachycaudus helichrysi*]

1. Antennal tubercles weakly developed .... A
   Antennal tubercles well developed .... B

Currant, Black

A	*Eriosoma (Schizoneura) ulmi*	I	*Nasonovia ribis-nigri*
B	*Aphis ribiensis*	J	*Nasonovia (Kakimia) houghtonensis similis*
C	*Aphis triglochinis*	K	*Nasonovia (Kakimia) cynosbati*
D	*Aphis neomexicana*		
E	*Aphis schneideri*	L	*Rhopalosiphoninus ribesinus*
F	*Aphis varians*		
G	*Cryptomyzus ribis*	M	*Hyperomyzus lactucae*
H	*Cryptomyzus galeopsidis*		[*Hyperomyzus (Neonasonovia) ribiellus*]

1. Terminal process much shorter than base of last antennal segment. (On roots) .... A

   Terminal process longer than base of last antennal segment. (On aerial parts) 2

2. Antennal tubercles weakly developed 3

   Antennal tubercles well developed 7

3.	Hairs on antennal segment III all shorter than diameter of segment at base. Last rostral segment with only 2 accessory hairs	4
	At least some of hairs on antennal segment III longer than basal diameter of segment. Last rostral segment with more than 3 accessory hairs	5

4.	Siphunculi shorter than terminal process of antenna, and also shorter than cauda; 1.5–2.0 times longer than their basal width. (Al. with antennal segments III–V swollen with numerous sensoria)	B
	Siphunculi longer than terminal process of antenna, and usually longer than cauda; 2.0–3.5 times longer than their basal width. (Al. with less sensoria on antennal segments III and IV and none on V, so that these segments appear no thicker than base of VI)	C

5.	Hairs on antennae rather numerous, long, fine, and wavy, the longest 2–4 times the basal diameter of antennal segment III	6
	Hairs on antennae rather sparse, erect, and straight, the longest less than twice the basal diameter of antennal segment III	D

6.	Usually large marginal tubercles on abdominal segments 2–5 as well as 1 and 7	E
	Marginal tubercles usually absent from abdominal segments 2–5	F

7.	Dorsal body hairs thick and long with knobbed apices, and many with tuberculate bases	8
	Dorsal body hairs pointed or with blunt apices, never with tuberculate bases	9

8.	Knobbed hairs on antennal segment III much shorter than those on head and antennal segments I and II. Siphunculi 3 or more times longer than cauda	G
	Knobbed hairs on antennal segment III of similar length to those on head and antennal segments I and II. Siphunculi less than 2.5 times longer than cauda	H

Currant, Grape

9.	Siphunculi tapering or cylindrical	10
	Siphunculi considerably swollen (clavate)	12

10.	Abdominal spiracles open, oval pores. Cauda with 7 hairs	I
	Abdominal spiracles partially covered by cowl-like opercula, so that they appear to be kidney- or crescent-shaped. Cauda usually with 5 hairs	11

11.	Siphunculi with transverse rows of small spinules over whole length, and less than 1.3 times longer than cauda	J
	Siphunculi usually somewhat imbricated or wrinkled but with rows of spinules, if present, only on distal half; more than 1.3 times longer than cauda	K

12.	Cauda dark, equilateral triangular in dorsal view	L
	Cauda pale, tongue-shaped, much longer than its basal width	M

Currant, Grape	see **Grape Vine**	

Currant, Red

	A	*Eriosoma (Schizoneura) ulmi*	G	*Rhopalosiphoninus ribesinus*
	B	*Aphis triglochinis*	H	*Hyperomyzus lactucae*
	C	*Aphis schneideri*	I	*Hyperomyzus (Hyperomyzella) rhinanthi* [*Nasonovia ribisnigri*]
	D	*Cryptomyzus ribis*		
	E	*Cryptomyzus galeopsidis*		
	F	*Nasonovia (Kakimia) cynosbati*		

1.	Terminal process much shorter than base of last antennal segment. (On roots)	A
	Terminal process longer than base of last antennal segment. (On aerial parts)	2
2.	Antennal tubercles weakly developed	3
	Antennal tubercles well developed	4

3. Antennal hairs shorter than basal diameter of
   antennal segment III .... B

   Antennal hairs long, fine, and wavy, the
   longest 2–4 times the basal diameter on
   antennal segment III C

4. Dorsal body hairs thick and long with knobbed
   apices, many with tuberculate bases 5
   Dorsal body hairs pointed or with blunt apices,
   never on tuberculate bases 6

5. Knobbed hairs on antennal segment III much
   shorter than those on head and antennal
   segments I and II. Siphunculi 3 or more
   times longer than cauda .... D

   Knobbed hairs on antennal segment III of
   similar length to those on head and antennal
   segments I and II. Siphunculi less than 2.5
   times longer than cauda .... E

6. Siphunculi not swollen. Spiracles partially
   covered by cowl-like opercula, so that they
   appear crescent-shaped .... F
   Siphunculi considerably swollen distally 7

7. Cauda equilateral triangular in dorsal view .... G
   Cauda tongue-shaped, much longer than its
   basal width 8

8. Dorsal abdomen unpigmented. Siphunculi
   dusky, cauda pale .... H
   Dorsal abdomen with a large black patch.
   Siphunculi and cauda dark I

Custard Apple

A	*Greenidea (Trichosiphon)* *anonae*	C	*Aphis citricola*
B	*Aphis gossypii*	D	*Toxoptera aurantii*

Daffodil

1.	Siphunculi with numerous long hairs	A
	Siphunculi without hairs	2
2.	Cauda paler than siphunculi	B
	Cauda dark	3
3.	Terminal process less than 3 times longer than base of last antennal segment. Stridulatory apparatus absent	C
	Terminal process more than 3.5 times longer than base of last antennal segment. Stridulatory apparatus present	D

Daffodil   *Myzus persicae* ⎤ see key to
            Aulacorthum solani ⎬ polyphagous Macrosiphini
            Macrosiphum euphorbiae ⎦

Dahlia
- A   *Aphis fabae* ⎤ see key to
    Aphis nasturtii ⎬ polyphagous *Aphis* spp.
    Aphis gossypii ⎦
- B   *Uroleucon (Uromelan) compositae*
- C   *Uroleucon ambrosiae*
- D   *Aulacorthum (Neomyzus) circumflexum*
- E   *Myzus persicae*
- F   *Aulacorthum solani*
    [*Brachycaudus helichrysi*]

1.	Antennal tubercles weakly developed	A
	Antennal tubercles well developed	2

2. Siphunculi black, with a subapical zone of polygonal reticulation   3

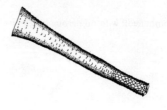

Siphunculi pale, or dark only at extreme apices, without polygonal reticulation   4

3. Cauda black   **B**
    Cauda pale   **C**

4. Dorsal abdomen with a large dark area, usually roughly horseshoe-shaped, and dark transverse bands on thoracic tergites   **D**
    Dorsal abdomen unpigmented   5

5. Inner faces of antennal tubercles convergent. Siphunculi slightly clavate   **E**

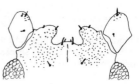

Inner faces of antennal tubercles parallel. Siphunculi tapering from base to flange   **F**

Damson
    A   *Hyalopterus pruni*
    B   *Phorodon humuli*
    C   *Brachycaudus helichrysi*
    D   *Brachycaudus (Appelia) prunicola*
    E   *Brachycaudus cardui*
    F   *Brachycaudus persicae*

1. Siphunculi very small and thin, shorter than cauda   **A**

Siphunculi longer than cauda   2

2. Cauda much longer than broad. Antennal tubercles each with a long, finger-like process   **B**

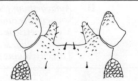

Cauda short and broad, rounded or helmet-shaped in dorsal view. Antennal tubercles weakly developed, without processes   3

Date

3. Siphunculi short, less than twice as long as cauda ... 4
   Siphunculi more than twice as long as cauda ... 5

4. Dorsal abdomen pale or dusky, without extensive dark pigmentation ... **C**
   Dorsal abdomen almost wholly black ... **D**

5. Last segment of rostrum more than 0.18 mm long. Dorsal abdomen variably pigmented but often quite pale. Siphunculi usually pale or dusky ... **E**
   Last segment of rostrum less than 0.175 mm long. Dorsal abdomen always extensively pigmented black. Siphunculi black ... **F**

Date

Cerataphis variabilis
Schizaphis rotundiventris
(or try key to aphids on **Palms**)

Dill

A	*Cavariella aegopodii*	**D**	*Hyadaphis coriandri*
B	*Dysaphis foeniculus*	**E**	*Myzus persicae*
C	*Semiaphis heraclei*	**F**	*Myzus ornatus*

1. Eighth abdominal tergite with posteriorly projecting supracaudal process ... **A**

   No supracaudal process ... 2

2. Cauda helmet-shaped in dorsal view, shorter than its basal width ... **B**

   Cauda tongue-shaped, much longer than its basal width ... 3

3. Siphunculi shorter than cauda. Antennal tubercles undeveloped ... 4
   Siphunculi much longer than cauda. Antennal tubercles well developed, their inner faces convergent ... 5

4. Siphunculi very small and flangeless, only about half as long as cauda or less, with aperture slanted towards the mid-line ... **C**

   Siphunculi two-thirds or more as long as cauda, with a flange ... **D**

86 Aphids on the world's crops

5. Siphunculi slightly clavate. Dorsal abdomen
   without dark ornamentation E

Siphunculi tapering from base to flange, slightly
curved. Dorsal abdomen with an intersegmental pattern of dark ornamentation F

Durian no aphids recorded

Egg Fruit (Egg Plant, Brinjal)

A { *Aphis fabae* group / *Aphis gossypii* } use key to polyphagous *Aphis* spp.

B { *Myzus persicae* / *Aulacorthum solani* / *Macrosiphum euphorbiae* } use key to polyphagous Macrosiphini

1. Antennal tubercles weakly developed A
   Antennal tubercles well developed B

Fennel
A *Cavariella aegopodii*   E *Hyadaphis coriandri*
B *Dysaphis apiifolia*   F *Aphis fabae*
C *Dysaphis foeniculus*   G *Myzus ascalonicus*
D *Hyadaphis foeniculi*   H *Myzus persicae*

1. Eighth abdominal tergite with posteriorly projecting supracaudal process A

   No supracaudal process 2

2. Cauda helmet-shaped in dorsal view, shorter
   than its basal width 3

   Cauda tongue-shaped, much longer than its
   basal width 4

3.	Siphunculi 2.5–3.5 times longer than their width at base. Hairs on frons very short and blunt	**B**
	Siphunculi short and broad-based, rarely more than twice their basal width. Hairs on frons long and pointed	**C**
4.	Siphunculi similar in length to cauda	5
	Siphunculi much longer than cauda	6
5.	Siphunculi about equal in length to cauda, or a little longer, and clavate	**D**
	Siphunculi a little shorter than cauda, swollen, but without any proximal constriction	**E**
6.	Cauda black with more than 10 hairs. Antennal tubercles weakly developed	**F**
	Cauda pale with usually 6 hairs. Antennal tubercles well developed	7
7.	Siphunculi rather small, shorter than antennal segment III, with minimum diameter of basal half less than diameter of middle part of hind tibia	**G**
	Siphunculi larger, a little longer than antennal segment III, and with minimum diameter of basal half greater than diameter of middle part of hind tibia	**H**

Fig
A *Reticulaphis fici* **C** *Aphis gossypii*
B *Greenidea ficicola* **D** *Toxoptera aurantii* [*Sitobion africanum*]

1.	Body flattened, oval, bearing superficial resemblance to whitefly pupa, with reduced antennae and legs hidden	**A**
	Body not aleyrodiform	2

2. Siphunculi with numerous long hairs. Cauda broadly rounded with an apical papilla   B

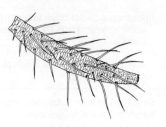

   Siphunculi without hairs. Cauda tongue-shaped   3

3. Cauda pale. No stridulatory apparatus   C
   Cauda black. Stridulatory apparatus present   D

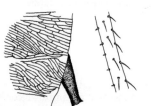

Filbert	see **Hazel Nut**

Flax, Linseed	*Aphis craccivora* *Aphis fabae* [*Acyrthosiphon ilka*] [*Acyrthosiphon mordvilkoi* Nevsky 1928] [*Linaphis lini* Zhang in Zhang & Zhong 1981]

Galangal	*Pentalonia nigronervosa*

Garlic	*Neotoxoptera formosana* [*Myzus ascalonicus*]

Ginger	[*Astegopteryx muiri*] [*Pentalonia nigronervosa*]

Gladiolus	A	*Dysaphis tulipae*	H	*Sitobion fragariae*
	B	*Aphis fabae*	I	*Sitobion akebiae*
	C	*Aphis gossypii*		[*Myzus (Sciamyzus) cymbalariae*]
	D	*Aulacorthum (Neomyzus) circumflexum*		[*Rhopalosiphoninus staphyleae tulipaellus*]
	E	*Myzus persicae*		
	F	*Aulacorthum solani*		[*Rhopalosiphum maidis*]
	G	*Macrosiphum euphorbiae*		[*Smynthurodes betae*]

Gladiolus

1.	Cauda short, helmet-shaped, about as long as its basal width in dorsal view		A
	Cauda elongate, tongue-shaped, much longer than its basal width		2

2.	Antennal tubercles weakly developed		3
	Antennal tubercles moderately to well developed, exceeding height of medial part of frons		4

3.	Cauda black with more than 10 hairs	B
	Cauda pale with 6–7 hairs	C

4.	Dorsal abdomen with a large dark patch, roughly horseshoe-shaped, and dark patches on thoracic tergites		D
	Dorsal abdomen either without pigmentation or with small segmental markings		5

5.	Inner faces of antennal tubercles convergent. Siphunculi slightly clavate		E
	Inner faces of antennal tubercles parallel or divergent. Siphunculi not clavate		6

6.	Inner faces of antennal tubercles parallel. Siphunculi dark only at extreme apices, and without polygonal reticulation		F
	Inner faces of antennal tubercles divergent. Siphunculi with a subapical zone of polygonal reticulation; either uniformly dark or pale with some distal darkening		7

7. Hairs on antennal segment III almost as long as basal diameter of the segment. Siphunculi mainly pale, sometimes darkened distally ... G

Hairs on antennal segment III short and inconspicuous, less than half as long as the basal diameter of the segment. Siphunculi uniformly pigmented ... 8

8. Proximal segments of antennae paler than distal segments. Siphunculi at least twice as long as cauda ... H
Antennae uniformly dark. Siphunculi less than twice as long as cauda ... I

Gooseberry

A *Eriosoma (Schizoneura) grossulariae*
B *Aphis grossulariae*
C *Aphis neomexicana*
D *Aphis varians*
E *Cryptomyzus ribis*
F *Cryptomyzus galeopsidis*
G *Hyperomyzus pallidus*
H *Nasonovia ribis-nigri*
I *Nasonovia (Kakimia) houghtonensis houghtonensis*
J *Nasonovia (Kakimia) cynosbati*
K *Nasonovia (Kakimia) brachycyclica*
[*Aphis schneideri*]
[*Hyperomyzus (Neonasonovia) picridis*]
[*Hyperomyzus (Neonasonovia) ribiellus*]

1. Terminal process much shorter than base of last antennal segment. (On roots) ... A

Terminal process longer than base of last antennal segment. (On aerial parts) ... 2

2. Antennal tubercles weakly developed ... 3

Antennal tubercles well developed ... 4

3. Lateral tubercles present and usually conspicuous on abdominal segments 2–6 as well as 1 and 7 ... B
Lateral tubercles usually absent from abdominal segments 2–6 ... 4

Gooseberry

4.	Siphunculi and cauda both pale. Hairs on antennae rather sparse, erect, and straight, the longest less than twice the basal diameter of antennal segment III	C
	Siphunculi pale, cauda dark. Hairs on antennae rather numerous, long, fine, and wavy, the longest 2–4 times the basal diameter of antennal segment III	D

5.	Dorsal body hairs thick and long, with knobbed apices, and many with tuberculate bases	6
	Dorsal body hairs pointed or with blunt apices, never on tuberculate bases	7

6.	Knobbed hairs on antennal segment III much shorter than those on head. Siphunculi 3 or more times longer than cauda	E
	Knobbed hairs on antennal segment III of similar length to those on head. Siphunculi less than 2.5 times longer than cauda	F

7.	Siphunculi considerably swollen (clavate)	G
	Siphunculi tapering from base to flange, or cylindrical on distal half, not swollen	8

8.	Siphunculi smooth, dark at apex with well-developed flange. Cauda usually with 7 hairs. Abdominal spiracles oval, open pores	H
	Siphunculi rough with rows of spinules or imbrication at least distally. Cauda usually with 5 hairs. Abdominal spiracles partially covered by cowl-like opercula so that they appear kidney- or crescent-shaped	9

9.	Siphunculi with transverse rows of spinules over entire length	I
	Siphunculi smooth or slightly imbricated but with transverse rows of spinules, if present, confined to distal half	10
10.	Dorsal abdomen pale or with rather indistinct sclerites	J
	Dorsal abdomen with a pattern of dark pigmentation	K

Gram (Black, Golden, Green)

Aphis craccivora
Acyrthosiphon gossypii

Grape Vine

A *Viteus vitifoliae*
B *Aploneura ampelinus*
C *Geoica lucifuga*
D *Aphis gossypii*
E *Aphis illinoisensis*

F *Aphis fabae*
[*Aphis craccivora*]
[*Aphis vitis* Scopoli 1763]
[*Aulacorthum solani*]
[*Macrosiphum euphorbiae*]
[*Prociphilus oleae*]

1.	Antennae 3-segmented	A
	Antennae 5- or 6-segmented	2
2.	Terminal process much shorter than base of last antennal segment. (On roots)	3
	Terminal process much longer than base of last antennal segment. (On aerial parts)	4
3.	Body spindle-shaped with sparse short hairs	B
	Body rounded, with numerous long hairs, including 1 or 2 especially long lateral hairs on each abdominal segment	C
4.	Cauda paler than siphunculi	D
	Cauda dark	5
5.	Hind tibia wholly black	E
	Hind tibia mainly pale, only black distally	F

Grasses

Many aphids occur on temperate and tropical grass crops. A key is given to the aphids on grasses of temperate pastures; this is intended mainly for aphids found colonizing species of *Poa* and *Festuca*. The aphid fauna of *Lolium* species is somewhat distinct, so there is a separate list and key for Ryegrass (q.v.). For tropical grasses, the large fauna on Bermuda grass (*Cynodon dactylon*) is listed and keyed, and there are individual lists, and keys where necessary, to the smaller number of aphid species likely to be found on the other tropical grasses of principal economic importance. However, few grass-feeding aphids are monophagous and if one key fails to give a satisfactory answer it may be worthwhile to try another one.

Grass, Bermuda

A	*Sipha (Rungsia) maydis*	M	*Tetraneura akinire*
B	*Schizaphis graminum*	N	*Tetraneura nigriabdominalis*
C	*Hysteroneura setariae*		
D	*Rhopalosiphum rufiabdominalis*	O	*Tetraneura radicicola/ yezoensis*
E	*Rhopalosiphum maidis*	P	*Tetraneura africana*
F	*Rhopalosiphum padi*	Q	*Asiphonella species*
G	*Sitobion graminis*	R	*Aploneura lentisci*
H	*Sitobion lambersi*	S	*Smynthurodes betae*
I	*Sitobion chanikiwiti*	T	*Geoica lucifuga*
J	*Sitobion avenae*	U	*Forda formicaria*
K	*Sitobion africanum*		[*Aploneura ampelina*]
L	*Sitobion miscanthi*		[*Forda hirsuta*]

1.	Terminal process longer than base of last antennal segment. (On aerial parts, except for D)	2
	Terminal process shorter than base of last antennal segment. (On roots)	13
2.	Dorsal surface heavily sclerotized with thick, spine-like hairs. Cauda broadly rounded	A
	Dorsum mainly unsclerotized with hairs fine if long. Cauda tongue-shaped	3
3.	Antennal tubercles low or undeveloped, not higher than medial part of frons. Siphunculi much shorter than the distance between their bases, without any polygonal reticulation	4
	Antennal tubercles sometimes rather low but always higher than medial part of frons, their inner faces divergent in dorsal view. Siphunculi about as long as the distance between their bases, with a subapical zone of polygonal reticulation	8
4.	Siphunculi mainly pale, except at extreme apices which are usually dark	B
	Siphunculi darker than body (if rather pale there is still a clear line of demarcation between body and bases of siphunculi)	5
5.	Cauda long and pale	C
	Cauda rather short and darker than body	6
6.	Antennae 5-segmented, bearing long fine hairs, many of which are over twice as long as diameter of antennal segment III. (On roots)	D
	Antennae 6-segmented, with shorter hairs. (On aerial parts)	7

7.	Body rather elongate. Terminal process less than 2.5 times longer than base of last antennal segment	E
	Body ovate. Terminal process more than 3 times longer than base of last antennal segment	F

8.	Cauda and siphunculi both black	G
	Cauda pale, or if dusky then much paler than siphunculi	9

9.	Terminal process more than 6 times longer than base of last antennal segment	10
	Terminal process less than 6 times longer than base of last antennal segment	11

10.	Cauda with only 2 long hairs, plus a few very short apical ones. Siphunculi darker than body, even at base	H
	Cauda with more than 2 long hairs. Siphunculi pale at base, dusky at apex	I

11.	Siphunculi less than 1.5 times longer than cauda. Second segment of hind tarsus more than 1.4 times longer than last segment of rostrum	J
	Siphunculi more than 1.5 times longer than cauda. Second segment of hind tarsus with claw less than 1.4 times longer than last rostral segment	12

12.	Dorsal abdominal pigmentation usually consisting of segmentally arranged and well-defined transverse black bars. Short haired; longest hairs on antennal segment III less than 20 μm long	K
	Dorsal abdominal pigmentation ill-defined and not segmentally divided. Longer haired; longest hairs on antennal segment III more than 20 μm long	L

13.	Siphunculi present as small, pigmented cones	14
	Siphunculi completely absent	17
14.	Last segment of rostrum short, less than 0.15 mm long and less than 1.8 times longer than hind tarsus	15
	Last segment of rostrum more than 0.15 mm long and more than 1.8 times longer than hind tarsus	16
15.	Eighth abdominal tergite with a pair of long stout hairs and a group of small fine hairs placed ventro-lateral of them on each side. Cauda usually with a few small hairs in addition to 2 long ones	M
	Eighth abdominal tergite with a pair of long stout hairs only, and cauda also usually with only 2 rather long hairs	N
16.	Body and antennae with numerous hairs of varying size and length	O
	Hairs on body and antennae all short, sparse, and inconspicuous	P
17.	Antennae 6-segmented with terminal process about one-half as long as base of last segment	Q
	Antennae usually 5-segmented, terminal process less than one-third of length of base of last segment	18
18.	Body spindle-shaped, about twice as long as broad. Antennae small, less than one-sixth of body length	R
	Body rather globose, less than 1.5 times longer than broad. Antennae more than one-fifth of body length	19
19.	Antennal segment II about twice as long as segment I and nearly as long as III	S
	Antennal segment II only a little longer than segment I and much shorter than III	20

20. Anus and anal plate displaced dorsally. Antennae less than one-quarter of body length, with segment III shorter than IV and V combined .... T

Anus and anal plate not displaced dorsally. Antennae more than one-third of body length, with segment III longer than IV and V combined .... U

Grass, Elephant
Melanaphis sacchari
Rhopalosiphum maidis
[*Sitobion chanikiwiti*]

Grass, Guinea
A *Sipha flava*
B *Sitobion lambersi*
C *Rhopalosiphum maidis*
D *Hysteroneura setariae*

1. Dorsal body hairs long, thick, and spine-like. Siphunculi very short truncated cones. Cauda knobbed .... A

Dorsal body hairs short. Siphunculi tubular. Cauda tongue-shaped 2

2. Antennal tubercles developed, their inner faces broadly divergent. Siphunculi with a subapical zone of polygonal reticulation .... B

Antennal tubercles weakly developed. Siphunculi without polygonal reticulation 3

3. Body rather elongate. Cauda dark C
   Body broadly ovate. Cauda pale .... D

Grass, Johnson
A *Anoecia* species
B *Tetraneura ulmi*
C *Tetraneura africana*
D *Rhopalosiphum maidis*
E *Melanaphis sacchari*
F *Schizaphis graminum*
G *Sitobion avenae*
[*Sipha (Rungsia) maydis*]

Grass, Kentucky Blue

1. Terminal process much shorter than base of last
   antennal segment (on roots) 2
   Terminal process much longer than base of last
   antennal segment (on aerial parts) 4

2. Antennae one-third or more as long as body.
   Tarsi 2-segmented .... A
   Antennae very short, one-fifth or less of
   body length. Tarsi 1-segmented 3

3. Wax pore-plates (ventro-laterally on each
   segment) each consisting of a ring of cells
   around an undivided central area. Last
   segment of rostrum 0.15–0.19 mm long, less
   than 2.4 times longer than hind tarsus .... B

   Wax pore-plates with central area divided.
   Last segment of rostrum 0.18–0.23 mm long,
   more than 2.3 times longer than hind tarsus C

4. Cauda dark. Terminal process less than 2.5
   times longer than base of last antennal
   segment .... D
   Cauda pale. Terminal process more than 3
   times longer than base of last antennal
   segment 5

5. Siphunculi shorter than cauda .... E
   Siphunculi longer than cauda 6

6. Siphunculi pale except at apices, without any
   polygonal reticulation .... F

   Siphunculi wholly dark, with a subapical zone
   of polygonal reticulation .... G

Grass, Kentucky Blue—
See key to aphids of temperate pasture grasses

Grass, Kikuyu
Tetraneura nigriabdominalis
Tetraneura yezoensis
[*Micromyzus (Kugegania) ageni*]
} for separation see p. 95, couplet 14

Grass, Rhodes   *Rhopalosiphum maidis*
                     Sitobion graminis

Grasses of Temperate Pastures (for Rye-Grass see separate key)

A	*Atheroides serrulatus*	U	*Baizongia pistaciae*
B	*Chaetosiphella berlesei*	V	*Geoica utricularia*
C	*Sipha Rungsia maydis*	W	*Geoica setulosa*
D	*Sipha glyceriae*	X	*Aploneura lentisci*
E	*Sipha flava*	Y	*Prociphilus erigeronensis*
F	*Cryptaphis poae*	Z	*Paracletus cimiciformis*
G	*Jacksonia papillata*	AA	*Forda formicaria*
H	*Sitobion fragariae*	BB	*Forda (Pentaphis)*
I	*Sitobion miscanthi*		*marginata* group
J	*Sitobion avenae*	CC	*Tetraneura ulmi*
K	*Metopolophium dirhodum*	DD	*Anoecia* species
L	*Metopolophium tenerum*		[*Aphis sambuci; Diuraphis*
M	*Metopolophium festucae*		*muehlei; Glabromyzus*
N	*Metopolophium sabihae*		*howardii; Metopolophium*
O	*Rhopalomyzus poae*		*festucae cerealium; Myzus*
P	*Utamphorophora*		*ascalonicus; Myzus*
	humboldti		*persicae; Rhopalosiphum*
Q	*Schizaphis graminum*		*insertum; Rhopalosiphum*
R	*Hysteroneura setariae*		*padiformis; Schizaphis*
S	*Rhopalosiphum maidis*		*borealis* Tambs-Lyche
T	*Rhopalosiphum padi*		1959; *Schizaphis*
			hypersiphonata]

1.	Terminal process either similar in length to, or longer than, base of last antennal segment. (On aerial parts)	2
	Terminal process much shorter (less than one-half) than base of last antennal segment. (On roots)	21

2.	Dorsal body hairs partly or mainly long and spine-like. Siphunculi either pores or very short truncated cones	3
	Dorsal body hairs if long, then fine and pointed, or with knobbed apices (but more often short and inconspicuous). Siphunculi tubular, much longer than their basal width	7

3.	Body very elongate (more than 3 times longer than broad), with long spine-like hairs only on head and posterior abdomen. Siphunculi merely pores in the heavily sclerotized dorsum	A
	Body ovate (less than twice as long as broad), with spine-like hairs on all segments. Siphunculi short, truncated cones	4

4.	Last segment of rostrum long and pointed, longer than second segment of hind tarsus	B
	Last segment of rostrum short and blunt, shorter than second segment of hind tarsus	5

Grasses of temperate pastures 99

5.	Cauda broadly rounded	C
	Cauda knobbed	6

6.	Dorsal cuticle wholly sclerotic with many small tubercles and denticles between the hairs	D
	Dorsal cuticle smoky with dark intersegmental markings, not strongly sclerotized nor adorned with tubercles or denticles	E

7.	Antennal tubercles moderately to well developed, exceeding height of middle of frons in dorsal view	8
	Antennal tubercles low or undeveloped, not higher than middle of frons in dorsal view	18

8.	Dorsal body hairs long and capitate. Dorsal abdomen wholly pigmented	F
	Body hairs rather short, not capitate. Dorsal abdomen pale or with dark markings	9

9.	Siphunculi of characteristic shape; narrowest in middle, with no flange, and aperture slanted towards mid-line. Cauda only a little longer than broad, with a basal constriction. Cuticle very rough with many ridges, tubercles, and denticles	G
	Siphunculi not so shaped. Cauda tongue-shaped, without a basal constriction. Cuticle fairly smooth	10

10.	Siphunculi tapering from base to flange, or cylindrical on distal half	11
	Siphunculi clearly clavate	17

11.	Siphuncili darker than body, with a subapical zone of polygonal reticulation	12
	Siphunculi pale, without polygonal reticulation	14

12. Proximal segments of antennae paler than distal segments. Siphunculi about twice as long as cauda or longer   H

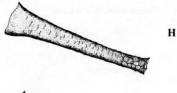

   Antennae uniformly dark. Siphunculi less than twice as long as cauda   13

13. Siphunculi more than 1.4 times longer than cauda. Second segment of hind tarsus less than 1.3 times longer than last segment of rostrum   I
    Siphunculi less than 1.4 times longer than cauda. Second segment of hind tarsus more than 1.25 times longer than last segment of rostrum   J

14. Each antennal segment pale at base and dusky at apex (base of VI paler than apex of V). Last segment of rostrum less than 0.71 times as long as second segment of hind tarsus. (Al. without dorsal abdominal pigmentation)   K
    Antennae progressively darker from base to apex. Last segment of rostrum more than 0.71 times as long as second segment of hind tarsus. (Al. with a segmental pattern of dark markings on dorsal abdomen)   15

15. Terminal process 2.0–2.8 times longer than base of last antennal segment   L
    Terminal process 2.9–3.9 times longer than base of last antennal segment   16

16. Siphunculi 0.32 mm or more in length, usually more than 1.33 times longer than cauda   M
    Siphunculi less than 0.32 mm in length, usually less than 1.33 times longer than cauda   N

17. Cauda as dark as siphunculi, triangular in dorsal view, less than 1.5 times longer than its basal width   O
    Cauda pale, tongue-shaped, about twice as long as its basal width in dorsal view   P

Grasses of temperate pastures

18. Siphunculi pale over most of length but often dark at extreme apex	Q
Siphunculi darker than body (if rather pale there is still a clear difference of pigmentation between body and base of siphunculi)	19
19. Cauda long and pale, siphunculi very dark	R
Cauda rather short and about as dark as siphunculi	20
20. Body rather elongate. Terminal process less than 2.5 times length of base of last antennal segment	S
Body ovate. Terminal process more than 3 times longer than base of last antennal segment	T
21. Siphunculi absent	22
Siphunculi present as pores on low cones	29
22. Anal plate displaced dorsally	23
Anal plate in normal ventral position	25
23. Prothorax almost parallel-sided in dorsal view. Antennal segment III shorter than segment V including terminal process. All body hairs pointed	U
Prothorax continuing curve of body in dorsal view. Antennal segment III longer than V including terminal process. Some body hairs spatulate	24

24. Anal plate with numerous short hairs

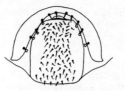

   Anal plate with two longitudinal rows of long
   hairs   W

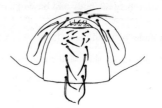

25. Body spindle-shaped, about twice as long as
    wide. Antennae very short, less than one-
    sixth of body length   X
    Body rather globose, less than 1.5 times
    longer than broad. Antennae more than one-
    fifth of body length   26

26. Wax pore-plates present on head, thorax, and
    abdomen, with an especially large one laterally
    on each abdominal segment. First segment of
    hind tarsus with only 2 hairs   Y

    No wax pore-plates. First segment of hind
    tarsus with 5 or more hairs   27

27. Body flattened dorso-ventrally. Antennae 6-
    segmented. Eyes usually with many facets   Z
    Body not flattened dorso-ventrally, rather
    globose. Antennae 5-segmented. Eyes usually
    with 3 facets   28

28. Primary sensorium on last antennal segment
    very large, occupying an area at least 4–5
    times larger than that occupied by the primary
    sensorium on IV. Antennal segment III longer
    than IV + V together   AA

    Primary sensorium on last antennal segment not
    much larger than that on segment IV. Antennal
    segment III not longer than IV + V together   BB

Greengage

29. Antennae very short, one-fifth of body length or less. Tarsi 1-segmented. Abdominal wax pore-plates present, each consisting of a ring of cells around an undivided central area ... **CC**

Total antenna length one-third or more of body length. Tarsi 2-segmented. No wax pore-plates ... **DD**

Greengage	use key to aphids on **Plum**

Ground Nut (Peanut)

A	*Prociphilus erigeronensis*	D	*Aphis gossypii*
B	*Myzus persicae*		[*Acyrthosiphan ilka/ bidentis*]
C	*Aphis craccivora*		

1. Terminal process much shorter than base of last antennal segment. (On roots with white wax) ... **A**

   Terminal process much longer than base of last antennal segment. (On aerial parts) ... 2

2. Antennal tubercles well developed, their inner faces convergent in dorsal view. Siphunculi pale, slightly clavate ... **B**

   Antennal tubercles weakly developed. Siphunculi dark and cylindrical or tapering ... 3

3. Dorsal abdomen with a solid black patch. Cauda and siphunculi both black ... **C**
   Dorsal abdomen unpigmented. Cauda paler than siphunculi ... **D**

Guarana	no aphids recorded

Guava

A	*Greenidea decaspermi*	D	*Aphis gossypii*
B	*Greenidea (Trichosiphon) formosana*	E	*Aphis citricola*
		F	*Toxoptera aurantii*
C	*Myzus persicae*		[*Aphis craccivora*]

1.	Siphunculi bearing numerous long hairs	2
	Siphunculi without hairs	3
2.	Siphunculi longer than hind tibiae, with a pattern of pale reticulation extending over almost their entire length	A
	Siphunculi shorter than hind tibiae, with pale reticulation only at base	B
3.	Antennal tubercles well developed, their inner faces convergent in dorsal view. Siphunculi pale, slightly clavate	C
	Antennal tubercles weakly developed. Siphunculi dark, not clavate	4
4.	Cauda paler than siphunculi, with usually 4–7 hairs	D
	Cauda and siphunculi both very dark, cauda with usually more than 7 hairs	5
5.	Terminal process of antenna less than 3 times longer than base of last segment. No stridulatory apparatus	E
	Terminal process more than 3.5 times longer than base of last antennal segment. Stridulatory apparatus present	F

Guinea Grass   see **Grass, Guinea**

Guinea Corn

Guinea Corn

A	*Anoecia corni*	I	*Schizaphis graminum*
B	*Forda orientalis*	J	*Sitobion leelamaniae*
C	*Sipha (Rungsia) maydis*	K	*Sitobion avenae*
D	*Sipha flava*	L	*Sitobion africanum*
E	*Rhopalosiphum maidis*	M	*Sitobion miscanthi*
F	*Rhopalosiphum padi*		[*Ceratovacuna lanigera*]
G	*Melanaphis sacchari*		[*Melanaphis formosana*]
H	*Hysteroneura setariae*		[*Tetraneura nigriabdominalis*]

1. Terminal process much shorter than base of last antennal segment. (On roots) — 2
   Terminal process longer than base of last antennal segment. (On aerial parts) — 3

2. Siphunculi present as pores on very low, hairy cones — A
   Siphunculi completely absent — B

3. Dorsal body hairs mainly long and spine-like — 4
   Dorsal body hairs short — 5

4. Cauda broadly rounded — C

   Cauda knobbed — D

5. Antennal tubercles weakly developed. Siphunculi without polygonal reticulation — 6

   Antennal tubercles developed, sometimes rather low but exceeding height of medial part of frons in dorsal view. Siphunculi with a sub-apical zone of polygonal reticulation — 10

6. Cauda darker than body — 7
   Cauda pale — 8

7. Body rather elongate. Terminal process less than 2.5 times longer than base of last antennal segment — E
   Body broadly oval. Terminal process more than 3 times longer than base of last antennal segment — F

8. Siphunculi shorter than cauda	G
Siphunculi longer than cauda	9
9. Siphunculi wholly dark	H
Siphunculi mainly pale, often dark at extreme apices	I
10. Siphunculi pale or a little dusky at apex only. Terminal process more than 6 times longer than base of last antennal segment	J
Siphunculi dark. Terminal process usually less than 6 times longer than base of last antennal segment	11
11. Siphunculi less than 1.4 times longer than cauda. Second segment of hind tarsus more than 1.3 times longer than last segment of rostrum	K
Siphunculi more than 1.4 times longer than cauda. Second segment of hind tarsus less than 1.25 times longer than last rostral segment	12
12. Dorsal abdominal pigmentation usually consisting of segmentally arranged and well-defined transverse black bars. Short haired; longest hairs on antennal segment III less than 20 μm long	L
Dorsal abdominal pigmentation ill-defined and not segmentally divided. Longer haired; longest hairs on antennal segment III more than 20 μm long	M

Hazel Nut, Cobnut, Filbert

A *Myzocallis coryli*
B *Pterocallis (Mesocallis) pteleae*
C *Pterocallis (Paratinocallis) corylicola*
D *Neochromaphis coryli*
E *Pterocallis heterophyllus*
F *Pterocallis montana*
G *Corylobium avellanae*
H *Macrosiphum (Neocorylobium) vandenboschi*
I *Macrosiphum (Neocorylobium) corylicola*
J *Macrosiphum (Neocorylobium) coryli*
   [*Illinoia corylina* (Davidson 1914)]

Hazel Nut, Cobnut, Filbert

1.	Adult viviparous females usually or always alate. Cauda knobbed, anal plate bilobed. Siphunculi small truncated cones		2
	Adult viviparous females in colonies mainly apterous. Cauda tongue-shaped, anal plate entire. Siphunculi long, tubular structures		7
2.	(Al.) Terminal process more than twice as long as base of last antennal segment		A
	(Al.) Terminal process similar in length to base of last antennal segment, or shorter		3
3.	(Al.) Terminal process about equal in length to base of last antennal segment		4
	(Al.) Terminal process less than 0.6 of length of base of last antennal segment		5
4.	(Al.) Antennal segment III dark, with usually more than 9 sensoria. Last segment of rostrum 1.2–1.4 times longer than second segment of hind tarsus. Abdomen with rather large lateral tubercles, each bearing a single small hair		B
	(Al.) Antennal segment III pale, sometimes dusky at apex, with usually 9 or less sensoria, one of them near the apex of the segment. Last rostral segment 0.9–1.0 times as long as second segment of hind tarsus. Abdomen with small flat marginal tubercles, each bearing 2–3 hairs		C
5.	(Al.) Wings with pattern of extensive dark markings. Terminal process very short, only about 0.1 of length of base of last segment		D
	(Al.) Wings without extensive pigmented areas. Terminal process 0.2–0.6 of the length of the base of last antennal segment		6
6.	(Al.) Terminal process 0.5–0.6 of length of base of last antennal segment. Spinal hairs on anterior abdominal tergites long and spine-like, on low tuberulate bases		E
	Terminal process 0.2–0.45 of length of base of last antennal segment. Dorsal abdominal hairs all very short		F
7.	Dorsal body hairs long and thick and slightly capitate, arising from large tubercles		G
	Dorsal body hairs shorter, not capitate, not borne on tubercles		8
8.	Dorsal cuticle strongly sclerotized and very wrinkled (although quite pale)		H
	Dorsal cuticle not sclerotic, quite smooth		9

9. Antennae with basal segments dark, more
distal segments becoming progressively paler.
Cauda dusky, about one-third of length of
siphunculi, which are dark at base I
Antennae with basal segments pale, more distal
segments progressively darker. Cauda very
pale, about half as long as siphunculi,
which are usually pale at base .... J

Hemp

Aphis fabae
Phorodon cannabis
[Aphis sativae Williams 1911]

Hop

A	Aphis fabae	
B	Aphis (Cerosipha) humuli	
C	Aphis gossypii	
D	Phorodon cannabis	
E	Phorodon humulifoliae	
F	Phorodon humuli humuli	
G	Phorodon humuli japonensis	
H	Rhopalosiphoninus staphyleae	
I	Myzus persicae	
J	Aulacorthum solani	
K	Macrosiphum euphorbiae	
L	Macrosiphum hamiltoni	

1. Antennal tubercles weakly developed 2
   Antennal tubercles well developed 4

2. Cauda black .... A
   Cauda pale or dusky 3

3. Antennae 5-segmented in both apt. and al. .... B
   Antennae 6-segmented (rarely 5-segmented in
   very small apt.) C

4. Antennal tubercles with finger-like forward
   projections, at least half as long as the first
   antennal segment 5
   Antennal tubercles without finger-like
   projections 8

5. Hairs on antennal tubercles, antennal segments
   I–III, and dorsal body with knobbed apices.
   Longest hairs on antennal segment III similar
   in length to basal diameter of segment .... D

   Hairs on antennal tubercles, antennae, and
   dorsal body with apices blunt or pointed.
   Longest hairs on antennal segment III less
   than half basal diameter of segment 6

Hop

6. Siphunculi with distal one-third strongly curved outwards, swollen on the curve and then narrowing abruptly, the narrow subapical section being completely smooth, the rest coarsely imbricated ... **E**

   Siphunculi with distal part slightly curved outwards but tapering gradually over most of length, there being no abrupt transition to a smooth narrow apical section ... 7

7. Finger-like projection on antennal tubercle reaching apex of antennal segment I or beyond, its length greater than that of antennal segment II. Last segment of rostrum similar in length to second segment of hind tarsus ... **F**

   Projection on antennal tubercles not reaching apex of antennal segment I, its length a little less than that of segment II. Last rostral segment about 1.5 times longer than second segment of hind tarsus ... **G**

8. Siphunculi markedly swollen, maximum diameter of swollen part at least 1.5 times minimum diameter of basal part. Dorsal abdomen with dark markings ... **H**

   Siphunculi tapering, cylindrical or slightly swollen. Dorsal abdomen without dark markings ... 9

9. Inner faces of antennal tubercles convergent in dorsal view. Terminal process rarely more than 4 times longer than base of antennal segment VI ... **I**

   Inner faces of antennal tubercles parallel or divergent in dorsal view. Terminal process usually more than 4 times longer than base of antennal segment VI ... 10

10. Inner faces of antennal tubercles parallel.
    Siphunculi without polygonal reticulation   **J**

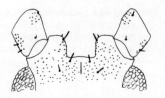

    Inner faces of antennal tubercles divergent.
    Siphunculi with a subapical zone of polygonal reticulation   **11**

11. Siphunculi cylindrical over most of length.
    Cauda usually a little more than half as long as siphunculi   **K**

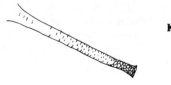

    Siphunculi slightly but clearly swollen on distal one-third proximal to reticulated subapical zone, which is less than 0.1 of total length of siphunculus. Cauda less than half as long as siphunculi   **L**

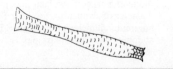

Horseradish   **A**   *Rhopalosiphoninus latysiphon*   **C**   *Macrosiphum euphorbiae* [*Aphis armoraciae*]
                    B   *Myzus persicae*

---

1. Siphunculi jet black, with middle part inflated like a balloon (subterranean)   **A**

    Siphunculi pale or dusky, not or only slightly swollen   **2**

Huckleberry

2. Inner faces of antennal tubercles convergent. Siphunculi slightly clavate, without any polygonal reticulation. Terminal process 3.5–4.5 times longer than base of antennal segment VI .... B

Inner faces of antennal tubercles divergent. Siphunculi cylindrical on distal part, with a subapical zone of polygonal reticulation. Terminal process 4.3–6.0 times longer than base of antennal segment VI C

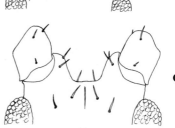

Huckleberry	no aphids recorded

Huckleberry, Garden

	A	*Smynthurodes betae*	
	B	*Aphis maidi radicis*	
	C	*Aphis craccivora* *Aphis fabae solanella* *Aphis nasturtii* *Aphis gossypii*	see key to polyphagous *Aphis* spp.
	D	*Myzus persicae* *Aulacorthum solani* *Macrosiphum euphorbiae*	see key to polyphagous Macrosiphini

1. Terminal process shorter than base of last antennal segment. Siphunculi absent .... A

   Terminal process longer than base of last antennal segment. Siphunculi present 2

2. Terminal process less than twice as long as base of last antennal segment. Cauda dark, triangular, less than 1.5 times longer than its basal width in dorsal view (subterranean) .... B

   Terminal process more than twice as long as base of last antennal segment. Cauda pale or dark, tongue-shaped, more than 1.5 times longer than its basal width in dorsal view. (On aerial parts) 3

3.	Antennal tubercles weakly developed	C
	Antennal tubercles well developed	D

Indigo
	A	Antalus albatus	D	Myzus persicae
	B	Aphis craccivora		[Anomalosiphon indigoferae
	C	Megoura pallipes		Ghosh, Ghosh & Raychaudhuri
				1971; Aphis indigoferae
				Shinji 1922]

1.	Siphunculi very short, broad, pigmented cones. Cauda shaped like an arrow-head in dorsal view. Anal plate bilobed	A
	Siphunculi tubular. Cauda tongue-shaped. Anal plate entire	2

2.	Abdomen with a solid black patch centred on tergites 4–5. Antennal tubercles weakly developed. Siphunculi tapering	B
	Dorsal abdomen without pigmentation. Antennal tubercles well developed. Siphunculi swollen in the middle or on distal part	3

3.	Siphunculi black. Inner faces of antennal tubercles divergent in dorsal view	C
	Siphunculi pale except at extreme apices. Inner faces of antennal tubercles convergent	D

Iris
	A	Dysaphis tulipae	F	Myzus ascalonicus
	B	Sitobion avenae	G	Myzus persicae
	C	Schizaphis (Paraschizaphis) scirpi	H	Macrosiphum euphorbiae
			I	Aulacorthum solani
	D	Aphis newtoni	J	Metopolophium dirhodum
	E	Aulacorthum (Neomyzus) circumflexum		

1.	Siphunculi wholly dark	2
	Siphunculi pale, at least basally	5

2.	Cauda helmet-shaped in dorsal view, no longer than broad at base	A
	Cauda tongue-shaped, much longer than its basal width	3

3.	Siphunculi with a subapical zone of polygonal reticulation. Lateral abdominal tubercles not developed	**B**
	Siphunculi without any polygonal reticulation. Lateral tubercles present on all abdominal segments	4
4.	Antennal segment III dark, with long fine hairs up to 3–4 times longer than the basal diameter of the segment. Siphunculi about twice as long as cauda	**C**
	Antennal segment III pale, with longest hairs 1–2 times longer than basal diameter of segment. Siphunculi similar in length to cauda	**D**
5.	Dorsal abdomen with a large black roughly horseshoe-shaped patch, and thorax with black patches	**E**
	Dorsal surface unpigmented	6
6.	Siphunculi slightly to moderately clavate	7
	Siphunculi tapering or cylindrical on distal half	8
7.	Siphunculi rather small, shorter than antennal segment III, the minimum diameter of their basal part less than the diameter of the middle part of hind tibia	**F**
	Siphunculi a little longer than antennal segment III and with minimum diameter of basal half greater than diameter of middle part of hind tibia	**G**

8. Siphunculi with a subapical zone of polygonal reticulation ... H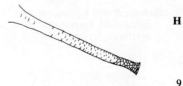

Siphunculi without any polygonal reticulation ... 9

9. Inner faces of antennal tubercles parallel in dorsal view. Siphunculi with a well-developed flange. Last segment of rostrum longer than second segment of hind tarsus ... I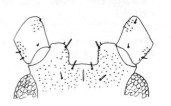

Inner faces of antennal tubercles divergent. Siphunculi with a small flange. Last segment of rostrum only 0.6–0.7 of length of second segment of hind tarsus ... J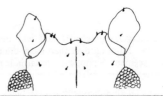

Johnson Grass	see **Grass, Johnson**
Jute, Gunny	no aphids recorded
Kale	as for **Cabbage**
Kapok	*Aphis gossypii* [*Toxoptera citricidus*]
Kikuyu Grass	see **Grass, Kikuyu**
Kiwi Fruit	*Toxoptera aurantii*
Kohlrabi	as for **Cabbage**
Kumquat	no aphids recorded

Lalang

Lalang	A *Hysteroneura setariae*	C	*Tetraneura nigriabdominalis*
	B *Anoecia fulviabdominalis*	D	*Tetraneura radicicola*

1. Terminal process more than 5 times longer than base of last antennal segment. Siphuncular tubular. Cauda pale long and finger-like. (On aerial parts) ... **A**
   Terminal process shorter than base of last antennal segment. Siphunculi as small cones. Cauda broadly rounded. (On roots) ... **2**

2. Antennae more than one-third of body length. Tarsi 2-segmented ... **B**
   Antennae less than one-fifth of body length. Tarsi 1-segmented ... **3**

3. Last segment of rostrum short, less than 0.15 mm long and less than 1.8 times length of hind tarsus. Anal plate with some very short hairs in addition to very long ones ... **C**
   Last segment of rostrum more than 0.20 mm long, more than 2.0 times length of hind tarsus. Anal plate with very long hairs only ... **D**

Lavender

Myzus ornatus
Myzus persicae
[*Eucarazzia elegans*]
(or try key to aphids on **Mint**)

Leek

Aphis fabae
Neotoxoptera formosana

Lemon

see **Citrus**

Lentil

Aphis craccivora
Aphis fabae

Lettuce

A *Pemphigus* species
B *Neotrama caudata*
C *Trama troglodytes*
D { *Aphis gossypii* / *Aphis fabae* / *Aphis citricola* } see key to polyphagous *Aphis* spp.
E *Uroleucon formosanus*
F *Uroleucon cichorii*
G *Uroleucon ambrosiae/ pseudoambrosiae*

H	*Nasonovia ribis-nigri*	L	*Macrosiphum euphorbiae*
I	*Aulacorthum (Neomyzus) circumflexum*		[*Macrosiphum constrictum* Patch 1923; *Myzus persicae; Myzus (Sciamyzus) cymbalariae; Prociphilus erigeronensis*]
J	*Aulacorthum solani*		
K	*Acyrthosiphon (Tlja) lactucae*		

1. Terminal process much shorter than base of last antennal segment. (On roots)   2
   Terminal process much longer than base of last antennal segment. (On aerial parts)   4

2. Second segment of hind tarsus of normal length, similar to the other tarsi. Body hairs very short and sparse   A

   Second segment of hind tarsus very long, much longer than the other tarsi, 0.7 or more of length of hind tibia. Body hairs numerous, long, and fine   3

3. Siphunculi present. Eyes with usually more than 12 facets   B
   Siphunculi absent. Eyes with only 3 facets   C

4. Antennal tubercles little developed   D

   Antennal tubercles well developed   5

5. Siphunculi wholly dark   6
   Siphunculi pale, at least at bases   8

6. Antennal segment III more than 1.5 times longer than segments IV and V together   E
   Antennal segment III about the same length as, or a little shorter than, segments IV and V together   7

Lettuce 117

7.	All dorsal abdominal hairs arising from small dark sclerites, and a large dark sclerite anterior to the base of each siphunculus	F
	Only some of dorsal abdominal hairs arising from small dark sclerites, and no sclerite immediately anterior to base of siphunculus	G

8.	Dorsal abdomen with dark markings	9
	Dorsal abdomen without dark markings	10

9.	Dorsal abdomen with paired small dark intersegmental markings, far apart between anterior segments but nearer mid-line between segment 4–5 and 5–6. Terminal process 6–9 times longer than base of antennal segment VI	H
	Dorsal abdomen with a large black, roughly horseshoe-shaped patch. Terminal process about 4 times longer than base of antennal segment VI	I

10.	Inner faces of antennal tubercles approximately parallel in dorsal view	J
	Inner faces of antennal tubercles divergent in dorsal view	11

11.	Siphunculi a little shorter than the distance between their bases, without subapical polygonal reticulation	K
	Siphunculi longer than the distance between their bases, with a subapical zone of polygonal reticulation	L

Lily

A	Dysaphis tulipae	H	Fimbriaphis lilii
B	Aphis fabae	I	Fimbriaphis scoliopi
C	Aphis gossypii	J	Macrosiphum euphorbiae
D	Aulacorthum (Neomyzus) circumflexum	K	Macrosiphum lilii
		L	Neotoxoptera oliveri
E	Indomegoura indica	M	Myzus persicae
F	Aulacorthum magnoliae		[Aphis ogilviei Theobald 1928]
G	Aulacorthum solani		[Myzus ascalonicus]

1. Cauda helmet-shaped, as short as its basal width in dorsal view — **A**

   Cauda tongue-shaped or triangular, clearly longer than its basal width — 2

2. Antennal tubercles weakly developed — 3

   Antennal tubercles well developed — 4

3. Cauda dark with more than 10 hairs — **B**
   Cauda paler than siphunculi, with 4–7 hairs — **C**

4. Dorsal abdomen with a large dark patch, usually roughly horseshoe-shaped, and dark transverse bars on thoracic tergites — **D**

   Dorsal surface without extensive dark markings — 5

5. Antennal segments III–VI and tibiae wholly dark, and femora black except at bases — 6
   At least segment III of antenna, tibiae, and femora, for the most part, pale or dusky — 7

Lily

6. Siphunculi very stout, wholly dark, cylindrical for basal two-thirds and then abruptly tapering to flange, with a subapical zone of polygonal reticulation. Inner faces of antennal tubercles divergent ..... **E**

   Siphunculi long and thin, slightly clavate. Inner faces of antennal tubercles convergent ..... **F**

7. Siphunculi cylindrical over most of length or tapering from base to apex 8
   Siphunculi slightly clavate 12

8. Siphunculi without any polygonal reticulation ..... **G**
   Siphunculi with a subapical zone of polygonal reticulation 9

9. Inner faces of antennal tubercles parallel or convergent in dorsal view. Cauda less than half as long as siphunculi 10
   Inner faces of antennal tubercles divergent. Cauda more than half as long as siphunculi 11

10. Siphunculi 2.7–3.3 times longer than cauda, which is pale. Terminal process more than 6.5 times longer than base of last antennal segment ..... **H**
    Siphunculi 3.2–3.5 times longer than cauda, which is dusky to dark. Terminal process less than 6 times longer than base of last antennal segment ..... **I**

11. Siphunculi pale at base, sometimes darkening towards apex. Femora and bases of tibiae normally pale ..... **J**
    Siphunculi wholly dark. Apices of femora and bases of tibiae ('knees') dark ..... **K**

12. Siphunculi shorter than antennal segment III. Cauda triangular in dorsal view, less than 1.5 times longer than its basal width. (Wing veins of al. with broad dusky borders) ..... **L**
    Siphunculi a little longer than antennal segment III. Cauda tongue-shaped in dorsal view, about twice as long as its basal width. (Wing veins of al. unbordered) ..... **M**

Lily, Day

A *Indomegoura indica*	C *Myzus persicae*
B *Myzus hemerocallis*	[*Aphis sambuci*]
	[*Rhopalosiphoninus staphyleae*]

1. Antennae, legs, and siphunculi black. Siphunculi very stout and of characteristic shape .... **A**

   Antennae, legs, and siphunculi mainly pale. Siphunculi thin **2**

2. Terminal process less than 3 times longer than base of last antennal segment. Siphunculi tapering or cylindrical on distal half, and coarsely imbricated. (Al. without a black dorsal abdominal patch) .... **B**

   Terminal process more than 3 times longer than base of last antennal segment. Siphunculi slightly clavate, moderately imbricated. (Al. with a black dorsal abdominal patch) .... **C**

Lily of the Valley

A *Aulacorthum speyeri*	C *Illinoia wahnaga*
B *Macrosiphum pechumani*	D *Myzus ascalonicus*
	E *Myzus persicae*

1. Dorsal thorax and abodmen with a distinctive and extensive pattern of black markings .... **A**

   Dorsal thorax and abdomen pale, without any extensive dark markings **2**

2. Siphunculi and cauda wholly black. Siphunculi cylindrical on distal half, not swollen .... **B**

   Siphunculi and cauda mainly pale. Siphunculi slightly or moderately clavate **3**

3. Inner faces of antennal tubercles divergent. Siphunculi with polygonal reticulation distal to the swollen part .... **C**

   Inner faces of antennal tubercles convergent or parallel in dorsal view. Siphunculi without any polygonal reticulation **4**

Lime, Citrus

4. Siphunculi rather small, shorter than antennal
segment III, with minimum diameter of
proximal part less than diameter of middle
part of hind tibia .... D

Siphunculi larger, a little longer than antennal
segment III, with minimum diameter of
proximal part greater than diameter of
middle part of hind tibia .... E

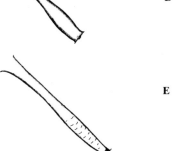

Lime, Citrus   see Citrus

Lime (Linden)

A   *Tiliphagus lycoposugus*
B   *Patchiella reaumuri*
C   *Longistigma caryae*
D   *Macrosiphum tiliae*
E   *Tiliaphis shinae*
F   *Tiliaphis shinjii*
G   *Eucallipterus tiliae*
      [*Tiliaphis coreana* Quednau 1979]
      [*Tiliaphis pseudoshinae* Quednau 1979]

1. Siphunculi absent or present only as inconspicuous pores. (Forming large 'leaf-nest' galls in spring) 2

Siphunculi present either as short truncated cones, broad dark hairy cones, or long tubes. (Not forming 'leaf-nest' galls) 3

2. Al. from gall with narrow elongate sensoria on antennal segments III–V, protruding as sharp transverse ridges and giving these segments a serrate outline .... A

Al. from gall with sensoria on antennal segments III–V fewer in number and arranged in a row on one side of the antenna, not protruding as sharp ridges .... B

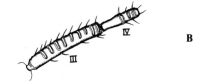

3. Very large aphid (body length more than 5 mm) with body and appendages covered in long hairs, and siphuncular pores on broad dark hairy cones. (Al. with dark pterostigma extending to tip of wing.) (Colonies on bark) C

Body length less than 5 mm. (On leaves) 4

4. Siphunculi very long, tubular, with subapical
zone of polygonal reticulation. Cauda tongue-
shaped, anal plate entire. (Colonies usually
with mainly apterous viviparae)   D
(Al.) Siphunculi in form of truncated cones,
about as long as, or only a little longer than,
their basal width. Cauda knobbed and anal
plate bilobed. All adult viviparae alate   5

5. Dorsal abdomen without dark markings   E
Dorsal abdomen with paired dark markings   6

6. Siphunculi pale, longer than their width at
base. Terminal process as long as, or a little
longer than, base of last antennal segment   F
Siphunculi shorter than their width at base
and usually dark. Terminal process a little
shorter than base of last antennal segment   G

Liquorice   *Aphis craccivora*

Litchi   *Toxoptera aurantii*

Loganberry   *Aphis idaei*
[*Illinoia davidsoni*]

Loofah   *Aphis gossypii*
[*Aulacorthum magnoliae*]

Loquat (Japanese Medlar)

A	*Nippolachnus himalayensis*	G	*Aphis gossypii*
B	*Nippolachnus piri*	H	*Aphis pomi*
C	*Tuberolachnus*	I	*Aphis citricola*
	(Tuberolachniella) sclerata		[*Nippolachnus*
D	*Pyrolachnus pyri*		*xitianmushanus* Zhang &
E	*Brachycaudus persicae*		Zhong 1982]
F	*Hyalomyzus eriobotryae*		[*Pyrolachnus macroconus*
			Zhang & Zhong 1982]

1. Terminal process shorter than base of last
antennal segment. Body densely hairy   2
Terminal process longer than base of last
antennal segment. Body sparsely haired   5

Lovage

2.	Eyes without a posterior process	3
	Eyes with a distinct posterior process (triommatidium)	4
3.	Legs black. Body oval	**A**
	Legs mainly pale. Body elongate	**B**
4.	Dorsal abdomen with a large broad central tubercle	**C**
	Dorsal abdomen without a central tubercle	**D**
5.	Dorsal abdomen black. Cauda shorter than its width at base in dorsal view	**E**
	Dorsal abdomen pale. Cauda longer than its basal width in dorsal view	6
6.	Antennal tubercles well developed, with inner faces parallel in dorsal view	**F**
	Antennal tubercles poorly developed	7
7.	Cauda paler than siphunculi, with 4–6 hairs	**G**
	Cauda dark, with usually more than 6 hairs	8
8.	Lateral tubercles present on abdominal segments 2–4. Cauda with rarely less than 14 hairs. Length of last segment of rostrum more than 130 μm	**H**
	Lateral tubercles absent from abdominal segments 2–4. Cauda with usually less than 12 hairs. Length of last segment of rostrum less than 120 μm	**I**

Lovage

A	*Myzus ornatus*		**C**	*Aphis fabae*
B	*Dysaphis apiifolia petroselini*		**D**	*Aphis citricola*

1.	Antennal tubercles well developed, their inner faces convergent in dorsal view. Siphunculi pale	**A**
	Antennal tubercles poorly developed. Siphunculi dark	2

2. Cauda helmet-shaped in dorsal view, no longer
   than its basal width B

   Cauda tongue-shaped, much longer than its
   basal width 3

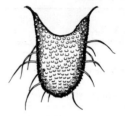

3. Abdomen with black markings, at least laterally
   on most segments, and usually black transverse
   bars on tergites 7 and 8 C
   Abdomen without any dark markings dorsally D

Lucerne (Alfalfa)

- A *Therioaphis trifolii*
- B *Nearctaphis bakeri*
- C *Aphis craccivora*
- D *Aphis gossypii*
- E *Acyrthosiphon pisum*
- F *Acyrthosiphon kondoi*
- G *Acyrthosiphon loti*
- H *Myzus ornatus*
- I *Myzus persicae*

[*Aphis alvata* Zhang 1981]
[*Aphis craccivora usuana* Zhang 1981]
[*Aphis loti* Kaltenbach 1862]
[*Aphis medicaginis*]
[*Aphis robiniae canavaliae* Zhang 1981]
[*Aulacorthum solani*]
[*Nearctaphis californica*]
[*Nearctaphis sensoriata*]

1. Cauda with a constriction and a knob-like
   apex; anal plate bilobed. Body hairs capitate,
   arising from pigmented tubercles A

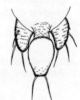

   Cauda tongue-shaped, rounded or triangular
   in dorsal view, anal plate entire. Body hairs
   not capitate, nor arising from pigmented
   tubercles 2

Lucerne (Alfalfa)

2.	Cauda short and broad, rounded or triangular in dorsal view, hardly longer than its basal width. Dorsal abdomen with many small dark hair-bearing sclerites, separate on tergites anterior to siphunculi but sometimes fused into larger patches or bars on posterior tergites		**B**
	Cauda tongue-shaped, much longer than its basal width. Dorsal abdomen either unpigmented or with a large central black patch		3

3.	Antennal tubercles weakly developed		4
	Antennal tubercles well developed		5

4.	Dorsal abdomen with a solid black patch. Cauda black		**C**
	Dorsal abdomen unpigmented. Cauda pale		**D**

5.	Inner faces of antennal tubercles divergent		6
	Inner faces of antennal tubercles convergent		8

6.	Articulation between antennal segments III and IV blackish pigmented. Base of antennal segment VI 0.25–0.4 mm long, 1.8–3 times longer than last segment of rostrum. Siphunculi very attenuate distally, their minimum diameter less than one-quarter of width of cauda at mid-length in dorsal view		**E**
	Articulation between antennal segments III and IV not blackish pigmented. Base of antennal segment VI 0.13–0.2 mm long, 1–1.5 times as long as last rostral segment. Siphunculi not so attenuate distally with minimum diameter more than one-quarter of width of cauda at mid-length in dorsal view		7

7.	Terminal process 4.3–6.0 times longer than base of last antennal segment, and usually 1.6–2.5 times longer than cauda. Siphunculi usually more than 1.7 times longer than cauda		**F**
	Terminal process 3.0–4.2 times longer than base of last antennal segment, and usually 1.1–1.6 times longer than cauda. Siphunculi usually less than 1.7 times longer than cauda		**G**

8. Siphunculi tapering from base to flange, with a slight outward curve distally. Dorsal abdomen with an intersegmental pattern of dark ornamentation

H

Siphunculi slightly clavate. Dorsal abdomen without dark ornamentation

I

Lupin

	A	*Aphis craccivora*	D	*Myzus persicae*
	B	*Aphis lupini*	E	*Macrosiphum zionensis*
	C	*Myzus ornatus*	F	*Macrosiphum albifrons*
				[*Macrosiphum euphorbiae*]

1. Antennal tubercles weakly developed — 2

    Antennal tubercles well developed — 3

2. Dorsal abdomen with an extensive black patch — A
    Dorsal abdomen without a black patch — B

3. Inner faces of antennal tubercles convergent. Siphunculi without polygonal reticulation. (Small to medium-sized aphids) — 4

    Inner faces of antennal tubercles divergent. Siphunculi with a subapical zone of polygonal reticulation. (Large aphids) — 5

4. Siphunculi tapering from base to flange, with a slight outward curve distally. Dorsal abdomen with an intersegmental pattern of dark ornamentation — C

    Sipunculi slightly clavate. Dorsal abdomen without dark ornamentation — D

Macadamia 127

5. Siphunculi black. Body elongate spindle-shaped ... **E**
   Siphunculi pale, sometimes darkening towards apex. Body ovate ... **F**

Macadamia Toxoptera aurantii

Maize

A	Sipha (Rungsia) maydis	P	Sitobion avenae
B	Sipha flava	Q	Sitobion africanum
C	Melanaphis sacchari	R	Sitobion miscanthi
D	Melanaphis formosana	S	Geoica utricularia
E	Schizaphis graminum	T	Geoica lucifuga
F	Hysteroneura setariae	U	Colopha ulmicola
G	Aphis gossypii	V	Pemphigus species
H	Aphis fabae	W	Anoecia species
I	Aphis maidi-radicis	X	Tetraneura ulmi
J	Aphis armoraciae	Y	Tetraneura yezoensis/ radicicola
K	Rhopalosiphum maidis		[Aphis craccivora]
L	Rhopalosiphum padi		[Metopolophium dirhodum]
M	Myzus persicae		[Metopolophium festucae cerealium]
N	Macrosiphum euphorbiae		
O	Sitobion howlandae		

1. Terminal process similar in length to, or longer than, base of last antennal segment ... 2
   Terminal process much shorter than (less than one-half) base of last antennal segment ... 19

2. Dorsal body hairs partly or mainly long and spine-like. Siphunculi very short, truncated cones, shorter than their basal diameter ... 3
   Dorsal body hairs if long then fine. Siphunculi tubular, longer than basal width ... 4

3. Cauda broadly rounded. Dorsal abdomen dark ... **A**

   Cauda with a knobbed apex. Dorsal abdomen pale or smoky, with small dark intersegmental markings ... **B**

4. Siphunculi a little shorter than cauda. Legs pale ... 5
   Siphunculi longer than cauda; if only a little longer than cauda, then the legs are mainly dark ... 6

5.	Cauda paler than siphunculi, which have a small flange only 1.2–1.3 times wider than the subapical constriction. Last segment of rostrum distinctly shorter than second segment of hind tarsus	C
	Cauda and siphunculi similarly pigmented, Siphuncular flange larger, 1.3–1.5 times wider than the subapical constriction. Last rostral segment as long as second segment of hind tarsus	D
6.	Antennal tubercles low or undeveloped, not higher than medial part of frons. Siphunculi much shorter than the distance between their bases	7
	Antennal tubercles sometimes rather low, but always higher than medial part of frons. Siphunculi similar in length to the distance between their bases, or longer	14
7.	Siphunculi pale over most of their length although often darkened at extreme apices	E
	Siphunculi darker than body (if rather pale there is still a clear difference of pigment between body and base of siphunculi)	8
8.	Cauda paler than siphunculi	9
	Cauda and siphunculi both similarly dark	10
9.	Terminal process more than 5 times longer than base of last antennal segment	F
	Terminal process less than 3.5 times longer than base of last antennal segment	G
10.	Dorsal cuticle with dark markings, usually including transverse bands on thorax and posterior to siphunculi, and smaller dark patches on anterior abdominal segments	11
	Dorsal cuticle of body without dark markings	13
11.	Cauda tongue-shaped, much longer than its basal width. Terminal process more than 2.5 times longer than base of last antennal segment. (On aerial parts)	H
	Cauda bluntly conical, about as long as its basal width. Terminal process less than twice as long as base of last antennal segment. (On roots)	12

Maize

12. Apt. usually with none, sometimes with 1–3 sensoria on antennal segment III, none on IV or V. Al. with 4–7 sensoria on antennal segment III, 0–3 on IV, and none on V ... **I**
    Apt. with 5–12 sensoria on antennal segment III, 1–6 on IV, 0–5 on V. Al. with 10–16 sensoria on III, 2–9 on IV, 0–6 on V ... **J**

13. Body rather elongate. Terminal process less than 2.5 times longer than base of last antennal segment ... **K**
    Body ovate. Terminal process more than 3 times longer than base of last antennal segment ... **L**

14. Inner faces of antennal tubercles convergent. Siphunculi slightly clavate, without any polygonal reticulation ... **M**
    Inner faces of antennal tubercles divergent. Siphunculi not clavate, with a subapical zone of polygonal reticulation ... 15

15. Hairs on antennal segment III almost as long as the basal diameter of the segment. Siphunculi always pale at base, sometimes darker apically ... **N**
    Hairs on antennal segment III short and inconspicuous, length less than half the basal diameter of the segment. Siphunculi usually wholly dark, sometimes pale at base ... 16

16. Dorsal abdomen without markings. Terminal process 7–9 times longer than base of antennal segment VI. Siphunculi pale at base becoming darker towards apex ... **O**
    Dorsal abdomen with markings, usually a segmental pattern of dark spots, patches, or bars which is sometimes rather faint. Terminal process 4–6 times longer than base of antennal segment VI. Siphunculi wholly dark ... 17

17. Siphunculi less than 1.5 times longer than cauda; second segment of hind tarsus more than 1.4 times longer than last segment of rostrum ... **P**
    Siphunculi more than 1.5 times longer than cauda; second segment of hind tarsus less than 1.4 times longer than last rostral segment ... 18

18.	Dorsal abdominal markings usually comprising segmentally arranged and well defined black bars, sometimes broken up into paired patches	Q
	Dorsal abdominal pigmentation ill-defined and not segmentally divided	R
19.	Siphunculi absent	20
	Siphunculi present as pores or small shallow cones	23
20.	Anal plate displaced dorsally	21
	Anal plate in normal ventral position	22
21.	Anal plate covered with numerous short hairs, but no long ones	S
	Anal plate with long hairs in irregular longitudinal rows, short hairs being confined to the region close to the anal aperture	T
22.	Tarsi 1-segmented	U
	Tarsi 2-segmented	V
23.	Total antennal length one-third or more of body length. Tarsi 2-segmented	W
	Antennae very short, one-fifth of body length or less. Tarsi 1-segmented	24
24.	Body and antennae with sparse, short hairs. Last rostral segment less than 0.2 mm long	X
	Body and antennae with numerous long hairs. Last rostral segment more than 0.2 mm long	Y

Mandarin Orange   see **Citrus**

Mango
- A *Greenidea mangiferae*
- B *Aphis gossypii*
- C *Toxoptera odinae*
- D *Toxoptera aurantii*
  [*Aphis citricola*]
  [*Aphis craccivora*]
  [*Aphis fabae solanella*]
  [*Macrosiphum centranthi*]
  [*Sinomegoura citricola*]

Manila Hemp

1.	Siphunculi with numerous very long hairs	A
	Siphunculi without hairs	2
2.	Cauda paler than siphunculi, with 4–7 hairs. No stridulatory apparatus	B
	Cauda dark, with more than 10 hairs. Stridulatory apparatus present	3
3.	Siphunculi clearly shorter than cauda. Terminal process less than 3 times longer than base of last antennal segment	C
	Siphunculi similar in length to cauda or longer. Terminal process more than 3.5 times longer than base of last antennal segment	D

Manila Hemp   *Pentalonia nigronervosa*

Marjoram, Sweet   *Myzus ornatus*

Medlar
- A   *Ovatus insitus*
- B   *Aphis pomi*
- C   *Rhopalosiphum insertum*
- [*Aphis craccivora*]

1.	Antennal tubercles well developed with inner faces convergent in dorsal view. Siphunculi pale, tapering from base to flange	A
	Antennal tubercles not developed	2
2.	Siphunculi wholly dark, tapering gradually from base to flange	B
	Siphunculi pale or a little dusky, only dark at apices, of characteristic shape	C

Melon

  A *Myzus persicae*
  B *Aphis craccivora*
  C *Aphis gossypii*

1. Antennal tubercles well developed, their inner faces convergent in dorsal view. Siphunculi pale, slightly clavate — **A**

   Antennal tubercles weakly developed. Siphunculi dark, tapering from base to flange — **2**

2. Dorsal abdomen with a large black patch. Cauda black — **B**
   Dorsal abdomen unpigmented. Cauda pale or a little dusky — **C**

Millet, African   no aphids recorded

Millet, Common

  A *Geoica utricularia*   E *Schizaphis graminum*
  B *Anoecia* species   F *Rhopalosiphum maidis*
  C *Tetraneura yezoensis*   G *Rhopalosiphum padi*
  D *Sitobion avenae*   [*Melanaphis sacchari*]

1. Terminal process shorter than base of last antennal segment. (On roots) — **2**
   Terminal process longer than base of last antennal segment. (On aerial parts) — **4**

2. Siphunculi absent. Anal plate displaced dorsally — **A**

   Siphunculi present as small shallow cones or pores — **3**

3. Tarsi 2-segmented. Antennae at least one-third as long as body. — **B**
   Tarsi 1-segmented. Antennae very short, less than one-fifth of body length — **C**

4. Siphunculi much darker than cauda, with a subapical zone of polygonal reticulation — **D**

   Siphunculi similar in pigmentation to cauda, although sometimes darker at extreme apices, and without any subapical reticulation — **5**

Millet, Great

5. Cauda pale, siphunculi pale with dark apices   E

    Cauda and siphunculi both darker than body   6

6. Body rather elongate, 9–14 times as long as the siphunculi. Terminal process less than 2.5 times longer than base of last antennal segment   F

    Body ovate, only 5.5–7.5 times as long as the siphunculi. Terminal process more than 3 times longer than base of last antennal segment   G

Millet, Great   see **Guinea Corn**

Millet, Pearl	**A**	*Forda orientalis*	**C**	*Rhopalosiphum maidis*
	B	*Sitobion leelamaniae*		[*Melanaphis sacchari*]
				[*Rhopalosiphum rufiabdominalis*]

1. Body globose. Siphunculi absent. Terminal process much shorter than base of last antennal segment. (On roots)   A

    Body elongate oval or spindle-shaped. Siphunculi evident. Terminal process longer than base of last antennal segment. (On aerial parts)   2

2. Terminal process of antenna very long, about 8 times longer than base of last segment. Siphunculi pale or dusky, about twice as long as cauda, with a subapical zone of polygonal reticulation   B

    Terminal process of antenna less than 2.5 times longer than base of last segment. Siphunculi dark, less than 1.5 times longer than cauda and not reticulated near apex   C

Mint	**A**	*Kaltenbachiella pallida*	**E**	*Aphis gossypii*
	B	*Eucarazzia elegans*	**F**	*Myzus ornatus*
	C	*Aphis menthae-radicis*	**G**	*Ovatus mentharius*
	D	*Aphis affinis*	**H**	*Ovatus crataegarius*

1. Terminal process much shorter than base of last antennal segment. Siphunculi absent	A
Terminal process longer than base of last antennal segment. Siphunculi evident	2
2. Siphunculi very considerably swollen, so that maximum diameter of distal part is more than twice the minimum diameter of the proximal part	B
Siphunculi not swollen	3
3. Antennal tubercles little developed	4
Antennal tubercles well developed	6
4. Cauda triangular in dorsal view, about as long as its basal width, and as dark as siphunculi (On roots)	C
Cauda tongue-shaped, longer than its basal width, paler than siphunculi	5
5. Terminal process usually less than twice as long as base of last antennal segment, shorter than antennal segment III and only 0.55–0.8 of its length when III is 0.24 mm or longer	D
Terminal process more than twice as long as base of last antennal segment, longer than antennal segment III in small specimens but only 0.85–1.1 of its length in large specimens with III 0.24–0.4 mm	E
6. Terminal process shorter than antennal segment III. Dorsal abdomen with a conspicuous intersegmental pattern of dark ornamentation	F
Terminal process longer than antennal segment III. Dorsal abdomen without conspicuous dark ornamentation	7
7. Inner side of antennal segment I and antennal tubercle each with a forwardly-directed process, and that on the antennal tubercle being longer than its basal width in dorsal view. Siphunculi attenuated and cylindrical on distal half, as long as the distance between their bases	G
Antennal segment I without a forwardly-directed process, and that on the antennal tubercle shorter than its basal width in dorsal view. Siphunculi tapering from base to flange, shorter than the distance between their bases	H

Mulberry

Mulberry *Aphis fabae*
Toxoptera aurantii
[*Uroleucon (Uromelan) compositae*]

Mung *Aphis craccivora*
Aphis gossypii

Mustard
A *Pemphigus populitransversus*
B *Lipaphis erysimi*
C *Aphis maidiradicis*
D *Brevicoryne brassicae*
E *Myzus persicae*
F *Macrosiphum euphorbiae*

1. Terminal process much shorter than base of last antennal segment. Siphunculi absent — **A**

   Terminal process longer than base of last antennal segment. Siphunculi present — **2**

2. Antennal tubercles little developed. Length of siphunculi not more than half width of head across eyes — **3**

   Antennal tubercles well developed. Length of siphunculi greater than width of head across eyes — **4**

3. Cauda tongue-shaped, clearly longer than its basal width in dorsal view. Dorsal abdomen with little or no pigmentation anterior to siphunculi — **B**

   Cauda short, triangular, no longer than its basal width in dorsal view. Dorsal abdomen with dark sclerites on segments anterior to siphunculi — **4**

4. Siphunculi tapering from base to flange. Terminal process less than twice as long as base of last antennal segment. (On roots) — **C**

   Siphunculi barrel-shaped. Terminal process more than twice as long as base of last antennal segment. (On aerial parts) — **D**

5. Inner faces of antennal tubercles convergent. Siphunculi slightly clavate, without any polygonal reticulation

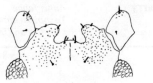

E

Inner faces of antennal tubercles divergent. Siphunculi with a subapical zone of polygonal reticulation

F

Nectarine *Myzus persicae*
(see also keys to aphids on **Peach** and **Almond**)

Nutmeg no aphids recorded

Oats

A	*Sipha maydis*	N	*Geoica utricularia*
B	*Sipha (Rungsia) flava*	O	*Forda (Pentaphis) marginata*
C	*Rhopalosiphum maidis*		[*Aphis maidi-radicis*]
D	*Rhopalosiphum padi*		[*Aulacorthum (Neomyzus) circumflexum*]
E	*Glabromyzus howardii*		[*Diuraphis noxia*]
F	*Sitobion fragariae*		[*Forda formicaria*]
G	*Sitobion miscanthi*		[*Geoica magnifica* (del Guercio 1913)]
H	*Sitobion avenae*		[*Metopolophium montanum*]
I	*Schizaphis graminum*		[*Rhopalosiphum rufiabdominalis*]
J	*Metopolophium dirhodum*		
K	*Metopolophium festucae cerealium*		
L	*Anoecia* species		
M	*Tetraneura ulmi*		

1. Terminal process similar in length to, or longer than, base of last antennal segment. (On aerial parts) 2
   Terminal process much shorter than base of last antennal segment. (On roots) 12

2. Dorsal body hairs partly or mainly long and spine-like. Siphunculi very short, truncated cones, shorter than their basal diameter   3

   Dorsal body hairs short. Siphunculi tubular, longer than their basal width 4

Oats 137

3.	Cauda broadly rounded		A
	Cauda with a knobbed apex		B

4.	Siphunculi darker than body (even if rather pale there is still a clear difference of pigmentation between body and bases of siphunculi)		5
	Siphunculi as pale as body, or dark only at extreme apices		10

5.	Siphunculi short, less than half as long as the distance between their bases	6
	Siphunculi more than half as long as the distance between their bases	7

6.	Body rather elongate. Terminal process less than 2.5 times longer than base of last antennal segment. Siphunculi less than 1.5 times longer than cauda		C
	Body ovate. Terminal process more than 3 times longer than base of last antennal segment. Siphunculi more than 1.5 times longer than cauda		D

7.	Siphunculi swollen in middle, without any polygonal reticulation		E
	Siphunculi tapering from base to flange, with a subapical zone of polygonal reticulation		8

8.	Proximal segments of antennae paler than distal segments. Siphunculi at least twice as long as cauda		F
	Antennae uniformly dark. Siphunculi less than twice as long as cauda		9

9.	Siphunculi more than 1.5 times longer than cauda. Second segment of hind tarsus less than 1.4 times longer than last segment of rostrum	G
	Siphunculi less than 1.5 times longer than cauda. Second segment of hind tarsus more than 1.4 times longer than last rostral segment	H

10.	Antennae shorter than distance from frons to bases of siphunculi. Siphunculi about half as long as the distance between their bases, and often dark at extreme apices	I
	Antennae longer than distance from frons to bases of siphunculi. Siphunculi more than three-quarters as long as the distance between their bases	11

11.	Each antennal segment pale at base and dusky at apex (base of VI paler than apex of V). Usually more than 8 caudal hairs. (Al. without dorsal abdominal pigmentation)	J
	Antennae progressively darker from base to apex (if base of VI is a little paler than apex of V, then apices of segments III and IV are not darkened). Usually less than 8 caudal hairs. (Al. with a pattern of transverse dark bars on dorsal abdomen)	K

12.	Siphunculi present as small cones or pores	13
	Siphunculi completely absent	14

13.	Total antennal length one-third or more of body length. Tarsi 2-segmented	L
	Antennae very short, less than one-fifth of body length. Tarsi 1-segmented	M

14.	Anus and anal plate displaced dorsally. Body with some hairs spatulate	N
	Anus and anal plate in normal ventral position. Body hairs all pointed	O

Oil Palm use key to aphids on **Palms**

Okra *Aphis gossypii*
Myzus persicae
[*Aphis albella* Nevsky 1951]

Olive

Olive		*Prociphilus oleae*

Onion	A	*Myzus ascalonicus*	D	*Neotoxoptera oliveri*
	B	*Myzus (Sciamyzus)*		[*Aphis fabae*]
		cymbalariae		[*Aphis gossypii*]
	C	*Neotoxoptera formosana*		[*Rhopalosiphum padi*]
				key as for **Chives**

Opium	see **Poppy**

Orange	see **Citrus**

Orchids	A	*Cerataphis orchidearum*	D	*Aulacorthum (Neomyzus)*
	B	*Aulacorthum solani*		*dendrobii*
	C	*Aulacorthum (Neomyzus)*	E	*Sitobion anselliae*
		circumflexum	F	*Sitobion luteum*
			G	*Sitobion indicum*

1. Body dorso-ventrally flattened, subcircular, wholly sclerotic with a distinctly crenulate margin and a pair of horns projecting forward from frons. Siphunculi are large round pores raised on shallow cones .... **A**

   Body ovate or spindle-shaped, mainly membranous. Siphunculi tubular 2

2. Siphunculi mainly pale, sometimes dark at extreme apices, without polygonal reticulation 3
   Siphunculi black, with a subapical zone of polygonal reticulation 5

3. Dorsal abdomen without pigmentation .... **B**
   Dorsal abdomen with a pattern of dark pigmentation 4

4. Dorsal abdomen with a large, roughly horseshoe-shaped black patch. Siphunculi tapering or cylindrical on distal half .... **C**

   Dorsal abdomen with a segmental pattern of paired dark sclerites, not joined across segments. Siphunculi swollen on distal half .... **D**

5. Dorsal abdomen without pigmentation anterior to siphunculi. Antennal segment III pale except at apex	E
Dorsal abdomen with a central dark patch anterior to siphunculi. Antennal segment III black except at very base	6
6. Dorsal abdominal patch broadly oval with only small marginal indentations. Cauda with 6–8 hairs	F
Dorsal abdominal patch very irregular in outline with large intersegmental marginal indentations. Cauda with 9–13 hairs	G

Palms

	A	Cerataphis formosana	E	Astegopteryx nipae
	B	Cerataphis palmae (= variabilis)	F	Astegopteryx rhapidis
	C	Cerataphis lataniae	G	Hysteroneura setariae
	D	Astegopteryx rappardi	H	Schizaphis rotundiventris (= cyperi)

1. Head with a pair of forwardly directed frontal horns. Siphunculi present merely as shallow cones or pores	2
Head without frontal horns. Siphunculi tubular	7

2. Body dorso-ventrally flattened, subcircular, wholly sclerotic with a distinct crenulate margin due to a continuous row of wax glands	3
Body elongate oval, not dorso-ventrally flattened, not wholly sclerotic and without a distinct crenulate margin	5

3. Frontal horns finger-like, rounded apically, very close together	A
Frontal horns pointed, well separated	4

4. Underside of head with at least one pair of thick, dagger-shaped hairs situated latero-ventral of bases of frontal horns. Cauda with 5–7, mainly long, hairs	B
Underside of head with only fine hairs. Cauda with 10–16 hairs of various sizes	C

Paprika 141

5.	Dorsal abdomen with segmental markings consisting of broad transverse bars, sometimes broken into paired, dark patches. Large lateral abdominal wax glands, segmentally arranged in often rather irregular groups	D
	Dorsal abdomen without segmental markings, wholly dusky. Wax glands either absent from segments anterior to siphunculi or all grouped in short rows with long axes parallel	6
6.	Groups of lateral wax glands usually well developed on all abdominal segments. Length of antennal segment III 1.0–1.5 times that of segment IV (when antennae are 5-segmented)	E
	Lateral wax glands reduced or absent on abdominal segments anterior to siphunculi. Length of antennal segment III 1.5–2.0 times that of segment IV (when antennae are 5-segmented)	F
7.	Cauda very pale, more than half the length of the siphunculi	G
	Cauda dusky or dark, less than half the length of the siphunculi	H

Paprika A *Aphis gossypii*
 B *Myzus persicae*
 C *Macrosiphum euphorbiae*
  key as for **Chili**

Parsley A *Cavariella aegopodii* D *Aphis fabae*
 B *Hyadaphis foeniculi* E *Rhopalosiphoninus latysiphon*
 C *Dysaphis apiifolia petroselini* F *Myzus persicae*
  G *Myzus ornatus*

1.	Eighth abdominal segment with a posteriorly projecting supracaudal process	A
	Eighth abdominal segment without a supracaudal process	2
2.	Antennal tubercles weakly developed	3
	Antennal tubercles well developed	5

3.	Siphunculi similar in length to cauda or a little longer, and clavate	B
	Siphunculi usually clearly longer than cauda, and tapering from base to flange	4
4.	Cauda helmet-shaped and no longer than its basal width in dorsal view	C
	Cauda tongue-shaped, much longer than its basal width in dorsal view	D
5.	Siphunculi jet black, narrow basally and bulbous on distal half	E
	Siphunculi pale, tapering or slightly clavate	6
6.	Siphunculi slightly clavate. Dorsal abdomen without dark ornamentation	F
	Siphunculi tapering from base to flange. Dorsal abdomen with an intersegmental pattern of dark ornamentation	G

Parsnip 143

Parsnip

A	*Cavariella aegopodii*	I	*Aphis gossypii*
B	*Cavariella theobaldi*	J	*Aphis citricola*
C	*Semiaphis* species	K	*Aphis armoraciae*
D	*Anuraphis subterranea*	L	*Aphis helianthi*
E	*Dysaphis crataegi kunzei*	M	*Aphis fabae*
F	*Dysaphis bonomii*	N	*Myzus ornatus*
G	*Hyadaphis foeniculi*	O	*Myzus persicae*
H	*Aphis decepta*	P	*Macrosiphum euphorbiae*

(Some of these species, especially **F**, **H**, and **L**, may only occur on wild parsnip.)

1. Eighth abdominal tergite with supracaudal process 2

   Eighth abdominal tergite without a supracaudal process 3

2. Siphunculi clavate. Terminal process less than 1.5 times longer than base of last antennal segment .... A

   Siphunculi tapering from base to flange. Terminal process more than twice as long as base of last antennal segment .... B

3. Siphunculi very small and flangeless, only about half of length of cauda, with aperture slanted towards mid-line C

   Siphunculi more than three-quarters of length of cauda if not much longer 4

4. Cauda short and broad, rounded or helmet-shaped, not as long as its basal width in dorsal view 5
   Cauda tongue-shaped, at least as long as its basal width in dorsal view 7

5. Siphunculi in form of truncated cones less than twice as long as their basal diameters, covered in densely packed rows of small spinules .... D

   Siphunculi tubular, more than 2.5 times longer than their basal diameters, with imbrication but without regular rows of small spinules 6

6. Siphunculi less than 3.5 times longer than their basal diameters. Dorsal abdominal pigmentation never in form of complete bars on tergites anterior to siphunculi	E
Siphunculi 4 or more times longer than their basal diameters. Dorsal abdomen often with complete transverse bars on all tergites	F
7. Antennal tubercles weakly developed	8
Antennal tubercles well developed	14
8. Siphunculi clavate	G
Siphunculi cylindrical or tapering	9
9. Cauda paler than siphunculi	10
Cauda and siphunculi similarly pigmented	11
10. Cauda with more than 10 hairs. (Al. with more than 40 sensoria on antennal segment III)	H
Cauda with 4–7 hairs. (Al. with less than 12 sensoria on antennal segment III)	I
11. Abdominal tergites 7 and 8 unpigmented	J
Abdominal tergites 7 and 8 with dark transverse bars or dashes	12
12. Antennal segment III (in apt.) with 5–12 sensoria. Hairs on hind femur all shorter than basal width of hind femur	K
Antennal segment III without sensoria. Hairs on hind femur long and fine, the longest exceeding the basal width of the hind femur	13
13. Terminal process 0.6–0.9 times as long as antennal segment III	L
Terminal process 1.0–1.6 times as long as antennal segment III	M
14. Dorsal abdomen with an intersegmental pattern of dark ornamentation	N
Dorsal abdomen without dark ornamentation	15

Pawpaw

15. Siphunculi slightly clavate, without polygonal reticulation. Antennal tubercles with inner faces convergent .... O

    Siphunculi not clavate, with a subapical zone of polygonal reticulation. Inner faces of antennal tubercles divergent .... P

Pawpaw	A	*Aphis gossypii*	C	*Macrosiphum euphorbiae*
	B	*Myzus persicae*		use key to aphids on **Chili**

Pea (Garden)	A	*Aphis fabae*	E	*Acyrthosiphon pisum*
	B	*Myzus persicae*		[*Acyrthosiphon pisivorum* Zhang 1980]
	C	*Macrosiphum euphorbiae*		[*Aulacorthum (Neomyzus) circumflexum*]
	D	*Aulacorthum solani*		

1. Cauda and siphunculi black. Antennal tubercles not developed .... A

   Cauda pale, siphunculi mainly pale, sometimes dark apically. Antennal tubercles well developed 2

2. Siphunculi slightly clavate. Inner faces of antennal tubercles convergent .... B

   Siphunculi not clavate. Inner faces of antennal tubercles parallel or divergent 3

3. Siphunculi with a subapical zone of polygonal reticulation. Antennae bearing hairs about as long as the basal diameter of antennal segment III C

   Siphunculi without polygonal reticulation. Antennae with only inconspicuous hairs, much shorter than the basal diameter of antennal segment III 4

4. Inner faces of antennal tubercles parallel.
   Siphunculi more than twice as long as cauda ... **D**

   Inner faces of antennal tubercles divergent.
   Siphunculi very attenuate distally, but less
   than twice as long as the very large cauda ... **E**

Peach

A	*Pterochloroides persicae*	J	*Myzus persicae*
B	*Hyalopterus pruni*		[*Aphis citricola;*
C	*Hyalopterus amygdali*		*Hysteroneura setariae;*
D	*Brachycaudus persicae*		*Macrosiphum euphorbiae;*
E	*Brachycaudus (Appelia) schwartzi*		*Myzus cerasi; Myzus yamatonis; Phorodon persifoliae* Shinji 1922;
F	*Brachycaudus helichrysi*		*Rhopalosiphum momo* Shinji 1927;
G	*Brachycaudus (Thuleaphis) amygdalinus*		*Rhopalosiphum nymphaeae;*
H	*Tuberocephalus momonis*		*Rhopalosiphum rufiabdominalis*]
I	*Myzus varians*		

---

1. Terminal process much shorter than base of
   last antennal segment. Siphunculi in form of
   dark, hairy cones ... **A**

   Terminal process longer than base of last
   antennal segment. Siphunculi tubular ... 2

2. Siphunculi shorter and much thinner than
   cauda ... 3
   Siphunculi longer than cauda ... 4

3. Siphunculi narrow, 3–4 times longer than
   their maximum diameter ... **B**

   Siphunculi rather broader, about twice as long
   as their maximum diameter ... **C**

4. Cauda short and broad, shorter than its
   basal width in dorsal view. Spiracles widely
   open, large and circular ... 5
   Cauda longer than its basal width in dorsal
   view. Spiracles smaller, oval or kidney-shaped ... 8

5. Siphunculi dark. Dorsal cuticle extensively
   sclerotized, black ... 6
   Siphunculi pale at least basally. Dorsal abdomen
   pale or with little sclerotization anterior to
   siphunculi ... 7

Peanut 147

6. Sclerotization of dorsal abdomen almost complete, extending laterally to spiracles and not segmentally divided. Anterior part of mesosternum with a pair of mammariform processes ... **D**

   Sclerotization of dorsal abdomen partly segmentally divided especially on anterior tergites, not extending laterally to spiracles. Mesosternum without mammariform processes ... **E**

7. Cauda helmet-shaped in dorsal view, almost as long as its basal width which is less than the length of the last rostral segment. Abdomen without any dark dorsal markings ... **F**

   Cauda very broadly rounded, much broader than long in dorsal view, its basal width greater than the length of the last rostral segment. Posterior abdominal tergites often with dark markings ... **G**

8. Siphunculi wholly dark, about 3 times as long as their basal diameter, very rough surfaced with spinules, small hairs, and coarse imbrication. Apex of cauda rather acutely pointed ... **H**

   Siphunculi pale, or dark only on distal half, about 4 times their basal diameter, imbricated but without spinules or hairs. Cauda with apex blunt in dorsal view ... 9

9. Siphunculi with basal half pale and distal half dark ... **I**

   Siphunculi wholly pale, except sometimes at extreme apex ... **J**

Peanut   see **Groundnut**

Pear   A *Aphanostigma piri*   F *Pterochloroides persicae*
              B *Eriosoma lanigerum*   G *Pyrolachnus pyri*
              C *Eriosoma (Schizoneura)*   H *Anuraphis subterranea*
                 lanuginosum   I *Anuraphis farfarae*
              D *Eriosoma (Schizoneura)*   J *Anuraphis pyrilaseri*
                 pyricola   K *Dysaphis pyri*
              E *Nippolachnus pyri*   L *Dysaphis reaumuri*

M	*Sappaphis piri*	V	*Aphis citricola*
N	*Melanaphis pyraria*		[*Anuraphis catonii;*
O	*Fimbriaphis gentneri*		*Aphanostigma iaksuiense;*
P	*Toxoptera aurantii*		*Aphis craccivora;*
Q	*Toxoptera citricidus*		*Brachycaudus almatinus*
R	*Schizaphis* species		Nevsky 1951;
S	*Rhopalosiphum insertum*		*Brachycaudus cardui;*
T	*Aphis gossypii*		*Brachycaudus persicae;*
U	*Aphis pomi*		*Dysaphis multisetosa* Basu 1969; *Dysaphis (Pomaphis) plantaginea; Longistigma xizangensis* Zhang 1981; *Macrosiphum rosae; Myzus persicae; Ovatus insitus; Nearctaphis bakeri; Sinomegoura pyri* Ghosh & Raychaudri 1968]

1. Antenna 3-segmented. Body small, pyriform, broadest anteriorly — **A**

   Antenna 5- or 6-segmented. Body ovate or elongate — 2

2. Terminal process much shorter than (less than half) base of last antennal segment — 3
   Terminal process at least two-thirds as long as base of last antennal segment, usually much longer — 7

3. Total length of hind leg much less than length of body — 4
   Total length of hind leg greater than length of body — 6

4. Distinct dorsal abdominal wax pore-plates present, consisting of groups of cells each enclosing a very small, narrow or sub-divided central area. (On roots, trunk, or branches) — **B**

   Dorsal abdominal wax pore-plates absent or present, if present each consisting of a ring of cells around a large undivided central area. (On roots only) — 5

5. Body length 2.0–2.7 mm. Length of last rostral segment 0.22–0.25 mm. Cauda with 2–4 hairs — **C**
   Body length 1.3–1.9 mm. Length of last rostral segment 0.12–0.16 mm. Cauda with 5 or more hairs — **D**

Pear 149

6.	Body narrowly spindle-shaped, more than twice as long as its maximum width. Body and appendages bearing long fine hairs more than twice as long as the diameter of the antennae	**E**
	Body ovate, less than twice as long as its maximum width. Body and appendages bearing thick hairs, shorter than the diameter of the antennae	7
7.	Abdomen with a double longitudinal row of large pigmented spinal tubercles. (Al. with maculate forewings)	**F**
	Abdomen without pigmented spinal tubercles. (Forewings of al. not maculate)	**G**
8.	Cauda helmet-shaped, triangular or rounded in dorsal view, not longer than its basal width	9
	Cauda tongue-shaped, clearly longer than its basal width in dorsal view	14
9.	Siphunculi short dark, truncated cones covered with transverse rows of densely packed small spinules. (Leaf galls containing fundatrix and al. spring migrants only)	10
	Siphunculi dark or pale, somewhat imbricated but without rows of spinules	12
10.	(Apt. fundatrix.) Terminal process longer than base of last antennal segment. Abdominal segments 1–4 each with a rather large, dark, marginal sclerite, bearing a large flat tubercle. (Al. spring migrant with spinal and marginal tubercles on abdominal tergites 6 and 7)	**H**
	(Apt. fundatrix.) Terminal process shorter than base of last antennal segment. Abdominal segments 1–4 with marginal tubercles not on large dark sclerites. (Al. spring migrant without spinal and marginal tubercles on abdominal segments 6 and 7)	11
11.	(Apt. fundatrix.) Abdominal segment 6 usually with marginal tubercles. (Al. spring migrant with terminal process 3.5–4.5 times longer than base of last antennal segment)	**I**
	(Apt. fundatrix.) Abdominal segment 6 without tubercles. (Al. spring migrant with terminal process 5–6 times longer than base of last antennal segment)	**J**

12.	Siphunculi pale or only a little dusky	13
	Siphunculi dark	K

13.	Antennae with sparse, short, blunt hairs. Siphunculi without hairs. (Al. with a solid dark patch on abdominal tergites 3–5)	L
	Antennae with numerous very fine long hairs. Siphunculi with a few long fine hairs. (Al. with dark transverse bars on all abdominal tergites)	M

14.	Siphunculi shorter than cauda. Dorsal abdomen with a solid dark patch	N
	Siphunculi longer than cauda. Dorsal abdomen mainly unpigmented	15

15.	Total antennal length greater than body length. Terminal process more than 5 times longer than base of last antennal segment	O
	Total antennal length less than body length. Terminal process less than 5 times longer than base of last antennal segment	16

16.	Stridulatory apparatus present. Siphunculi less than 1.5 times longer than cauda	17
	Stridulatory apparatus absent. Siphunculi usually more than 1.5 times longer than cauda	18

17.	Cauda usually with less than 20 hairs. Hairs on antennal segment III shorter than diameter of segment at base. Body length generally less than 2 mm	P
	Cauda with more than 20 hairs. Hairs on antennal segment III longer than basal diameter of segment. Body length generally greater than 2 mm	Q

Pear, Avocado 151

18.	Lateral tubercle on abdominal segment 7 positioned posteriodorsally to spiracle. Terminal process more than 3 times longer than base of last antennal segment	19
	Lateral tubercle on abdominal segment 7 positioned posterioventrally to spiracle. Terminal process usually less than 3 times longer than base of last antennal segment	20
19.	Siphunculi tapering from base to flange	R
	Siphunculi cylindrical over most of length, slightly swollen distally, with a constriction just proximal to flange	S
20.	Cauda paler than siphunculi, with usually 4–6 hairs	T
	Cauda dark, usually with more than 6 hairs	21
21.	Lateral tubercles present on abdominal segments 2–4. Cauda usually with more than 13 hairs. Length of last segment of rostrum more than 130 μm	U
	Lateral tubercles absent from abdominal segments 2–4. Cauda usually with less than 12 hairs. Length of last rostral segment less than 120 μm	V

Pear, Avocado   see **Avocado Pear**

Pecan
- A *Phylloxera* species
- B *Longistigma caryae*
- C *Melanocallis caryaefoliae*
- D *Monelliopsis pecanis*
- E *Monellia caryella*
  [*Monellia medina* Bissell 1978]
  [*Monelliopsis nigropunctata*]
  [*Protopterocallis pergandei* Bissell 1978]

1.	Antennae 3-segmented. Body pear-shaped, broadest anteriorly	A
	Antennae 6-segmented. Body elongate or oval	2

2.	Large, densely hairy aphid (body length over 4 mm) on bark	B
	Smaller aphids (body length less than 3 mm), adults sparsely haired and all alate. (On leaves)	3

3.	(Al.) Siphunculi black, truncate, similar in length to antennal segment I. Thorax, middle, and hind femora very dark. Large, dark paired spinal and lateral tubercles on anterior abdominal segments	C
	(Al.) Siphunculi pale, very small, and tuberculate. Thorax, middle, and hind femora mainly pale or dusky. Abdominal tubercles small	4

4.	(Al.) Abdominal tubercles variably pigmented, but the lateral tubercle on segment 5 always forming a dark spot immediately anterior to each siphunculus. Sides of head, thorax, and wings never with broad, dark bands. Basal part of antennal segment III hardly swollen — siphunculus	D
	(Al.) Lateral tubercles on abdominal segment 5 not deeply pigmented. Summer and autumn (but not spring) generations with a broad, dark band around front and sides of head, along sides of thorax and anterior abdominal segments; and a broad dark leading edge to forewing. Basal part of antennal segment III clearly swollen — siphunculus	E

Pepper, Black   *Aphis gossypii*

Pepper (Green or Red)   see **Chili**

Pigeon Pea	A	*Cervaphis rappardi indica*	D	*Aphis gossypii*
	B	*Aphis craccivora*	E	*Sitobion nigrinectaria*
	C	*Aphis citricola*	F	*Megoura lespedezae*

1.	Body with large, branched, lateral hair-bearing processes	A
	Body without large lateral processes	3
2.	Cauda dark	3
	Cauda paler than siphunculi	4

Pineapple 153

3.	Dorsal abdomen with a solid black patch	B
	Dorsal abdomen unpigmented	C
4.	Siphunculi shorter than head width across eyes. Terminal process less than 3.5 times longer than base of last antennal segment. Cauda less than 2.5 times longer than last rostral segment	D
	Siphunculi longer than head width across eyes. Terminal process more than 3.5 times longer than base of last antennal segment. Cauda more than 3 times longer than last rostral segment	5
5.	Siphunculi more than 1.5 times longer than cauda, with a subapical zone of polygonal reticulation	E
	Siphunculi less than 1.5 times longer than cauda, without any polygonal reticulation	F

Pineapple   No aphids recorded

Pinks   see **Carnation**

Pistachio
- A *Smynthurodes betae*
- B *Paracletus cimiciformis*
- C *Slavum* species
- D *Geoica utricularia*
- E *Forda (Pentaphis) hirsuta*

Key to apterous viviparae in galls (but *not* fundatrices):

1.	Body with numerous long hairs. Antennal segment II elongate, much longer than I, with long fine hairs, the longest of them exceeding the basal diameter of the segment	A
	Body hairs sparse, or short if numerous. Antennal segment II similar in length to I, with hairs shorter than its basal diameter	2
2.	Terminal process very short, less than one-quarter as long as base of last segment	3
	Terminal process more than one-quarter as long as base of last antennal segment	4
3.	Antennae 6-segmented	B
	Antennae 5-segmented	C
4.	Primary sensoria on antennal segment V completely fringed with hairs. Dorsal abdominal wax gland cells rather loosely associated	D
	Primary sensoria on antennal segment V without a complete fringe of hairs. Dorsal abdominal wax gland cells fully associated into plates	E

Plum

	A	Pterochloroides persicae	J Aphis citricola
	B	Hyalopterus pruni	[Aphis chetansapa Hottes
	C	Brachycaudus helichrysi	& Frison 1931]
	D	Brachycaudus (Appelia)	[Aphis longicauda Baker 1920]
		prunicola	[Brachycaudus almatinus
	E	Brachycaudus cardui	Nevsky 1951]
	F	Brachycaudus persicae	[Brachycaudus divaricatae
	G	Phorodon humuli	Shaposhnikov 1956]
	H	Hysteroneura setariae	[Myzus persicae]
	I	Rhopalosiphum	[Rhopalosiphum padi]
		nymphaeae	[Schizaphis longicaudata]

1. Terminal process much shorter than base of last antennal segment .... A

   Terminal process much longer than base of last antennal segment 2

2. Siphunculi very small and thin, shorter than cauda .... B

   Siphunculi longer than cauda 3

3. Cauda rounded or helmet-shaped in dorsal view, not longer than its basal width 4

   Cauda tongue-shaped, much longer than its basal width in dorsal view 7

4. Siphunculi short, less than twice as long as cauda 5
   Siphunculi more than twice as long as cauda 6

5. Dorsal abdomen pale or dusky, without extensive sclerotization C
   Dorsal abdomen mostly sclerotized, black .... D

6. Last segment of rostrum more than 0.18 mm long. Dorsal abdomen variably pigmented, often quite pale. Siphunculi usually pale or dusky .... E
   Last segment of rostrum less than 0.175 mm long. Dorsal abdomen always extensively sclerotized, black. Siphunculi black .... F

Pomegranate

7. Siphunculi pale. Antennal tubercles well
    developed with long finger-like processes G

    Siphunculi dark. Antennal tubercles weakly
    developed, without processes 8

8. Cauda very pale and conspicuous, at least
    one-tenth of body length H
    Cauda dusky or dark, black if more than
    one-tenth of body length 9

9. Siphunculi more than twice as long as cauda,
    cylindrical over most of length, slightly
    swollen subapically with a constriction just
    proximal to flange I

    Siphunculi less than twice as long as cauda,
    tapering from base to flange J

Pomegranate A *Aphis punicae*
  B *Aphis gossypii*
      [*Aphis citricola*]
      [*Aphis craccivora*]
      [*Aulacorthum (Neomyzus) circumflexum*]
      [*Myzus ornatus*]
      [*Myzus persicae*]

1. Siphunculi often rather pale, usually less
    than 1.5 times longer than cauda which has
    7–9 hairs. (Al. often with one or more
    sensoria on antennal segment IV) A
    Siphunculi dark, usually more than 1.5 times
    longer than cauda which has 4–7 hairs.
    (Al. without sensoria on antennal segment IV) B

Poppy A *Aphis fabae*
  B *Myzus persicae*
  C *Acyrthosiphon ilka/ bidentis*
      [*Aphis gossypii*]
      [*Lipaphis erysimi*]
      [*Rhopalosiphum padi*]
      [*Rhopalosiphum rufiabdominalis*]

1.	Siphunculi and cauda black	A
	Siphunculi and cauda pale	2

2.	Inner faces of antennal tubercles convergent. Siphunculi slightly clavate	B
	Inner faces of antennal tubercles divergent. Siphunculi tapering from base to flange	C

Potato

- A *Smynthurodes betae*
- B *Pemphigus* species
- C *Rhopalosiphum rufiabdominalis*
- D *Aphis fabae*
- E *Aphis nasturtii*
- F *Aphis frangulae/ gossypii*
- G *Rhopalosiphoninus latysiphon*
- H *Aulacorthum (Neomyzus) circumflexum*
- I *Aulacorthum solani*
  Macrosiphum euphorbiae
  Myzus ascalonicus   see key to
  Myzus ornatus   polyphagous
  Myzus persicae   Macrosiphini
  [*Acyrthosiphon malvae*]

1.	Terminal process much shorter than base of last antennal segment. Siphunculi absent	2
	Terminal process much longer than base of last antennal segment. Siphunculi present	3

2.	Body and appendages with numerous hairs. Abdomen without wax pore plates	A
	Body and appendages with very sparse, short hairs. Abdomen with wax pore-plates on posterior segments	B

3.	Antennal tubercles weakly developed, not projecting beyond medial part of frons in dorsal view	4
	Antennal tubercles well developed	7

Potato 157

4.	Antennal hairs very long, the longest more than 3 times longer than the basal diameter of antennal segment III	C
	Antennal hairs rarely more than twice as long as basal diameter of antennal segment III	5
5.	Cauda black with more than 10 hairs	D
	Cauda paler than siphunculi, with less than 10 hairs	6
6.	Siphunculi pale or dusky. Some hairs on hind femur long and fine, the longest of them exceeding the diameter of the femur at base (illustr.). (Al. with 8–12 sensoria on antennal segment III, 2–4 on segment IV)	E
	Siphunculi dark. Hairs on hind femur all rather short. (Al. with 3–7 sensoria on antennal segment III, none on segment IV)	F
7.	Siphunculi jet black with distal part very swollen	G
	Siphunculi pale or dusky, tapering, cylindrical, or only slightly swollen	8
8.	Dorsal abdomen with a large, roughly horseshoe-shaped dark patch	H
	Dorsal abdomen unpigmented	I

Potato, Sweet   see **Sweet Potato**

Prune   as for **Plum**

Pumpkin
A *Smynthurodes betae*
B *Aphis craccivora*
C *Aphis maidiradicis*
D *Aphis fabae*
E *Aphis gossypii*
F *Myzus persicae*
G *Macrosiphum euphorbiae*
[*Acyrthosiphon vasiljevi* Mordvilko 1915; *Aulacorthum magnoliae*; *Brachycaudus helichrysi*]

---

1. Terminal process much shorter than base of last antennal segment. Siphunculi absent — **A**

   Terminal process longer than base of last antennal segment. Siphunculi evident — 2

---

2. Antennal tubercles absent or weakly developed — 3

   Antennal tubercles well developed — 5

---

3. Dorsal abdomen with a large black patch centred on abdominal tergites 4–5 — **B**
   Dorsal abdomen unpigmented or with a little broken pigmentation anterior to siphunculi — 4

---

4. Cauda bluntly conical, about as long as its basal width. Terminal process less than twice as long as base of last antennal segment — **C**

   Cauda tongue-shaped, much longer than its basal width. Terminal process more than 2.5 times longer than base of last antennal segment — 5

---

5. Cauda black. Dorsal abdomen with dark markings, at least on posterior segments — **D**
   Cauda pale. Dorsal abdomen without dark markings — **E**

Pyrethrum 159

6. Inner faces of antennal tubercles convergent.
   Siphunculi slightly clavate, without any
   polygonal reticulation .... **F**

   Inner faces of antennal tubercles divergent.
   Siphunculi with a subapical zone of polygonal
   reticulation .... **G**

Pyrethrum   no aphids recorded

Quince   A   *Watabura nishiyae*   J   *Myzus ornatus*
             B   *Pterochloroides persicae*   K   *Myzus persicae*
             C   *Anuraphis farfarae*   L   *Ovatus insitus* (or *crataegarius*)
             D   *Dysaphis pyri*   [*Eriosoma lanigerum;*
             E   *Rhopalosiphum insertum*   *Eriosoma (Schizoneura)*
             F   *Aphis craccivora*   *lanuginosum; Eriosoma*
             G   *Aphis pomi*   *(Schizoneura) pyricola;*
             H   *Aphis citricola*   *Nearctaphis bakeri;*
             I   *Aulacorthum solani*   *Nearctaphis crataegifoliae*]

---

1. Siphunculi absent. Antennae very short, 4-
   segmented (illustr.). Tarsi with only one claw
   developed, the other much reduced .... **A**

   Siphunculi present as tubes or dark hairy
   cones. Antennae 5- or 6-segmented. Tarsi
   with two similar-sized claws 2

---

2. Terminal process of antenna very short.
   Siphunculi in form of dark hairy cones .... **B**

   Terminal process longer than base of last
   antennal segment. Siphunculi tubular 3

---

3. Cauda short and broad, helmet-shaped, or
   rounded in dorsal view, no longer than its
   basal width 4
   Cauda tongue-shaped, clearly longer than its
   basal width in dorsal view 5

4.	Siphunculi short, truncated cones, about twice as long as basal diameter, with transverse rows of closely packed spinules	C
	Siphunculi 3 or more times longer than their basal diameter, with irregular imbrications but without rows of spinules	D
5.	Antennal tubercles weakly developed, not projecting beyond medial part of frons in dorsal view	6
	Antennal tubercles well developed	9
6.	Siphunculi cylindrical over much of length, slightly swollen distally, with a constriction just proximal to flange	E
	Siphunculi tapering from base to flange	7
7.	Dorsal abdomen with a large black patch	F
	Dorsal abdomen unpigmented	8
8.	Lateral tubercles present on abdominal segments 2–4. Cauda with usually more than 13 hairs. Length of last segment of rostrum more than 130 μm	G
	Lateral tubercles absent from abdominal segments 2–4. Cauda usually with less than 12 hairs. Length of last rostral segment less than 120 μm	H
9.	Inner faces of antennal tubercles parallel. Antennal segment III with a sensorium near its base	I
	Inner faces of antennal tubercles convergent. Antennal segment III without any sensoria	10
10.	Dorsal abdomen with an intersegmental pattern of dark ornamentation	J
	Dorsal abdomen without dark ornamentation	11
11.	Siphunculi slightly clavate. Antennal segment VI pale, with terminal process 2.5–4.5 times longer than base	K
	Siphunculi tapering from base to flange. Antennal segment VI dark, with terminal process 5–8 times longer than base	L

Quinine

Quinine	A	*Toxoptera odinae*	C	*Aphis gossypii*
	B	*Toxoptera aurantii*	D	*Aphis citricola*

1. Siphunculi clearly shorter than cauda	A
Siphunculi similar in length to, or longer than, cauda	2

2. Terminal process more than 3.5 times longer than base of last antennal segment. Stridulatory apparatus present	B
Terminal process less than 3 times longer than base of last antennal segment. No stridulatory apparatus	3

3. Cauda paler than siphunculi	C
Cauda as dark as siphunculi	D

Radish

Radish	A	*Myzus persicae*
	B	*Brevicoryne brassicae*
	C	*Lipaphis erysimi*
		[*Akkaia sikkimensis* Agarwala & Raychaudhuri 1977]

use first part of key to aphids on **Cabbage**

Raffia

Raffia	*Cerataphis palmae*

Ragi

Ragi	A	*Sitobion leelamaniae*	E	*Tetraneura nigriabdominalis*
	B	*Hysteroneura setariae*	F	*Tetraneura yezoensis*
	C	*Rhopalosiphum maidis*		[*Schizaphis graminum;*
	D	*Rhopalosiphum padi*		*Sitobion avenae;*
				Sitobion miscanthi]

1. Terminal process longer than base of last antennal segment. Siphunculi tubular	2
Terminal process shorter than base of last antennal segment. Siphunculi in form of small, shallow cones	5

2.	Antennal tubercles moderately developed, with inner faces broadly divergent. Siphunculi with a subapical zone of polygonal reticulation	A
	Antennal tubercles weakly developed, not projecting beyond medial part of frons in dorsal view. Siphunculi without polygonal reticulation	3
3.	Siphunculi black, cauda long and pale	B
	Siphunculi and cauda both dark	4
4.	Body rather elongate. Terminal process less than 2.5 times longer than base of last antennal segment	C
	Body broadly oval. Terminal process more than 3 times longer than base of last antennal segment	D
5.	Last segment of rostrum short, less than 0.15 mm long and less than 1.8 times longer than second segment of hind tarsus	E
	Last segment of rostrum more than 0.2 mm long and more than 1.8 times longer than second segment of hind tarsus	F

Rape   as for **Mustard**

Raspberry, American Red

    A *Aphis rubicola*
    B *Illinoia rubicola*
    C *Amphorophora agathonica*
       [*Aphis rubifolii*]

1.	Small aphid (body length less than 2 mm), with siphunculi tapering from base to flange	A
	Large aphid (body length more than 3 mm), with siphunculi swollen on distal half	2
2.	Siphunculi dark, at least distally, with a subapical zone of polygonal reticulation distal to swollen part	B
	Siphunculi pale, without subapical reticulation	C

Raspberry, Black (Blackcap)

A	*Amphorophora sensoriata*	**C**	*Amphorophora rubitoxica*
B	*Amphorophora rubicumberlandi*		[*Aphis idaei; Aphis rubicola*]

1.	Antennae wholly dusky, with sensoria on segment III extending at least three-quarters of length of segment	**A**
	Antennae mainly pale with only apices of segments III–V dusky or dark, sensoria on III not extending more than half way towards apex of segment	**2**
2.	Siphunculi pale. Last segment of rostrum shorter than second segment of hind tarsus	**B**
	Siphunculi dusky. Last segment of rostrum longer than second segment of hind tarsus	**C**

Raspberry, European

A	*Aphis idaei*	**E**	*Macrosiphum euphorbiae*
B	*Aphis gossypii*		[*Macrosiphum funestum;*
C	*Myzus ornatus*		*Matsumuraja hirakurensis*
D	*Amphorophora idaei*		Sorin 1971;
			Rhopalosiphum insertum]

1.	Antennal tubercles weakly developed	**2**
	Antennal tubercles well developed	**3**
2.	Siphunculi pale or dusky, thinner distally than middle part of hind tibia, and more than 2.5 times longer than cauda	**A**
	Siphunculi dark, thicker distally than the middle part of the hind tibia, and less than 2.5 times longer than cauda	**B**
3.	Inner faces of antennal tubercles convergent. Small aphid, with an intersegmental pattern of dark ornamentation	**C**
	Inner faces of antennal tubercles divergent. Larger aphid, without dorsal abdominal pigmentation	**4**
4.	Siphunculi swollen on distal part, without any polygonal reticulation	**D**
	Siphunculi not swollen, with a subapical zone of polygonal reticulation	**E**

Rhododendron

A large aphid fauna occurs on *Rhododendron* species in South East Asia; in some cases these are only keyed to genus, with the species described from *Rhododendron* in each genus listed separately in square brackets.

A	*Vesiculaphis* species	G	*Chaetomyzus rhododendri*
	[*V. caricis* (Fullaway 1910)]	H	*Illinoia rhokalaza*
	[*V. kongoensis* Takahashi 1965]	I	*Illinoia rhododendri*
	[*V. grandis* Basu 1964]	J	*Illinoia azaleae*
	[*V. rhododendri* A.K. Ghosh & Raychaudhuri 1972]	K	*Illinoia lambersi*
		L	*Neoacyrthosiphon* species
B	*Toxoptera odinae*		[*N. rhododendri* M.R. Ghosh, A.K. Ghosh, & Raychaudhuri 1971]
C	*Aphis citricola*		[*N. taiheisanum* (Takahashi 1935)]
D	*Indiaphis* species		[*N. (Pseudoacyrthosiphon) holsti* (Takahashi 1935)]
	[*I. crassicornis* Basu 1969]		[*N. (Pseudoacyrthosiphon) takahashii* A.K. Ghosh 1969]
	[*I. rostrata* A.K. Ghosh & Raychaudhuri 1972]		
E	*Sinomegoura rhododendri*		
F	*Indomasonaphis indica*	M	*Ericolophium itoe*

---

1. Siphunculi conspicuously warty, usually swollen in middle, often with a slight S-curve .... **A**

    Siphunculi smooth or imbricated, not warty; cylindrical or clavate **2**

---

2. Antennal tubercles weakly developed or undeveloped **3**

    Antennal tubercles moderately to well developed, projecting above medial part of frons in dorsal view **5**

---

3. Siphunculi shorter than cauda. Stridulatory apparatus present .... **B**

    Siphunculi longer than cauda. Stridulatory apparatus absent **4**

---

4. Cauda dark and more than 1.5 times its basal width in dorsal view. Siphunculi dark, more than 2.5 times longer than their basal width .... **C**

    Cauda pale, less than 1.5 times its basal width in dorsal view. Siphunculi pale or dusky, very stout, less than twice as long as their basal width .... **D**

---

5. Siphunculi similar in length to, or shorter than, cauda .... **E**

    Siphunculi much longer than cauda **6**

6.	Siphunculi clavate (if only slightly clavate, then there is a subapical zone of polygonal reticulation)	7
	Siphunculi not clavate, without any polygonal reticulation	12
7.	Dorsal abdominal hairs with tuberculate bases	8
	Dorsal abdominal hairs without raised bases	9
8.	Cauda with numerous hairs (about 50)	F
	Cauda with 5–6 hairs	G
9.	Siphunculi wholly dark	H
	Siphunculi pale, at least at base	10
10.	Hairs on antennal segment III inconspicuous, very short and blunt, the longest less than half the basal diameter of the segment	11
	Hairs on antennal segment III erect and conspicuous, the longest of them about equal to the basal diameter of the segment	I
11.	Last segment of rostrum less than 0.14 mm long. Length of second segment of hind tarsus more than 1.5 times the maximum diameter of the swollen part of the siphunculus	J
	Last segment of rostrum more than 0.14 mm long. Length of second segment of hind tarsus less than 1.5 times the maximum diameter of the swollen part of the siphunculus	K
12.	Dorsal abdominal hairs with papillate or tuberculate bases	L
	Dorsal abdominal hairs without raised bases	M

Rhubarb   A   *Dysaphis radicola*
                   B   *Aphis fabae*
                   C   *Macrosiphum euphorbiae*
                         [*Aphis rheicola* Nevsky 1951]
                         [*Xerophilaphis rhei* Nevsky 1951]

1.	Cauda helmet-shaped in dorsal view, shorter than its basal width	A
	Cauda tongue-shaped, much longer than its basal width	2

2. Antennal tubercles weakly developed, cauda
    black, siphunculi black without polygonal
    reticulation   B

    Antennal tubercles well developed, cauda pale,
    siphunculi at least basally pale, with a sub-
    apical zone of polygonal reticulation   C

Rice

A	Sipha glyceriae		N	Anoecia corni
B	Sipha flava		O	Anoecia fulviabdominalis
C	Schizaphis graminum		P	Tetraneura basui
D	Sitobion graminis		Q	Tetraneura akinere
E	Sitobion miscanthi/ akebiae		R	Tetraneura nigriabdominalis
F	Sitobion avenae		S	Tetraneura ulmi
G	Hysteroneura setariae		T	Tetraneura radicicola/ yezoensis
H	Rhopalosiphum rufiabdominalis			[Chaetogeoica polychaeta Raychaudhuri, Pal & Ghosh 1978; Diuraphis noxia; Forda orientalis; Geoica utricularia; Melanaphis sacchari; Metopolophium dirhodum; Prociphilus sp.]
I	Rhopalosiphum nymphaeae			
J	Rhopalosiphum padi			
K	Paracletus cimiciformis			
L	Geoica lucifuga			
M	Geoica setulosa			

1. Body hairs long and spine-like 2
    Body hairs short or, if long, fine 3

2. Terminal process about equal in length to base
    of last antennal segment. Dorsal cuticle
    heavily sclerotized, with numerous small
    tubercles or denticles between the hairs   A
    Terminal process about twice as long as base
    of last antennal segment. Dorsal cuticle pale
    or smoky with dark intersegmental markings,
    not adorned with small tubercles or denticles   B

3. Terminal process longer than base of last
    antennal segment   4
    Terminal process much shorter than base of
    last antennal segment   11

4. Siphunculi mainly pale, often dark at extreme
    apices   C

    Siphunculi wholly darker than body colour,
    so that there is a clear line of demarcation
    between body and bases of siphunculi   5

5.	Siphunculi with a subapical zone of polygonal reticulation. Antennal tubercles with inner faces divergent in dorsal view, sometimes low but higher than medial part of frons	6
	Siphunculi without polygonal reticulation. Antennal tubercles undeveloped or weakly developed, not higher than medial part of frons in dorsal view	8
6.	Cauda dark	D
	Cauda pale	7
7.	Siphunculi more than 1.4 times longer than cauda. Second segment of hind tarsus less than 1.3 times longer than last segment of rostrum	E
	Siphunculi less than 1.4 times longer than cauda. Second segment of hind tarsus more than 1.25 times longer than last segment of rostrum	F
8.	Cauda long and pale	G
	Cauda dark and rather short	9
9.	Antennae 5-segmented, bearing long, fine hairs, many of them over twice as long as basal diameter of antennal segment III	H
	Antennae with shorter hairs and usually 6-segmented	10
10.	Siphunculi usually more than twice as long as cauda and slightly but distinctly clavate. Last rostral segment 140–190 μm long, about 1.5 times longer than base of antennal segment VI and about equal in length to antennal segment V	I
	Siphunculi usually less than twice as long as cauda, cylindrical or with slight thickening subapically, but not distinctly clavate. Last rostral segment 90–130 μm long, 1.1–1.25 times as long as base of antennal segment VI and 0.6–0.8 of length of antennal segment V	J
11.	Siphunculi absent	12
	Siphunculi present as small pores or shallow cones	14
12.	Body flattened dorso-ventrally. Antennae 6-segmented. Eyes often with many facets. Body hairs all pointed	K
	Body globose. Antennae 5-segmented. Eyes always with 3 facets. Some body hairs spatulate	13

Aphids on the world's crops

13. Each abdominal segment with 1 or 2 erect
    and pointed lateral hairs on each side **L**

    Lateral abdominal hairs similar to dorsal
    hairs, i.e. of variable length, often curved
    and spatulate **M**

14. Tarsi 2-segmented. Total antennal length
    one-third or more of body length 15
    Tarsi 1-segmented. Antennae very short, one-
    fifth of body length or less 16

15. Large, flat, round lateral tubercles placed
    near spiracles on abdominal segments 1–7,
    including segments 5 and 6 (anterio-ventral
    and posterioventral of siphunculi) **N**

    Large flat round tubercles placed near
    spiracles on abdominal segments 1–4 and 7,
    but absent from segments 5 and 6 **O**

16. Last segment of rostrum short, about as long
    as its basal width and less than 1.8 times
    longer than hind tarsus without claw 17
    Last rostral segment long, much longer than
    its basal width and more than 1.8 times
    longer than hind tarsus without claw 19

17. Penultimate antennal segment with only
    about 4 long hairs **P**

    Penultimate antennal segment with more than
    10 hairs 18

18. Eighth abdominal tergite with a pair of long stout hairs and a group of small fine hairs placed ventro-lateral of them on each side. Cauda usually with a few small hairs in addition to 2 long ones .... Q

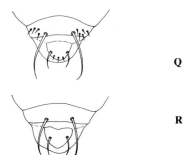

Eighth abdominal tergite with a pair of long stout hairs only, and cauda also usually with only two rather long hairs .... R

19. Abdominal wax pore-plates typically comprising a ring of cells around a large undivided central area. Body and appendages only sparsely hairy .... S

All abdominal wax glands small and arranged singly. Body and appendages with numerous fine hairs of varying length .... T

Rose

A	*Maculolachnus submacula*	
B	*Maculolachnus sijpkensi*	
C	*Longicaudus trirhodus*	
D	*Chaetosiphon* species (see separate key below)	
E	*Pseudaphis abyssinica*	
F	*Myzaphis bucktoni*	
G	*Myzaphis rosarum*	
H	*Myzaphis turanica*	
I	*Pseudocercidis rosae*	
J	*Wahlgreniella nervata*	
K	*Macrosiphum euphorbiae*	
L	*Macrosiphum rosae*	
M	*Macrosiphum mordvilkoi*	
N	*Eomacrosiphum nigromaculosum*	
O	*Macrosiphum pallidum*	
P	*Macrosiphum pachysiphon*	
Q	*Sitobion rosaeiformis*	
R	*Sitobion fragariae*	
S	*Sitobion rubiphila*	
T	*Sitobion ibarae*	
U	*Myzus japonensis*	
V	*Metopolophium alpinum*	
W	*Metopolophium dirhodum*	
X	*Metopolophium montanum*	
Y	*Rhodobium porosum*	
Z	*Fimbriaphis fimbriata*	

[*Acyrthosiphon pseudodirhodum; Acyrthosiphon tutigula; Amphorophora catharinae* Nevsky 1928; *Aphiduromyzus rosae* Uramov & Ibranova 1967; *Aphis fabae; Aphis gossypii; Aulacorthum solani; Bipersona hottesi* Knowlton & Smith 1936; *Capitophorus corambus* Hottes & Frison 1931; *Macrosiphum centranthi; Macrosiphum floridae* Ashmead 1882; *Myzaphis avariolosa* David, Rajasingh & Narayanan 1971; *Myzus distinctus* Nevsky 1929; *Myzus ornatus; Myzus persicae; Metopolophium chandrani* (David & Narayanan 1968); *Placoaphis siphunculata* Richards 1961; *Rhopalosiphum insertum; Schizoneura rosaefoliae* Shinji 1939; *Toxoptera aurantii*]

1. Terminal process very short, one-quarter or less of the length of the base of the last antennal segment. Siphunculi merely pores on flat dark discs ... 2

   Terminal process longer than base of last antennal segment. Siphunculi tubular ... 3

2. Dorsal abdominal hairs arising from small dark sclerites ... A

   Dorsal abdominal hairs not arising from small dark sclerites ... B

3. Siphunculi as short cones, much shorter than cauda ... C

   Siphunculi similar in length to, or longer than, cauda ... 4

4. Dorsal body hairs distinctly capitate, often long and with tuberculate bases ... D

   Dorsal body hairs with blunt or pointed apices, usually short, never with tuberculate bases ... 5

5. Antennal tubercles weakly developed or absent (but medial part of frons may project prominently) ... 6
   Antennal tubercles moderately to well developed, projecting anteriorly beyond medial part of frons in dorsal view ... 9

6. Medial part of frons flat or slightly convex. Siphunculi short, similar in length to cauda. Dorsal cuticle not pitted ... E

   Medial part of frons projecting prominently in dorsal view. Siphunculi clearly longer than cauda. Dorsal cuticle pitted all over ... 7

Rose

7. Projecting medial part of frons square-sided or rounded, bearing 2–4 hairs, which are usually shorter than the basal diameter of antennal segment III. Dorsal abdomen with two broad longitudinal dark bands converging at about the level of the siphunculi. Siphunculi cylindrical or very slightly swollen distally ... 8

    Projecting medial part of frons rounded, bearing 4 hairs which are as long as, or longer than, the basal diameter of antennal segment III. Dorsal abdomen without clear dark markings. Siphunculi distinctly clavate ... F

8. Projecting medial part of frons square-sided, bearing 2 hairs rarely more than half as long as the basal diameter of antennal segment III ... G

    Projecting medial part of frons rounded, bearing 4 hairs usually more than half as long as basal diameter of antennal segment III ... H

9. Siphunculi not broadening out at base. Cauda short, about as long as its basal width in dorsal view ... I

    Siphunculi broadening out at base. Cauda tongue-shaped, clearly longer than its basal width in dorsal view ... 10

10. Siphunculi distinctly clavate ... J

    Siphunculi cylindrical or tapering on distal half ... 11

11. Siphunculi with a subapical zone of polygonal reticulation, usually dark or at least dusky on distal part ... 12
    Siphunculi without a subapical zone of polygonal reticulation, usually pale ... 21

12. Siphunculi pale at least over basal half,
    sometimes dusky towards apex .... **K**

    Siphunculi wholly darker than body colour,
    except sometimes at extreme base 13

13. Hairs on antennal segment III conspicuous,
    length of longest more than half diameter
    of segment at base. Dorsal abdomen usually
    pale, unsclerotized 14
    Hairs on antennal segment III short, length
    one-half or less of diameter of segment at
    base. Dorsal abdomen sclerotized, often
    smoky or with dark markings 18

14. Front of head black or very dark, like the
    siphunculi 15
    Front of head pale or dusky, paler than
    siphunculi 16

15. Antennal segment III with a cluster of
    sensoria restricted to basal part .... **L**
    Antennal segment III with sensoria extend-
    ing over most of length of segment .... **M**

16. Compound (raspberry-like) lateral tubercles
    present on pronotum and often on abdominal
    segments 1–7. Basal parts of siphunculi
    very heavily and densely imbricated .... **N**

    Lateral tubercles always simple if present.
    Basal parts of siphunculi only lightly
    imbricated 17

17. Antennal segments III–V black. Siphunculi
    thin and cylindrical distally, mainly dark but
    often pale at base .... **O**

    Antennal segments III–V pale or only dark
    apically. Siphunculi thick, tapering gradually
    from base to flange, always wholly black .... **P**

18.	Femora pale, rarely a little darker at apices	19
	Femora with at least apical third dark brown or black	20

19.	Body narrowly spindle-shaped, at least 3 times longer than broad. Terminal process more than 4.5 times longer than base of last antennal segment	Q
	Body rather broadly spindle-shaped, less than 2.5 times longer than broad. Terminal process less than 4.5 times longer than base of last antennal segment	R

20.	Cauda one-eighth to one-tenth of body length, clearly less than half as long as siphunculi, which have inconspicuous reticulation confined to distal one-tenth. Hairs on antennae and dorsal head minute, virtually invisible in unmounted specimens. Femora with numerous spinules	S
	Cauda long, one-fifth to one-sixth of body length, at least half as long as siphunculi, which have reticulation extending over about the distal one-fifth. Hairs on antennae and dorsal head one-quarter to one-half of basal diameter of antennal segment III. Femora without spinules	T

21.	Inner faces of antennal tubercles with distinctly convergent projections in dorsal view	U
	Inner faces of antennal tubercles parallel or divergent in dorsal view	22

22.	Antennal tubercles with inner faces smooth, and divergent in dorsal view	23
	Antennal tubercles with inner faces approximately parallel in dorsal view, and roughened or spinulose	25

23. Siphunculi more than twice as long as cauda. Total antennal length about the same as body length (not in fundatrix)	V
Siphunculi less than 1.8 times longer than cauda. Total antennal length less than 0.8 of body length	24
24. Last segment of rostrum less than 0.7 of length of second segment of hind tarsus. (Al. without distinct segmental markings on dorsum)	W
Last segment of rostrum more than 0.8 of length of second segment of hind tarsus. (Al. with broad transverse dark bands on dorsum)	X
25. Antennal segment III with about 10 circular sensoria on basal half. Sensoriated part of segment III, the apices of segments IV and V and the whole of VI usually dark. (Al. with little if any dorsal abdominal pigmentation)	Y
Antennal segment III without sensoria. Antennal segments uniformly pale, except for dark base to VI. (Al. with a dark dorsal abdominal patch)	Z

Key to *Chaetosiphon* species on roses (apterous viviparae only):

1.	Siphunculi bearing several capitate hairs (subgenus *Chaetosiphon*)	2
	Siphunculi without capitate hairs (subgenus *Pentatrichopus*)	3
2.	Siphunculi less than twice as long as cauda. Hairs on antennal segments IV and V similar in length to those on segment III	*Chaetosiphon chaetosiphon*
	Siphunculi more than 2.5 times longer than cauda. Hairs on antennal segments IV and V short, less than half as long as basal diameters of segments	*Chaetosiphon gracilicorne*
3.	Terminal process less than twice length of base of last antennal segment. Hairs on antennal segment III blunt or pointed, not capitate	4
	Terminal process more than twice length of base of last antennal segment (except in fundatrix). At least some of hairs on antennal segment III distinctly capitate	5
4.	Siphunculi less than 1.5 times longer than cauda. Dorsal abdomen with dusky transverse bands	*C. (Pentatrichopus) glaber*
	Siphunculi more than twice as long as cauda. Dorsal abdomen pale.	*C. (Pentatrichopus) coreanum*
5.	Head smooth dorsally. Siphunculi less than 2.5 times longer than cauda	*C. (Pentatrichopus) tetrarhodum*
	Head spinulose dorsally. Siphunculi more than 2.5 times longer than cauda	6
6.	Abdomen with two longitudinal lateral rows of long capitate hairs on each side (i.e. most segments with 2 lateral hairs on each side), as well as 4 medio-dorsal rows	*C. (Pentatrichopus) thomasi*
	Abdomen with one longitudinal lateral row of long capitate hairs on each side (i.e. most segments with only one lateral hair on each side), as well as usually 4 medio-dorsal rows	*C. (Pentatrichopus) fragaefolii*

Rubber   no aphids recorded

Rudbeckia

- A *Uroleucon leonardi*
- B *Uroleucon ambrosiae*
- C *Uroleucon rudbeckiae*
- D *Macrosiphum rudbeckiarum*
- E *Macrosiphum cockerelli*
  [*Aphis armoraciae;*
  Macrosiphum euphorbiae;
  Myzus persicae]

1. Siphunculi uniformly dark — 2
   Siphunculi pale, or pale at base becoming darker apically — 3

2. Lateral tubercles present on most abdominal marginal sclerites — A
   Lateral abdominal tubercles absent — B

3. Siphunculi less than twice as long as cauda, and usually dark distally — C
   Siphunculi long and thin, more than twice as long as cauda, and usually wholly pale — 4

4. Antennal segment III with 2–3 sensoria in apt., 8–10 sensoria in al. — D
   Antennal segment III with 4–10 sensoria in apt., 16–33 sensoria in al. — E

Rye

- A *Forda formicaria*
- B *Sipha (Rungsia) maydis*
- C *Schizaphis graminum*
- D *Metopolophium dirhodum*
- E *Rhopalosiphum maidis*
- F *Rhopalosiphum padi*
- G *Sitobion fragariae*
- H *Sitobion miscanthi*
- I *Sitobion avenae*
  [*Diuraphis noxia; Forda marginata* group;
  Metopolophium festucae cerealium; Myzus persicae;
  Schizaphis jaroslavi]

1. Terminal process shorter than base of last antennal segment. Antenna 5-segmented, with very large sensorium on segment V. Siphunculi absent. (On roots) — A
   Terminal process longer than base of last segment. Antenna 5- or 6-segmented. Siphunculi present. (On aerial parts) — 2

2. Dorsal abdomen with extensive dark-brown sclerotic area which includes the bases of the inconspicuous, conical siphunculi. Cauda broadly rounded. Body hairs long and spine-like — B
   Dorsal abdomen not extensively pigmented. Siphunculi evident, tubular. Cauda tongue-shaped. Body hairs short and sparse — 3

3.	Siphunculi as pale as body, or dark only at extreme apex	4
	Siphunculi darker than body; even if rather pale there is still a clear difference of pigmentation between body and base of siphunculi	5

4.	Antennae shorter than distance from frons to bases of siphunculi. Siphunculi about half as long as the distance between their bases, usually with dark apices	C
	Antennae longer than distance from frons to bases of siphunculi. Siphunculi more than three-quarters as long as the distance between their bases	D

5.	Cauda darker than body. Siphunculi short, less than half as long as the distance between their bases	6
	Cauda pale. Siphunculi at least two-thirds as long as the distance between their bases, with a subapical zone of polygonal reticulation	7

6.	Body rather elongate. Terminal process less than 2.5 times longer than base of last antennal segment	E
	Body ovate. Terminal process more than 3 times longer than base of last antennal segment	F

7.	Proximal segments of antennae paler than distal segments. Siphunculi more than twice as long as cauda	G
	Antennae uniformly dark. Siphunculi less than twice as long as cauda	8

8.	Siphunculi more than 1.5 times longer than cauda. Second segment of hind tarsus less than 1.4 times longer than last segment of rostrum	H
	Siphunculi less than 1.5 times longer than cauda. Second segment of hind tarsus more than 1.4 times longer than last rostral segment	I

Rye-Grass

A	*Atheroides serrulatus*	J	*Metopolophium festucae*
B	*Sipha (Rungsia) maydis*	K	*Metopolophium dirhodum*
C	*Diuraphis noxius*	L	*Anoecia* species
D	*Diuraphis (Holcaphis) frequens*	M	*Tetraneura ulmi*
		N	*Aploneura lentisci*
E	*Sitobion fragariae*	O	*Forda formicaria*
F	*Rhopalosiphum rufiabdominalis*		[*Forda marginata* group; *Metopolophium festucae cerealium*; *Myzus persicae*; *Utamphorophora humboldti*]
G	*Rhopalosiphum maidis*		
H	*Rhopalosiphum padi*		
I	*Schizaphis graminum*		

Rye-Grass

1.	Terminal process either similar in length to or longer than base of last antennal segment	2
	Terminal process much shorter than (less than half) base of last antennal segment	12

2.	Dorsal body hairs partly or mainly long and spine-like. Cauda broadly rounded	3
	Dorsal body hairs fine if long. Cauda tongue-shaped	4

3.	Body very elongate, more than 3 times longer than broad, with long, spine-like hairs only on head and posterior abdomen. Siphunculi merely pores in the heavily sclerotized tergum	A
	Body ovate, less than twice as long as broad, with spine-like hairs on all segments. Siphunculi short, truncated cones	B

4.	Siphunculi very small rounded cones, much shorter than cauda	5
	Siphunculi tubular, as long as or longer than cauda	6

5.	Eighth abdominal tergite with posteriorly directed supracaudal process. Terminal process more than 1.5 times longer than base of last antennal segment	C
	No supracaudal process. Terminal process less than 1.5 times longer than base of last antennal segment	D

6.	Siphunculi darker than body colour (if rather pale there is still a clear difference of pigmentation between body and bases of siphunculi)	7
	Siphunculi as pale as body, or only dark at extreme apices	10

7.	Cauda pale. Siphunculi about as long as the distance between their bases, with a subapical zone of polygonal reticulation	E
	Cauda dark. Siphunculi less than half as long as the distance between their bases	8

8.	Antennae 5-segmented, bearing long, fine hairs, many of which are over twice as long as the diameter of antennal segment III	**F**
	Antennae 6-segmented, with shorter hairs, mostly shorter than the diameter of antennal segment III	9
9.	Body rather elongate. Terminal process less than 2.5 times longer than base of last antennal segment	**G**
	Body ovate. Terminal process more than 2.5 times longer than base of last antennal segment	**H**
10.	Antennae shorter than distance from frons to bases of siphunculi. Siphunculi about half as long as the distance between their bases, and usually with dark apices	**I**
	Antennae longer than distance from frons to bases of siphunculi. Siphunculi more than three-quarters as long as the distance between their bases	11
11.	Antenna progressively darker from base to apex. Usually less than 8 caudal hairs. (Al. with a pattern of dark transverse bars on dorsal abdomen)	**J**
	Each antennal segment pale at base and dusky at apex (base of VI paler than apex of V). Usually with more than 8 caudal hairs. (Al. without dorsal abdominal pigmentation)	**K**
12.	Siphunculi present as pores or very small cones	13
	Siphunculi completely absent	14
13.	Total antennal length one-third or more of body length. Tarsi 2-segmented	**L**
	Antennae only one-fifth of body length or less. Tarsi 1-segmented	**M**
14.	Body spindle-shaped, at least twice as long as broad. (Wax-covered in life)	**N**
	Body globose, less than 1.5 times longer than broad. (Without wax in life)	**O**

Safflower

A *Aphis (Protaphis) anuraphoides*
B *Aphis craccivora*
C *Aphis gossypii*
D *Myzus persicae*
E *Aulacorthum solani*
F *Uroleucon (Uromelan) jaceae*
G *Uroleucon (Uromelan) compositae*
[*Uroleucon (Uromelan) gobonis*]

Saffron

1.	Terminal process similar in length to base of last antennal segment		A
	Terminal process much longer than base of last antennal segment		2
2.	Antennal tubercles weakly developed		3
	Antennal tubercles well developed		4
3.	Cauda dark. Dorsal abdomen with an extensive black patch		B
	Cauda paler than siphunculi. Dorsal abdomen unpigmented		C
4.	Siphunculi and cauda pale		5
	Siphunculi and cauda blackish		6
5.	Inner faces of antennal tubercles convergent. Antennal segment III without a sensorium. Siphunculi slightly clavate		D
	Inner faces of antennal tubercles parallel. Antennal segment III with a single sensorium near its base. Siphunculi tapering or cylindrical		E
6.	Tibiae wholly dark. Femur with basal half to three-fifths very pale, distal part contrasting black, with an abrupt transition between. Sensoria usually confined to basal three-fifths of antennal segment III		F
	Tibia dark at both ends, but with a paler middle section. Femur with about the basal third pale, distal part dark, with a gradual transition between. Sensoria usually extending over at least two-thirds of antennal segment III		G

Saffron   A   *Dysaphis tulipae*

B   { *Aulacorthum solani*
Myzus ascalonicus
Myzus persicae }   see key to polyphagous Macrosiphini

[*Aphis gossypii*]

180   *Aphids on the world's crops*

1. Cauda short, no longer than its basal width. Siphunculi dark. Antennal tubercles not developed	A
Cauda tongue-shaped, much longer than its basal width. Siphunculi pale. Antennal tubercles well developed	B

Sage   A   *Eucarazzia elegans*   C   *Aphis passeriniana*
            B   *Aphis salviae*   D   *Aphis gossypii*

1. Siphunculi very swollen on distal half. Cauda short, no longer than its basal width	A
Siphunculi tapering from base to flange. Cauda tongue-shaped	2
2. Dorsal abdomen with extensive dark pigmentation	B
Dorsal abdomen pale	3
3. Last segment of rostrum more than 1.7 times longer than second segment of hind tarsus	C
Last segment of rostrum less than 1.4 times longer than second segment of hind tarsus	D

Sago Palms   *Cerataphis palmae*
                  (or see key to aphids on **Palms**)

Salsify   *Brachycaudus (Appelia)*   *Smynthurodes betae*
               tragopogonis   [*Prociphilus erigeronensis*]
               Aphis armoraciae

Salsify, Black A *Brachycaudus (Appelia) tragopogonis*
                    B *Aphis fabae*

1. Cauda short and broad, shorter than its basal width in dorsal view	A
Cauda tongue-shaped, much longer than its basal width	B

Sesame

Sesame	A	*Aphis gossypii*
	B	*Myzus persicae*

1. Antennal tubercles weakly developed. Siphunculi dark, tapering from base to flange ... A

   Antennal tubercles well developed, their inner faces convergent in dorsal view. Siphunculi pale, slightly clavate ... B

Shallot   as for **Chives**

Silk Cotton Tree   see **Kapok**

Sloe

Hyalopterus pruni
Brachycaudus helichrysi
Brachycaudus (Appelia) prunicola
Brachycaudus cardui
Brachycaudus persicae
Phorodon humuli

Rhopalosiphum nymphaeae
[*Brachycaudus divaricatae* Shaposhnikov 1956]
[*Myzus persicae*]
[*Rhopalosiphum padi*]
use key to aphids on **Plum**

Sop, Sweet or **Sour**   see **Custard Apple**

Sorghum   see **Guinea Corn**

Soybean

	A	*Acyrthosiphon pisum*	D	*Aphis gossypii*
	B	*Aphis craccivora*		[*Aulacorthum solani*]
	C	*Aphis glycines*		[*Hyadaphis coriandri*]
				[*Macrosiphum euphorbiae*]
				[*Myzus persicae*]

1. Large aphid, antennal tubercles very well
   developed, siphunculi pale and very attenuate   A
   Medium to small aphids, antennal tubercles
   weakly developed, siphunculi dark   2

2. Dorsal abdomen with a large black patch,
   cauda dark   B
   Dorsal abdomen unpigmented, cauda paler than
   siphunculi   3

3. Cauda with 8–10 hairs; very pale, contrasting
   with the very dark siphunculi   C
   Cauda with 4–7 hairs; often dusky, not much
   paler than siphunculi   D

Spinach   *Pemphigus* species   *Macrosiphum euphorbiae*
                 Aphis gossypii   *Aulacorthum solani*
                 Aphis fabae   [*Hayhurstia atriplicis*]
                 Myzus persicae   use key to aphids on **Beets**

Stock (Gilliflower)
                 Myzus persicae
                 Brevicoryne brassicae
                 Lipaphis erysimi
                 [*Aphis gossypii*]
                 use key to aphids on **Cabbage**

Strawberry   A *Chaetosiphon*   P *Fimbriaphis wakibae*
                     (Pentatrichopus) minor   [*Amphorophora*
                 B *Chaetosiphon (P.) jacobi*   *agathonica*; *Aphis*
                 C *Chaetosiphon (P.) thomasi*   *(Cerosipha) ichigicola*
                 D *Chaetosiphon (P.)*   Shinji 1924; *Aphis*
                     fragaefolii   *maidiradicis*; *Aphis*
                 E *Aphis forbesi*   *nasturtii*; *Aphis ruborum*;
                 F *Aphis gossypii*   *Hyperomyzus*
                 G *Acyrthosiphon rogersii*   (*Hyperomyzella*) *rhinanthi*;
                 H *Sitobion fragariae*   *Macrosiphum pallidum*;
                 I *Macrosiphum rosae*   *Myzaphis rosarum*; *Myzus*
                 J *Macrosiphum euphorbiae*   (*Sciamyzus*) *cymbalariae*;
                 K *Myzus ascalonicus*   *Ovatus valuliae* (Robinson
                 L *Myzus ornatus*   1974); *Pemphigus*
                 M *Aulacorthum solani*   *bursarius*(?);
                 N *Rhodobium porosum*   *Trichosiphonaphis*
                 O *Fimbriaphis fimbriata*   *polygoni* (van der Goot 1917)]

1. Dorsal surface of body adorned with long,
   capitate hairs with tuberculate bases   2

   Body hairs not long and capitate, nor with
   tuberculate bases   5

Strawberry

2.	Hairs on abdominal tergites anterior to siphunculi usually all minute; rarely with rather long and capitate spinal hairs on some of these tergites, but never with lateral ones		A
	At least 4 long capitate hairs (2 spinal, 2 lateral) on each abdominal segment anterior to siphunculi		3

3.	Dorsal abdomen dark brown		B
	Dorsal abdomen pale		4

4.	Abdomen with two longitudinal lateral rows of long capitate hairs on each side (i.e. most segments with 2 lateral hairs on each side), as well as 4 medio-dorsal rows		C
	Abdomen with one longitudinal lateral row of long capitate hairs on each side (i.e. most segments with only one lateral hair on each side), as well as usually 4 medio-dorsal rows		D

5.	Antennal tubercles absent or weakly developed		6
	Antennal tubercles moderately to well developed, always projecting beyond medial part of frons in dorsal view		7

6.	Cauda and siphunculi equally dark. Abdominal tergites 7 and 8 often with narrow transverse bands. Last segment of rostrum longer than cauda and more than 1.5 times longer than second segment of hind tarsus		E
	Cauda paler than siphunculi. Abdominal tergites 7 and 8 without pigmentation. Last segment of rostrum shorter than cauda and less than 1.5 times longer than second segment of hind tarsus		F

7.	Inner faces of antennal tubercles divergent and smooth		8
	Inner faces of antennal tubercles parallel or convergent, and spinulose or imbricated		11

8.	Siphunculi wholly pale, without any polygonal reticulation		G
	Siphunculi dark or pale, with a subapical zone of polygonal reticulation		9

9.	Hairs on antennal segment III very short, less than half as long as basal diameter of segment. Body rather broadly spindle-shaped. Siphunculi dark	H

	Hairs on antennal segment III longer and more erect, the longest nearly as long as the basal diameter of the segment. Body more narrowly spindle-shaped. Siphunculi either wholly black or mainly pale	10

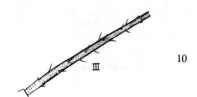

10.	Front of head and siphunculi black	I
	Front of head dusky or pale, siphunculi mainly pale, sometimes dusky on distal part	J

11.	Siphunculi clavate, with minimum diameter of basal part less than diameter of middle part of hind tibia	K

	Siphunculi tapering or cylindrical on distal half, without a constriction on basal half	12

12.	Dorsal abdomen with an intersegmental pattern of dark ornamentation. Inner faces of antennal tubercles distinctly convergent. Terminal process 1.8–2.5 times longer than base of antennal segment VI	L

	Dorsal abdomen without dark ornamentation. Inner faces of antennal tubercles parallel or very slightly convergent. Terminal process more than 3 times longer than base of antennal segment VI	13

13.	Siphunculi dusky or dark at extreme apex and with a large flange	M

	Siphunculi pale at apex with only a small flange	14

Sugar Beet

14. Antennal segment III with about 10 circular sensoria on basal two-thirds. Sensoriated part of segment III, apices of segments IV and V and the whole of VI often dark. (Al. without dorsal abdominal pigmentation)	N
Antennal segment III usually without sensoria, never with more than 4. Antennae mainly pale, sometimes with darker V and base of VI. (Al. with a dark dorsal abdominal patch)	15
15. Dorsal abdomen always pale and membranous. Last segment of rostrum less than 0.12 mm long. Siphunculi less than 1.8 times longer than cauda	O
Dorsal abdomen sclerotic, usually with an unclearly defined darker brown mid-dorsal region of variable extent. Last segment of rostrum usually more than 0.12 mm long. Siphunculi 1.8 or more times longer than cauda	P

Sugar Beet   see **Beets**

Sugar Cane
- A *Ceratovacuna lanigera*
- B *Geoica lucifuga*
- C *Forda orientalis*
- D *Tetraneura javensis*
- E *Tetraneura nigriabdominalis*
- F *Sipha flava*
- G *Rhopalosiphum maidis*
- H *Melanaphis sacchari*
- I *Sitobion miscanthi*
- J *Hysteroneura setariae*
  [*Ceratovacuna perglandulosa* Basu, Ghosh, & Raychaudhuri 1975; *Pseudoregma panicola*; *Tetraneura kalimpongensis* Raychaudhuri, Pal & Ghosh 1978]

1. Terminal process shorter than base of last antennal segment	2
Terminal process longer than base of last antennal segment	6
2. Front of head with a pair of horns projecting forward. (Colonies on leaves)	A
Front of head without horns. (On roots)	3
3. Siphunculi completely absent	4
Siphunculi present as small dark cones	5

4.	Anal plate displaced dorsally. Dorsal abdominal hairs mainly with spatulate or fan-shaped apices	B
	Anal plate in normal position. Dorsal abdominal hairs all small and pointed	C
5.	Abdominal wax pore-plates typically with one large cell incompletely surrounded by many smaller cells	D
	Abdominal wax pore-plates small, typically composed of one cell or a few similar-sized cells	E
6.	Dorsal body hairs long and spine-like. Cauda with a constriction and a knob-like apex	F
	Dorsal hairs small. Cauda tapering, tongue-shaped	7
7.	Cauda and siphunculi short and dark, neither more than one-tenth of body length	8
	Cauda pale, more than one-tenth of body length. Siphunculi dark, one-seventh to one-quarter of body length	9
8.	Body rather elongate. Siphunculi a little longer than cauda. Terminal process less than 2.5 times longer than base of last antennal segment	G
	Body ovate. Siphunculi a little shorter than cauda. Terminal process more than 3 times longer than base of last antennal segment	H
9.	Body broadly spindle-shaped. Siphunculi with a subapical zone of polygonal reticulation	I
	Body broadly ovate. Siphunculi without polygonal reticulation	J

Sumac

A Melaphidina species—see separate key below
B *Toxoptera odinae*
C *Aphis gossypii*
D *Aulacorthum magnoliae*
E *Myzus persicae*
F *Juncomyzus rhois*
G *Glabromyzus rhois*
  [*Glabromyzus howardii*]
  [*Glabromyzus rhusifoliae* (Richards 1973)]
  [*Glabromyzus schlingeri* Hille Ris Lambers 1966]

Sumac 187

1.	Terminal process much shorter than base of last antennal segment (gall formers)	A
	Terminal process longer than base of last antennal segment	2
2.	Siphunculi shorter than cauda. Stridulatory apparatus present	B
	Siphunculi longer than cauda. No stridulatory apparatus	3
3.	Siphunculi tapering continuously from base to flange	C
	Siphunculi swollen or cylindrical on distal half to two-thirds	4
4.	Antennal tubercles well developed, with inner faces convergent or parallel	5
	Antennal tubercles weakly developed, with inner faces broadly divergent	6
5.	Antennae and legs mainly black. Inner faces of antennal tubercles approximately parallel in dorsal view	D
	Antennae and legs mainly pale. Inner faces of antennal tubercles convergent	E
6.	Front of head, antennal segments I and II, siphunculi, and cauda all dark brown to black. Cauda only a little longer than its basal width in dorsal view	F
	Siphunculi dark but front of head, antennal segments I and II, and cauda all pale. Cauda at least 1.5 times its basal width in dorsal view	G

Key to alate viviparae of Melaphidina from galls on **Sumac** (based mainly on Tsai & Tang 1946):

1. Stigma of forewing extending in a curve around tip of wing 2

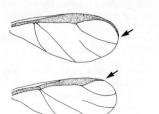

   Stigma of forewing not curved around tip of wing 3

2. Antennae 6-segmented; each of segments III–VI with a single very large sensorium occupying most of its surface area, in which are scattered many small islands of sclerotized cuticle. Gall elliptical, reddish, with many fine longitudinal ridges

   Schlechtendalia elongallis (Tsai & Tang 1946)

   Antennae 5-segmented; segments III–VI with at least partial separation of sensoriated surface into smaller units. Gall elliptical or angulate with projections

   Schlechtendalia chinensis (Bell)

3. Antennae 6-segmented 4
   Antennae 5-segmented 6

4. Each of antennal segments III–VI with a single large oblong sensorium occupying most of its surface area. Gall spindle-shaped

   Kaburagia rhusicola Takagi

   Antennal segments III–VI with separate, transversely elongate sensoria 5

5. Antennal sensoria on segments III–VI numerous (minimum of 8 on segment III) some of them forming complete rings around the antenna. Galls sac-like, rather thin walled

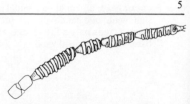

   Melaphis rhois (Fitch) (part)

   Antennal sensoria on segments III–VI fewer in number (maximum of 5 on segment III), never completely ringing the antenna. Galls red, flattened, irregularly branched from base

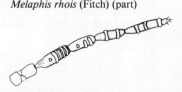

   Floraphis meitanensis Tsai & Tang 1946

Swede

6.	Anntenal segment III with more than 10 narrow, ring-shaped sensoria	*Melaphis rhois* (Fitch) (part)	
	Antennal segment III with less than 10 broad or narrow ring-shaped sensoria		7

7. Antennal segments IV and V each with a single, large sensorium occupying most of its surface area, in which are scattered small, rounded islands of cuticle. Gall pale yellowish brown

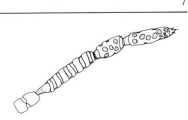

Nurudea ibofushi Matsumura

Antennal segments IV and V with separated, ring-shaped sensoria   8

8. Ring-shaped sensoria on antennal segments IV and V (especially the primary sensoria at the apices of these segments) in the form of broad bands. Gall not prominently reddish, with surface densely pubescent

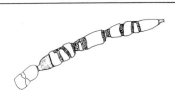

Nurudea shiraii (Matsumura)

Ring-shaped sensoria on antennal segments IV and V narrower and more numerous. Gall rosy red   *Nurudea yanoniella* (Matsumura)

Swede
- A   *Myzus persicae*
- B   *Brevicoryne brassicae*
- C   *Lipaphis erysimi*
  [*Myzus ascalonicus*]
  use first couplets of key to aphids on **Cabbage**

Sweet Potato
- A   *Geopemphigus floccosus*
- B   ⎧ *Aulacorthum solani* ⎫
        ⎨ *Myzus persicae*   ⎬ use key to polyphagous Macrosiphini
        ⎩ *Macrosiphum euphorbiae* ⎭
- C   *Aphis gossypii*
- D   *Aphis craccivora*

---

1. Terminal process shorter than base of last antennal segment. Siphunculi absent. (On roots)   A

Terminal process longer than base of last antennal segment. Siphunculi evident, tubular. (On aerial parts)   2

2. Antennal tubercles well developed. Siphunculi mainly pale	B
Antennal tubercles weakly developed. Siphunculi wholly dusky or dark	3

3. Dorsal abdomen pale	C
Dorsal abdomen with an extensive black patch	D

Sweet Sop see **Custard Apple**

Sweet William as for **Carnation**

Tangerine see **Citrus**

Taro *Aphis gossypii*
Pentalonia nigronervosa
Patchiella reaumuri
[*Hydronaphis colocasiae*
  Raychaudhuri, Raha, &
  Raychaudhuri 1977]

Taro, Giant *Pentalonia nigronervosa*

Tea A *Toxoptera aurantii*
B *Aphis gossypii*

1. Stridulatory apparatus present. Terminal process more than 3.5 times longer than base of last antennal segment. Cauda black with more than 10 hairs A

   Stridulatory apparatus absent. Terminal process less than 3.5 times longer than base of last antennal segment. Cauda pale or dusky with 4–7 hairs B

Thyme *Kaltenbachiella pallida*
[*Aphis serpylli* Koch]
(or try key to aphids on **Mint**)

Tobacco

	A	*Smynthurodese betae*	
	B	*Rhopalosiphum rufiabdominalis*	
	C	*Aphis fabae* group *Aphis nasturtii* *Aphis gossypii*	see key to polyphagous *Aphis* spp.
	D	*Nasonovia ribis-nigri*	
	E	*Aulacorthum solani* *Myzus persicae* *Macrosiphum euphorbiae*	see key to polyphagous Macrosiphini

---

1. Terminal process much shorter than base of last antennal segment. Siphunculi absent. Body and appendages densely hairy ... **A**

    Terminal process much longer than base of last antennal segment. Siphunculi evident, tubular. Body and appendages sparsely hairy ... **2**

---

2. Antennal tubercles absent or weakly developed, not projecting beyond medial part of frons in dorsal view ... **3**

    Antennal tubercles well developed ... **4**

---

3. Antennal hairs very long, the longest more than 3 times the basal diameter of antennal segment III. Terminal process of antenna characteristically curved and 2.5–3.5 times longer than cauda. (On roots) ... **B**

    Antennal hairs rarely more than twice as long as basal diameter of antennal segment III. Terminal process usually straight and not more than twice as long as cauda ... **C**

---

4. Terminal process 6.5–10 times longer than base of antennal segment VI. Dorsal abdomen with paired intersegmental markings, far apart between anterior segments but nearer mid-line between segments 4–5 and 5–6 ... **D**

    Terminal process 3.5–6.5 times longer than base of antennal segment VI. Dorsal abdomen without dark markings ... **E**

Tomato

A   *Smynthurodes betae*

B $\left\{\begin{array}{l}\textit{Aphis craccivora}\\ \textit{Aphis fabae}\\ \textit{Aphis gossypii}\end{array}\right\}$ see key to polyphagous *Aphis* spp.

C $\left\{\begin{array}{l}\textit{Myzus persicae}\\ \textit{Aulacorthum solani}\\ \textit{Macrosiphum euphorbiae}\end{array}\right\}$ see key to polyphagous Macrosiphini

[*Brachyunguis plotnikovi* (Nevsky 1928); *Prociphilus erigeronensis*; *Rhopalosiphum rufiabdominalis*]

---

1. Terminal process much shorter than base of last antennal segment. Siphunculi absent. Body and appendages densely haired   **A**

    Terminal process longer than base of last antennal segment. Siphunculi evident, tubular. Body and appendages sparsely hairy   2

2. Siphunculi wholly dark. Antennal tubercles weakly developed   **B**

    Siphunculi pale, at least basally. Antennal tubercles well developed   **C**

Tree Tomato

A   *Myzus persicae*

B $\left\{\begin{array}{l}\textit{Aphis gossypii}\\ \textit{Aphis fabae}\\ \textit{Aphis citricola}\end{array}\right\}$ see key to polyphagous *Aphis* spp.

---

1. Antennal tubercles well developed with inner faces convergent in dorsal view. Siphunculi pale and slightly clavate   **A**

    Antennal tubercles weakly developed. Siphunculi dark, tapering   **B**

Tulip

Tulip
- **A** *Rhopalosiphoninus latysiphon*
- **B** *Rhopalosiphoninus staphyleae*
- **C** *Dysaphis tulipae*
- **D** *Sitobion akebiae*
- **E** *Aphis fabae*
- **F** *Aphis gossypii*
- **G** *Aulacorthum (Neomyzus) circumflexum*
- **H** { *Myzus ascalonicus*; *Myzus persicae*; *Aulacorthum solani*; *Macrosiphum euphorbiae* } use key to polyphagous Macrosiphini
- [*Myzus (Sciamyzus) cymbalariae*; *Neotoxoptera oliveri*]

1.	Siphunculi mainly or wholly dark	2
	Siphunculi pale, at least on basal half	7

2.	Siphunculi markedly clavate	3
	Siphunculi tapering or cylindrical on distal half	4

3.	Dorsal abdomen with a large, black, sclerotic shield. Siphunculi jet black, with a narrow cylindrical basal part less than one-quarter of maximum diameter of inflated part	A
	Dorsal abdomen with a broken pattern of sclerotic markings. Siphunculi dark but with inflated section paler, basal part not cylindrical, at least one third of the maximum diameter of the inflated part	B

4.	Cauda helmet-shaped in dorsal view, no longer than its basal width	C
	Cauda tongue-shaped, longer than its basal width	5

5. Body broadly spindle-shaped. Siphunculi with
   a subapical zone of polygonal reticulation   D

   Body ovate. Siphunculi without polygonal
   reticulation   6

6. Cauda black, with more than 10 hairs. Dorsal
   abdomen with small dark sclerites at least
   laterally on segments anterior to siphunculi,
   and transverse dark bands on tergites 7 and
   8   E

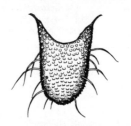

   Cauda paler than siphunculi, with 4–7 hairs.
   Dorsal abdomen without dark markings   F

7. Dorsal abdomen with a large, usually U-shaped
   dark patch, and thorax with dark transverse
   patches   G

   Dorsal abdomen and thorax without dark
   markings   H

Tung Oil   *Myzus persicae*

Turnip   A   *Smynthurodes betae*   C   *Brevicoryne brassicae*
               B   *Pemphigus*   D   *Lipaphis erysimi*
                   populitransversus   E   *Myzus persicae*
                                                 key as for **Cabbage**

Vaccinium   see also **Blueberry**
               A   *Ericaphis latifrons*   G   *Acyrthosiphon knechteli*
               B   *Aphis vaccinii*   H   *Aulacorthum rufum*
               C   *Aphis gossypii*   I   *Aulacorthum vaccinii*
               D   *Illinoia* species—see   J   *Aulacorthum flavum*
                   Blueberry   K   *Fimbriaphis scammelli/*
               E   *Wahlgreniella vaccinii*   *pernettyae*
               F   *Acyrthosiphon*   [*Macrosiphum nasonovi*
                   brachysiphon   Mordvilko 1918]

Vaccinium

1.	Antennal tubercles weakly developed, not or hardly projecting anteriorly beyond medial part of frons in dorsal view		2
	Antennal tubercles well developed, with inner faces parallel or divergent in dorsal view		4

2.	Siphunculi pale or dusky, curved outwards subapically and with a large flange		**A**
	Siphunculi dark, not curved outwards apically and with a small flange		3

3.	Cauda black with more than 10 hairs		**B**
	Cauda paler than siphunculi with 4–7 hairs		**C**

4.	Siphunculi swollen on distal half, but with a constricted or cylindrical subapical part with polygonal reticulation		**D**
	Siphunculi swollen, cylindrical, or tapering on distal half, without a cylindrical reticulated subapical part		5

5.	Siphunculi distinctly swollen on distal half, so that the maximum diameter of the swollen part is about 1.5 times greater than the minimum diameter of the proximal part		**E**
	Siphunculi tapering, cylindrical, or very slightly swollen on distal half		6

6.	Siphunculi similar in length to cauda		**F**
	Siphunculi at least 1.5 times longer than cauda		7

7.	Total antennal length clearly greater than body length; terminal process 3.7–4.8 times longer than base of antennal segment VI		8
	Total antennal length similar to or a little less than body length; terminal process 2.5–3.5 times longer than base of antennal segment VI		9

8. Siphunculi pale, tapering gradually from base to flange, less than twice as long as the long pointed cauda   G

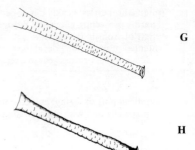

Siphunculi dusky with dark apices, cylindrical over most of length or slightly swollen on distal half, more than twice as long as the blunt cauda   H

9. Dorsal abdomen black   I
   Dorsal abdomen pale   10

10. Medial frontal prominence not evident. Antennal segment III with a single sensorium near its base. (Al. with extensive dark dorsal abdominal pigmentation, including transverse bars on tergites 1 and 2)   J

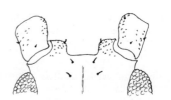

Frons with a distinct medial prominence, giving a W-shaped outline in dorsal view. Antennal segment III without a single sensorium. (Al. with a black dorsal abdominal patch on tergites 3–5, but no transverse bars on tergites 1 and 2)   K

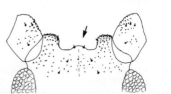

Vanilla   *Cerataphis orchidearum*

Velvet Bean   see **Bean, Velvet**

Vetch
- A *Aphis craccae*
- B *Aphis craccivora*
- C *Aphis fabae*
- D *Aphis nasturtii*
- E *Megoura viciae*
- F *Myzus ornatus*
- G *Aulacorthum solani*
- H *Acyrthosiphon pisum*

---

1. Antennal tubercles weakly developed   2
   Antennal tubercles well developed   5

2. Dorsal abdomen with extensive black pigmentation on tergites anterior to siphunculi   3
   Dorsal abdomen without pigmentation or with only scattered dark markings on tergites anterior to siphunculi   4

Walnut

3.	Siphunculi shorter than cauda, or up to 1.1 times longer	A
	Siphunculi clearly (at least 1.3 times) longer than cauda	B
4.	Cauda black with more than 10 hairs	C
	Cauda pale with less than 10 hairs	D
5.	Siphunculi black, swollen in middle, similar in length to or a little shorter than the black cauda	E
	Siphunculi pale or dusky, sometimes darker at extreme apices, cylindrical or tapering, much longer than the pale cauda	6
6.	Dorsal abdomen with an intersegmental pattern of dark ornamentation. Antennae only 0.6–0.8 of body length, with terminal process 1.7–2.5 times longer than base of last segment	F
	Dorsal abdomen without dark ornamentation. Antennae longer than body, with terminal process 3–5 times longer than base of last segment	7
7.	Inner faces of antennal tubercles approximately parallel. Siphunculi more than twice as long as cauda	G
	Inner faces of antennal tubercles divergent. Siphunculi very attenuate distally, less than twice as long as the very large cauda	H

Walnut
A *Callaphis juglandis* C *Chromaphis hirsutustibis*
B *Callaphis nepalensis* D *Chromaphis juglandicola*

1.	(Al.) Body length more than 3.5 mm. Dorsal abdomen with broad dark transverse bands on most tergites	2
	(Al.) Body length less than 3 mm. Dorsal abdomen without dark markings, or with bars or paired dark patches on tergites 4–5 only	3

2. (Al.) Knob of cauda globular, about as long as its greatest width. Abdominal tergites 1 and 2 with broken pigmentation	A
(Al.) Knob of cauda elongate, about twice as long as its greatest width. Abdominal tergites 1 and 2 with solid transverse dark bands like the more posterior tergites	B
3. (Al.) Hairs on hind tibia very long, the longest more than 3.5 times the middle diameter of the tibia. Wings with dark spots at ends of veins	C
(Al.) Hairs on hind tibia shorter, less than twice the middle diameter of the tibia. Wings without dark spots at ends of veins	D

Water Cress

A	Aphis nasturtii	D	Myzus persicae
B	Lipaphis erysimi		[Brevicoryne barbareae
C	Myzus ascalonicus		Nevsky 1929]

1. Antennal tubercles weakly developed, broadly divergent in dorsal view	2
Antennal tubercles well developed, their inner faces parallel or convergent in dorsal view	3

2. Antennal segment III shorter than segments IV and V together. Siphunculi tapering from base to flange	A
Antennal segment III about equal in length to, or longer than, IV and V together. Siphunculi slightly swollen proximal to flange	B

3. Siphunculi shorter than antennal segment III. Minimum diameter of basal part of siphunculus less than middle diameter of hind tibia	C
Siphunculi longer than antennal segment III. Minimum diameter of basal part of siphunculus greater than middle diameter of hind tibia	D

Water Melon

A	Myzus persicae
B	Aphis craccivora
C	Aphis gossypii
	As for **Melon**

Wheat

Wheat	A	*Sipha flava*	W	*Baizongia pistaceae*
	B	*Sipha (Rungsia) maydis*	X	*Geoica utricularia*
	C	*Sipha (Rungsia) elegans*	Y	*Geoica lucifuga*
	D	*Diuraphis noxia*	Z	*Aploneura lentisci*
	E	*Diuraphis (Holcaphis) tritici*	AA	*Paracletus cimiciformis*
	F	*Rhopalosiphum rufiabdominalis*	BB	*Pemphigus* species
			CC	*Forda formicaria*
	G	*Rhopalosiphum maydis*	DD	*Forda marginata*
	H	*Rhopalosiphum padi*	EE	*Forda hirsuta*
	I	*Glabromyzus howardii*	FF	*Forda orientalis*
	J	*Hysteroneura setariae*		[*Aphis armoraciae;*
	K	*Sitobion graminis*		*Colopha nirecola*
	L	*Sitobion fragariae*		(Matsumura 1917);
	M	*Sitobion miscanthi/ akebiae*		*Colopha ulmicola; Laingia psammae* Theobald 1922;
	N	*Sitobion avenae*		*Melanaphis pyraria;*
	O	*Schizaphis graminum*		*Metopolophium*
	P	*Metopolophium dirhodum*		*longicaudatum; Myzus persicae; Prociphilus*
	Q	*Metopolophium festucae cerealium*		*erigeronensis; Rhopalosiphum*
	R	*Anoecia* species		*padiformis; Schizaphis*
	S	*Tetraneura nigriabdominalis*		*hypersiphonata* A.N. Basu 1969; *Sipha glyceriae;*
	T	*Tetraneura yezoensis*		*Sitobion scabripes* L.K.
	U	*Tetraneura ulmi*		Ghosh 1972; *Smynthurodes*
	V	*Tetraneura africana*		*betae.*]

1. Terminal process similar in length to, or longer than, base of last antennal segment — 2
   Terminal process much shorter than base of last antennal segment — 18

2. Dorsal body hairs partly or mainly long and spine-like. Cauda either with a knob-like apex or broadly rounded — 3
   Dorsal body hairs short, or fine if long. Cauda tongue-shaped — 5

3. Cauda with a knob-like apex. Terminal process 2–2.8 times longer than base of last antennal segment — A

   Cauda broadly rounded in dorsal view. Terminal process 1.3–2.2 times longer than base of last antennal segment — 4

4. Dorsal abdomen uniformly shiny dark brown to black. Terminal process 1.3–1.8 times longer than base of last antennal segment — B
   Dorsal abdomen yellowish brown with some dark markings pleurally, leaving pale sides and a pale dorsal stripe. Terminal process 1.6–2.2 times longer than base of last antennal segment — C

5.	Siphunculi very small, much shorter than cauda; either papilliform or short rounded cones no longer than their basal width	6
	Siphunculi tubular, similar in length to or longer than cauda	7
6.	Eighth abdominal tergite with a posteriorly projecting supracaudal process. Terminal process 1.6–2.3 times longer than base of last antennal segment	D
	Eighth abdominal tergite without a supracaudal process. Terminal process less than 1.5 times longer than base of last antennal segment	E
7.	Siphunculi darker than body (if rather pale basally, there is still a clear difference of pigmentation between body and bases of siphunculi)	8
	Siphunculi as pale as body, or dark only at extreme apices	16
8.	Siphunculi less than half as long as the distance between their bases	9
	Siphunculi more than half as long as the distance between their bases	11
9.	Antennae usually 5-segmented, bearing long fine hairs, many of which are over twice as long as the diameter of antennal segment III. Terminal process of antenna characteristically curved	F
	Antennae 6-segmented, with hairs mostly shorter than diameter of antennal segment III. Terminal process of antenna straight	10
10.	Body rather elongate. Terminal process less than 2.5 times longer than base of last antennal segment. Siphunculi less than 1.5 times longer than cauda	G
	Body ovate. Terminal process more than 3 times longer than base of last antennal segment. Siphunculi more than 1.5 times longer than cauda	H

Wheat

11.	Siphunculi clavate	I
	Siphunculi tapering	12
12.	Siphunculi without polygonal reticulation. Body broadly ovate	J
	Siphunculi with a subapical zone of polygonal reticulation. Body broadly spindle-shaped	13
13.	Cauda and siphunculi both dark	K
	Cauda pale or if dusky much paler than siphunculi	14
14.	Proximal segments of antennae paler than distal segments. Siphunculi at least twice as long as cauda	L
	Antennae uniformly dark. Siphunculi less than twice as long as cauda	15
15.	Siphunculi more than 1.4 times longer than cauda. Second segment of hind tarsus less than 1.3 times longer than last segment of rostrum	M
	Siphunculi less than 1.4 times longer than cauda. Second segment of hind tarsus more than 1.25 times longer than last rostral segment	N
16.	Antennae shorter than the distance from front of head to bases of siphunculi. Siphunculi about half as long as the distance between their bases (and usually with dark apices)	O
	Antennae longer than the distance from front of head to bases of siphunculi. Siphunculi more than three-quarters as long as the distance between their bases	17
17.	Each antennal segment pale at base and darker at apex (base of VI paler than apex of V). Cauda with usually more than 8 hairs. (Al. without dorsal abdominal pigmentation)	P
	Antennae progressively darker from base to apex (if base of VI is a little paler than apex of V, then apices of III and IV are not darkened). Cauda with usually less than 8 hairs. (Al. with a pattern of transverse dark bars on dorsal abdomen)	Q
18.	Siphunculi present as small cones, or merely as pores	19
	Siphunculi completely absent	23

19. Total antenna length one-third or more of body length. Tarsi 2-segmented	R
Antennae very short, total length less than one-fifth of body length. Tarsi 1-segmented	20

20. Last segment of rostrum less than 0.15 mm long, and less than 1.8 times longer than hind tarsus. Disc of anal plate with some very small hairs in addition to very large ones	S
Last segment of rostrum more than 0.15 mm long, more than 1.8 times longer than hind tarsus. Disc of anal plate with large hairs only	21

21. Last segment of rostrum 0.20–0.26 mm long. Antennae and dorsal abdomen densely covered in small hairs. All abdominal wax glands small, and arranged singly	T
Last segment of rostrum 0.15–0.23 mm long. Antennae and dorsal abdomen sparsely hairy. Abdominal wax glands mainly consisting of multi-faceted plates	22

22. Last segment of rostrum 0.15–0.19 mm long, less than 2.4 times longer than hind tarsus. Abdominal wax pore-plates typically consisting of a ring of facets around a large undivided central area	U
Last segment of rostrum 0.18–0.23 mm long, more than 2.3 times longer than hind tarsus. Abdominal wax pore-plates with central area divided into facets	V

23. Anal plate displaced dorsally	24
Anal plate in normal ventral position	26

24. Prothorax almost parallel sided in dorsal view. Antennal segment III shorter than segment V including terminal process. All body hairs pointed   **W**

    Prothorax continuing curve of body in dorsal view. Antennal segment III longer than V including terminal process. At least some body hairs spatulate   25

25. Anal plate covered with numerous short hairs, but no long ones   **X**

    Anal plate with long hairs in irregular longitudinal rows, short hairs being restricted to the region close to the anal aperture   **Y**

26. Body spindle-shaped. Antennae very short, less than one-eighth of body length   **Z**
    Body rounded or oval in dorsal view. Antennae at least one-fifth of body length   27

27. Body flattened dorso-ventrally. Antennae 6-segmented. Eyes usually with many facets   **AA**
    Body not flattened dorso-ventrally. Antennae 5- or 6-segmented. Eyes usually with 3 facets   28

28. Base of last antennal segment clearly longer than penultimate segment. Large wax pore-plates often present, at least on posterior abdominal segments   **BB**
    Base of last antennal segment similar in length to penultimate segment. Wax plates absent   29

29. Sensorium on antennal segment V very large, occupying an area at least 4–5 times larger than that occupied by the sensorium on segment IV CC

Sensorium on antennal segment V not much larger than that on IV 30

30. Hairs on antennal segment III very short and blunt, less than one-quarter of diameter of segment and barely visible under a binocular microscope. Body hairs also mainly very short and inconspicuous DD
Hairs on antennal segment III longer, over half as long as diameter of the segment 31

31. Hairs on antennal segment III erect, thick, and bristle-like, mainly of length less than the diameter of the segment EE
Hairs on antennal segment III often longer than diameter of segment, with very finely pointed apices FF

Yam   *Aphis gossypii*
Aulacorthum magnoliae
[*Rhopalosiphum maidis;*
Toxoptera citricidus]

Keys to Polyphagous Aphids

Three keys are provided to assist in the recognition of apterous viviparae of certain widely distributed and polyphagous aphid species which will be commonly encountered on crops in many parts of the world. The first two keys are to 5 polyphagous *Aphis* species and 5 polyphagous species of Macrosiphini. These two groups may for present purposes be separated by the following couplet:

Antennal tubercles low or undeveloped, not exceeding height of medial part of frons in dorsal view. Siphunculi usually darker than general body colour, even at their bases. Lateral tubercles present at least on abdominal segments 1 and 7 polyphagous *Aphis* spp.

Keys to polyphagous aphids

    Antennal tubercles well developed. Siphunculi pale, at least on basal half, sometimes darker distally. Lateral tubercles absent from abdominal segments 1 and 7, and rare on other segments   polyphagous Macrosiphini spp.

The third key is to aphids widely distributed on various tropical trees and shrubs.

Key to polyphagous *Aphis* spp. (apterous viviparae only)

1. Dorsal abdomen with an extensive black patch centred on tergites 4–5. Cauda black, tongue-shaped, rather pointed and with 4–7 hairs   *Aphis craccivora*
   Dorsal abdomen of preserved specimens unpigmented or with scattered black markings. Cauda pale or dark, but if black it usually bears more than 7 hairs   2

2. Cauda pale, sometimes dusky in large specimens, but then clearly much paler than siphunculi   3
   Cauda dark, like the siphunculi   4

3. Siphunculi uniformly dark. Hairs on hind femur all shorter than diameter of femur at its base

   Aphis gossypii

   Siphunculi if dark then paler at base than apex. Hairs on hind femur mainly longer, some of them about as long as, or longer than, diameter of femur at its base

   Aphis nasturtii

4. Dorsal abdomen with small dark sclerites at least laterally on segments anterior to siphunculi, and transverse dark bands on tergites 7 and 8   *Aphis fabae*
   Dorsal abdomen without any dark markings   *Aphis citricola*

Key to polyphagous Macrosiphini spp. (apterous viviparae only)

1. Dorsal abdomen with an intersegmental pattern of dark ornamentation. Terminal process less than 2.5 times longer than base of last antennal segment. Siphunculi tapering, with a shallow S-curve, and coarsely imbricated. Small, oval aphid (less than 2 mm long)   *Myzus ornatus*

   Dorsal abdomen without clear dark intersegmental markings. Terminal process more than 2.5 times longer than base of last antennal segment. Siphunculi tapering, cylindrical, or clavate, not coarsely imbricated. Small, medium, or large aphids   2

2. Inner faces of antennal tubercles clearly divergent. Siphunculi long, with a subapical zone of polygonal reticulation extending over about distal one-sixth. Cauda long, one-seventh to one-fifth of body length. Rather large, spindle-shaped aphid

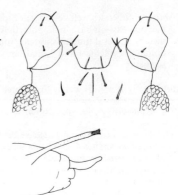

Macrosiphum euphorbiae

Inner faces of antennal tubercles convergent or approximately parallel. Siphunculi without polygonal reticulation. Cauda shorter, less than one-eighth of body length. Small to medium-sized oval aphids 3

3. Siphunculi tapering gradually from base to flange, with no sign of a swelling on distal half. Antennal segment III with a single sensorium near the base

Aulacorthum solani

Siphunculi slightly to moderately swollen on distal half. Antennal segment III without any sensoria. 4

4. Siphunculi rather small, shorter than antennal segment III, with minimum diameter of basal half less than diameter of middle part of hind tibia. Inner faces of antennal tubercles approximately parallel

Myzus ascalonicus

Siphunculi longer than antennal segment III, with minimum diameter of basal half greater than diameter of middle part of hind tibia. Inner faces of antennal tubercles convergent

Myzus persicae

Keys to polyphagous aphids

Key to aphids polyphagous on tropical trees and shrubs (apterous viviparae only)

1. Terminal process much shorter than base of last antennal segment. Siphunculi on flat, dark, hairy cones — *Longistigma caryae*
   Terminal process at least twice as long as last antennal segment. Siphunculi tubular — 2

2. Antennal tubercles low or undeveloped, not exceeding height of medial part of frons in dorsal view — 3
   Antennal tubercles well developed — 6

3. Siphunculi much shorter than cauda — *Toxoptera odinae*
   Siphunculi longer than or at least as long as cauda — 4

4. Terminal process 3.5–5.0 times longer than base of last antennal segment. Cauda with 10–26 hairs. Stridulatory apparatus present

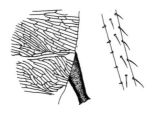

Toxoptera aurantii

   Terminal process 2.0–3.1 times longer than base of last antennal segment. Cauda with 4–12 hairs. No stridulatory apparatus — 5

5. Cauda black, with 6–12 hairs, usually 8–9. Hairs on hind femur long and fine, the longest exceeding the diameter of the hind femur at its base

Aphis citricola

   Cauda pale or dusky, with 4–8 hairs, usually 5–6. Hairs on hind femur all shorter than diameter of femur at its base

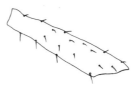

Aphis gossypii

6. Siphunculi tapering over most of length, about equal in length to or shorter than the very long, dark cauda. Legs and siphunculi with about the same degree of pigmentation

Sinomegoura citricola

Siphunculi cylindrical over most of length or slightly clavate on distal half, and more than 1.5 times longer than cauda. Legs black except at bases of femora, much darker than siphunculi

Aulacorthum magnoliae

C. *The Aphids*

Introduction to the section

This section provides a systematic review of the aphid species in the foregoing keys. The species are given in alphabetical order of their genera, with a brief introductory account for each genus providing references to any taxonomic accounts available for the genus in different parts of the world. The purposes of the accounts for each species are two-fold. First, information is provided on the appearance in life, habits, etc. of the aphid in order to help to confirm whether a correct identification has been obtained by using the keys. Secondly, aspects of the biology relevant to the aphid's economic importance are summarized with selected references to the literature.

Knowledge of the 450 or so species covered by this book varies from a vast amount of information on some of the best-known pest aphids such as *Myzus persicae* and *Acyrthosiphon pisum*, each of which could fill several volumes on their own, to a single record of the species and its host plant, with no other biological information at all. Obviously, the best-known species are generally those of greatest economic importance. However, it is always possible for a little-known aphid to suddenly become a major pest; e.g. *Acyrthosiphon kondoi* in the last decade. This is one of the reasons why in this book we have tried not to neglect the little-known species.

With regard to references to the available literature, it was a matter of choice between giving a large number of references with little or no indication of their relative value or relevance to the reader, or of being highly selective in citing only good and, where possible, up-to-date sources of further information, together with those pieces of original work which seem to us of special significance. We decided that the latter alternative would be most useful, particularly to those with limited library facilities. If the *Review of Applied Entomology* is also consulted, then the reader should be able to obtain without much difficulty all the essential information available about any species.

Size ranges of apterous and alate viviparous morphs of each species are given where sufficient material was available to feel confident that the extremes had been included. Diploid chromosome numbers ($2n$) are given where available. The original references for these may generally be found in the reviews by Kuznetsova and Shaposhnikov (1973) or Blackman (1980a), except where otherwise indicated; previously unpublished chromosome numbers are indicated by an asterisk (*).

Systematic treatment of genera (in alphabetical order)

Acyrthosiphon Mordvilko
Aphidinae: Macrosiphini

A largely palaearctic genus of about 100 species living without host alternation on various dicotyledons but particularly Leguminosae, Rosaceae, and Euphorbiaceae. They are medium-sized to rather large aphids; usually green but sometimes brownish, pink, or yellow; short-haired; with generally rather long antennae, legs, siphunculi, and cauda. Eastop (1971) gave keys for the identification of the species then known, and there are accounts of the Japanese species by Miyazaki (1971, pp. 52–55) and of the Indian species by Raychaudhuri, Ghosh, and Basu (1978, pp. 115–131).

Acyrthosiphon brachysiphon Hille Ris Lambers

Waxy green, found only on undersides of leaves and young shoots of *Vaccinium uliginosum*. Distribution is mainly boreal (Greenland, Iceland, Baffin Island, Sweden) but also recorded from the Swiss Alps. Monoecious life cycle, with only three generations a year in the Arctic, the fundatrices occurring in July and apterous males and oviparae in August.

Acyrthosiphon gossypii Mordvilko
(= *sesbaniae* David)

Plate 100

Appearance in life: Green; adults as well as larvae lightly dusted with fine wax (unlike *A. pisum*, in which adult apterae are conspicuous in colonies because they have no wax bloom). Apterae 2.5–3.8 mm, alatae 2.1–3.5 mm.

Host plants: Mostly Leguminosae of the tribe Phaseoleae (*Dolichos, Phaseolus, Vigna*, but not known from *Glycine*), plus a few species in related tribes, e.g. *Vicia faba* and *Alhagi camelorum*. It is a pest of cotton in central Asia and also occurs on wild Malvaceae there and in the Middle East. More rarely it is found on other plants including Cruciferae and Zygophyllaceae. The host plant transfer experiments of David (1956), and discrepant relative abundance on different plants in different localities, suggest that populations with particular host preferences occur.

Virus transmission: Recorded as a vector of pea leaf roll and pea mosaic to broad beans in Iran and Sudan, respectively.

Distribution: Central Asia, Mediterranean, Middle East and North Africa (India, Turkey, Iraq, Iran, Israel, Egypt, Sudan, Upper Volta, Algeria, Sicily, and perhaps as far east as Korea and Japan).

Biology: Anholocyclic over much of its range, but recorded as producing sexuales and overwintering in the egg stage on *Alhagi camelorum* in central Asia.

Starý and Gonzalez (1978) give an account of the parasites of *A. gossypii* in central Asia. Müller (1975b) has reviewed the literature concerning this species. $2n = 6$.

Acyrthosiphon ilka Mordvilko

Green, on *Papaver somniferum* in the Mediterranean and Middle East. Other Papaveraceae and several plants in other families—for example Cruciferae (*Isatis, Erysimum*), Linaceae (*Linum*), Thymelacaceae (*Daphne*), Compositae (*Lampsana*)—are also colonized. There is some doubt as to whether or not *A. bidentis* Eastop, living on a similar range of host plants in North and East Africa, is synonymous with *A. ilka* (see Leclant and Remaudière, 1974). Biology is little known.

Acyrthosiphon knechteli (Börner)

Green or pink aphids, found only on shoot apices and undersides of leaves of *Vaccinium uliginosum*. Recorded from North, West, and Central Europe (Sweden, Germany, Poland, Czechoslovakia, Switzerland). Biology is little known.

Acyrthosiphon kondoi Shinji
'Blue Alfalfa Aphid', 'Blue-Green Aphid'   Plate 102

Appearance in life: Bluish green, on stems and leaves of host plants. Alatae have a darker brown thorax than alatae of *A. pisum*. Apterae 2.1–2.9 mm, alatae 1.5–2.8 mm.

Host plants: Leguminosae, mainly of the tribes Trifoleae (*Medicago, Melilotus, Trifolium*), and Loteae (*Dorycnium, Lotus*) but also on *Astragalus* (Galegeae) and *Lens* (Viceae). An important pest on lucerne.

Distribution: Now a widely-distributed species. Origin unknown but probably from somewhere in temperate or warm temperate Asia; known from Japan, Korea, India, Pakistan, Afghanistan, and Iran. Introduced to both California and New Zealand in 1974–75 and has now spread through much of the U.S.A. and to Argentina, Chile, eastern Australia, and (in 1980) South Africa.

Biology: Introduced populations are apparently all anholocyclic. The sexual morphs have not been described and the life cycle in Asia is unknown. Gonzales *et al.* (1979) have reared parasites from *A. kondoi* collected in Japan; the control of *A. kondoi* in eastern Australia by native and introduced parasitoids and fungi is discussed by Lehane (1982). For work on resistance to *A. kondoi* in lucerne see Nielson *et al.* (1976), Okamoto (1974), Lance (1980), and references in latest issues of *Review of Applied Entomology*. $2n = 10$.

Acyrthosiphon (Tlja) lactucae (Passerini)
(= *barri* Essig)

Appearance in life: Pale yellowish green or pink, covered with a pale grey wax bloom. Alatae very pale. Apterae 1.7–2.9 mm, alatae 1.8–2.9 mm.

Host plants: On stems, and rarely on undersides of leaves of *Lactuca* species (*sativa, scariola, virosa*), and probably specific to this plant genus.

Virus transmission: A vector of lettuce mosaic virus.

Distribution: Europe, the Middle East, and (by introduction) throughout North America.

Biology: Monoecious holocyclic on *Lactuca* species. Males are alate. $2n = 16^*$. (*A. scariolae* Nevsky, previously synonymized with *A. lactucae*, has $2n = 18^*$.)

Acyrthosiphon loti Theobald

Green with faint greyish transverse lines, or rarely pink. Found commonly throughout most of Europe on certain wild Leguminosae, especially *Lotus*, but may occur sporadically on lucerne (Meier, 1958). The holocycle has been observed on *Lotus* species, *Hippocrepis comosa*, and *Onobrychis sativa*. Both apterous and alate males are recorded. $2n = 10$.

Acyrthosiphon pisum (Harris)   'Pea Aphid'   Plate 101
(= *destructor, pisi, onobrychidis*)

A. pisum is a complex of races and subspecies with different host plant ranges and preferences, some recognizable by their morphology, but the extent of interbreeding is largely unknown.

Appearance in life: Rather large, green, or pink aphids with slender appendages forming colonies on young growth and developing pods of many herbaceous and some shrubby Leguminosae. Immatures very lightly dusted with wax. Apterae 2.5–4.4 mm, alatae 2.3–4.3 mm.

Host plants: Mostly Leguminosae of the tribes Genisteae (*Cytisus, Genista, Sarothamnus, Spartium*), Trifoleae (*Medicago, Melilotus, Ononis, Trifolium, Trigonella*), Fabeae (*Lathyrus, Lens, Pisum, Vicia*), Hedysareae (*Hippocrepis, Onobrychis*), and also colonizing a few members of other tribes, e.g. *Lotus* (Loteae), *Astragalus* (Galegeae), and *Glycine* (Phaseoleae). Under dry conditions sometimes found on *Capsella bursapastoris*. There is much evidence that, at least in Europe, many populations occur with particular host-plant preferences; e.g. pea-feeding populations are distinguishable from those

Acyrthosiphon

colonizing *Medicago* (Müller 1980). One form which lives only on *Ononis* is probably a distinct species.

Virus transmission: A vector of more than 30 virus diseases, including non-persistent viruses of beans, peas, beet, clover, cucurbits, Narcissus, and Cruciferae, and the persistent viruses pea enation mosaic and pea leaf-roll.

Distribution: Origininally probably palaearctic, now almost world-wide, although a number of the earlier records of *A. pisum* from outside Europe apply to species such as *Macrosiphum euphorbiae* and *Aulacorthum solani*. C.I.E. Distribution Map **23** (revised 1982).

Biology: Holocyclic on various leguminous hosts in temperate regions; in warmer climates there is presumably facultative anholocycly. Both apterous and alate males are known. Cook (1963) gives an account of the ecology of *A. pisum* in North America. Starý and Gonzalez (1978) discuss the parasite complex of *A. pisum* in central Asia. Mackay et al. (1983) studied sexual morph production. Eastop (1971) gives references to a selection of the enormous literature on this species. Probably relatively few genotypes have been distributed outside Europe and hence the biology and host-plant relations of introduced populations may show significant differences (Blackman, 1981). $2n = 8$.

Acyrthosiphon pseudodirhodum (Patch)

Pale yellowish white, on *Spiraea latifolia* and perhaps other *Spiraea* species, and on wild *Rosa*, but not so far recorded from cultivated roses. In north-western U.S.A. and Canada (including Manitoba). Life cycle unknown.

Acyrthosiphon rogersii (Theobald)   Plate 99

This is one of a group of morphologically similar species with different host-plant associations, and is often regarded as a subspecies of *A. malvae* (Mosley) (= *pelargonii* Kaltenbach).

Appearance in life: Various shades of green, often pale yellow-green and shining. Often forming rather large colonies on the petioles of young strawberry leaves. Apterae 1.5–2.7 mm, alatae 1.8–2.7 mm.

Host plants: A. rogersii lives only on *Fragaria*. However, another member of the *malvae* group, *A. geranii* Kaltenbach, may sometimes be found on *Fragaria* but will also feed on *Malva, Geranium,* and *Erodium*; the two species are only really separable by their host-plant preferences.

Virus transmission: Vector of strawberry mottle virus.

Distribution: Northern and western Europe.

Biology: Holocyclic on *Fragaria*. Both apterous and alate males are recorded. Host-plant relationships and hybridization within the *A. malvae* group are discussed by Prior and Stroyan (1964) and Müller (1972). $2n = 10$.

Acyrthosiphon rubi Narzikulov

Shiny green aphid recorded from *Rubus* species (*caesius, ellipticus*) in Tadjikistan and northern India. This species may have polyphagous tendencies since it has also been found on several species of Compositae and Polygonaceae, including buckwheat (David and Hameed, 1975).

Acyrthosiphon tutigula (Hottes)

Pale yellow-green. Primary host is wild *Rosa*, with a migration to *Corydalis*. (The generic position of this host-alternating species is uncertain.) Not so far found on cultivated roses. Only recorded from Colorado, U.S.A.

Aiceona Takahashi   Anoeciinae

Rather large (2–3 mm) hairy aphids, in life brownish to purplish black and densely dusted with white wax. The 14 recognized species occur in Eastern and South Eastern Asia and are associated mostly with Lauraceae, but one is described from Juglandaceae. The genus is unusual in that the sexuales are alate and may be produced early in the year (Takahashi, 1960, p. 6). Siphunculi are well developed in the female morphs but curiously absent from males. The alatae of some species have largely or partially pigmented wing membranes.

Neither the systematic position of the genus nor the taxonomy of the world fauna is entirely satisfactory. Two species have been recorded from camphor. It is possible that species which normally live on wild Lauraceae sometimes colonize camphor. Accounts of *Aiceona* are available for India (M.R. Ghosh and Raychaudhuri, 1973), for north-east India (Raychaudhuri, Pal, and A.K. Ghosh 1980), and for Japan (Takahashi, 1960).

Aiceona actinodaphni Takahashi

Recorded from *Cinnamomum camphora* by Tseng and Tao (1938), as well as from 4 species of *Litsea* in China. However, it is not certain that they (only) had the *Actinodaphne*-feeding species.

Aiceona japonica Takahashi

According to the original description from Japan (Takahashi, 1960) it is 'rarely found on *Cinnamomum camphora*' in addition to 'a tree of Lauraceae'.

Allocotaphis Börner Aphidinae: Myzini

A monotypic genus for a hairy aphid with elongate siphunculi and a short cauda.

Allocotaphis quaestionis (Börner)

Rather large green aphids with a waxy bloom, rolling (but not discolouring) the leaves of apple in spring. The second generation are all alatae; green, with rust-brown thorax, a black dorsal abdominal patch, and very numerous tuberculate sensoria on antennal segments III–V, migrating for the summer to *Senecio doronicum*. The gynoparae returning to apple initially give rise to single-parent colonies on the undersides of the leaves (Hille Ris Lambers and Wildbolz, 1958). Distribution is apparently boreo-alpine, in Carpathians, Caucasus, and Alps.

Aloephagus Essig Pemphiginae: Fordini

A monotypic genus probably of African origin. The absence of siphunculi, short, 5-segmented antennae, elongate second rostral segment and characteristic habitat and appearance in life make confusion with other aphids unlikely. Hille Ris Lambers (1954a) described *Aloephagus* and discussed its taxonomic position.

Aloephagus myersi Essig Aloe Aphid Plate 150

Appearance in life: Densely wax-dusted. Apterae 1.8–2.5 mm, alatae *c.* 2.3 mm.

Host plants: Aloe species, living under the leaf bases. Host alternation from *Pistacia* is suspected (Hille Ris Lambers, personal communication).

Distribution: Widespread in Africa south of the Sahara, where it is probably native, but originally described from California, U.S.A., and also found in glasshouses in Europe.

Biology: Biology little studied; probably alternates to *Pistacia* in Africa, and is anholocyclic elsewhere. Attended by ants. $2n = 22^*$.

Amphorophora Buckton Aphidinae: Macrosiphini

A genus of about 27 species, mostly Nearctic, but a few eastern and western Palaearctic. Thirteen of the species live on *Rubus*. They are medium-sized to rather large aphids, usually greenish and often rather pale, with long appendages bearing rather short to medium length hairs. The siphunculi are rather long, pale or dusky and slightly to moderately swollen on the distal half. Most species on *Rubus* are found in loose colonies on young shoots or solitarily on leaves. As far as is known, the biology of most *Amphorophora* involves overwintering as eggs, a third parthenogenetic generation containing many alatae (which may also occur more sparsely in later generations), and production of apterous or alate males and oviparae in autumn. Host alternation does not occur.

Amphorophora agathonica Hottes
Large American Raspberry Aphid   Plate 124

Appearance in life: Apterae pale green, rather large, with siphunculi pale at base, dusky towards apex. Alatae with pale brown head and thorax. Apterae 2.4–4.5 mm, alatae 2.8–4.2 mm (but much smaller adults have been reared in culture).

Host plant: North American red raspberry, *Rubus idaeus* ssp. *strigosus*.

Virus transmission: A principal vector of raspberry mosaic in North America, but apparently not a vector of raspberry leaf curl or vein chlorosis.

Distribution: Widespread in North America north of about latitude 38°N, including Alaska and Nova Scotia.

Biology: Typical for the genus, with alate males. Population studies were made by Kennedy and Schaefers (1974), host-plant resistance to *A. agathonica* was studied by Kennedy and Schaefers (1973), and genetic variation in colonizing ability is discussed by Converse *et al.* (1971). Prior to 1971, the name rubi was used for this species in the North American literature, as it was thought to be the same as the species on European red raspberry. $2n = 14$.

Amphorophora idaei (Börner)   Large European Raspberry Aphid   Plate 125

Appearance in life: Adult apterae rather large, pale green, rather shiny, with pale siphunculi. Alatae with pale brown head and thorax. Apterae 2.6–4.1 mm, alatae 2.5–4.0 mm.

Host plants: European red raspberry, *Rubus idaeus* s.str., and cultivated varieties derived from it.

Virus transmission: A vector of raspberry mosaic diseases, and of raspberry leaf spot.

Distribution: Europe.

Biology: Monoecious holocyclic on raspberry, with alate males. Rautapää

(1967) gives an account of the biology of *A. idaei* in Finland, and Briggs (1965) studied the genetics of host-plant resistance-breaking mechanisms. In this and most other work prior to 1977, this aphid is referred to as *A. rubi* (Kaltenbach); Blackman *et al.* (1977) discriminate between *A. idaei* and *A. rubi* and discuss the confusion between them and the North American *A. agathonica*. $2n = 18$.

Amphorophora rubi (Kaltenbach)   Large Blackberry Aphid

Adult apterae medium to rather large, shiny, pale yellowish-green to green aphids on wild and cultivated blackberries (*Rubus fruticosus* agg.) in Europe. Monoecious holocyclic on *Rubus* with alate males; some anholocyclic overwintering may also occur. References to *A. rubi* on raspberry in the literature apply to *A. idaei* (Börner) in Europe and mostly to *A. agathonica* Hottes in North America. $2n = 20$.

Amphorophora rubicumberlandi Knowlton and Allen

Medium to rather large aphids with conspicuously swollen siphunculi, living often in large colonies on the canes of wild and cultivated black raspberries (*Rubus occidentalis, R. leucodermis*) in north-western U.S.A. (Oregon, Washington). Biology is little known.

Amphorophora rubitoxica Knowlton   Plate 126

Adult apterae medium to rather large, rather shiny, deep green, or pale yellowish green with a median dorsal deeper green stripe, the siphunculi conspicuously swollen and darker on distal half. On young shoots and under leaves of *Rubus* species including *laciniatus, occidentalis, palmatus, parviflorus, procerus, ursinus,* and *vitifolius,* in western North America. Life cycle and biology are little known. It is parasitized by *Monoctonus paulensis* (Aphidiidae) in California (Calvert and van den Bosch, 1972). $2n = 30$.

Amphorophora sensoriata Mason   Plate 123

Very pale, slightly bluish-green aphids in colonies on stems of black raspberries (*Rubus occidentalis*), falling off easily when disturbed. In north-eastern North America, as far south and west as Kansas. Life cycle and biology are little known. $2n = 72$.

Anoecia Koch   Anoeciinae

A distinctive holarctic genus of uncertain taxonomic position, consisting of about 20 species associated with *Cornus* and/or the subterranean parts of Gramineae or Cyperaceae, where they are usually attended by ants. Mostly palaearctic and extending into the oriental region, but several species are known only from North America and specimens have also been seen from South America. Some species alternate between *Cornus* and Gramineae, while others are entirely subterranean on Gramineae or Cyperaceae, overwintering either as eggs or anholocyclically. Adult apterae of most species on grass roots are medium-sized greenish grey or grey in colour, with a sclerotic dorsal abdominal 'plate'. They have compound eyes and are fairly mobile when disturbed. Immatures are paler, white or cream in colour. Alatae have a dark dorsal abdominal patch and a large dark pterostigma providing a conspicuous dark spot on the wing. The species of *Anoecia* are very difficult to separate. Zwölfer (1958) gives an account of the taxonomy and biology of the Central European species. Chakrabarti *et al.* (1982) key the Indian species.

Anoecia corni (F.) — Dogwood Aphid   Plate 129

This is probably the most widely distributed species of *Anoecia*, occurring on the roots of numerous grasses and some cereals, and overwintering either anholocyclically on grass roots or as eggs on *Cornus sanguinea*; in Europe, Central and South East Asia, India, and North America. Adult apterae are pale greenish grey with sclerotized parts dark grey. Alatae are grey and black with a conspicuous black pterostigma on the wing. Knechtel and Manolache (1943) provide an account of this species. $2n = $ 6, 7, or 8.

Anoecia fulviabdominalis (Sasaki)

Grey-green aphids on roots of rice, and sometimes other Gramineae (*Agropyron, Triticum, Hordeum, Imperata, Echinochloa, Digitaria*) in South East Asia (Japan, Korea). Heteroecious, alate sexuparae flying in autumn to *Cornus* species (*controversa, brachypoda*), although anholocyclic overwintering may also occur. Tanaka (1961) gives an account (in Japanese) of its biology, summarized in Yano *et al.* (1983).

Antalus Adams   Drepanosiphinae: Lizeriini

One of the few genera endemic to Africa, comprising three species associated with Leguminosae of the tribes Galegeae (*Indigofera*) and Hedysareae (*Aeschynomene, Humularia*). Small, frail, wax-producing predominantly yellow or green aphids.

Antalus albatus Adams

Very small (body length 1.5 mm) aphids, the apterae with reddish-brown head and prothorax, dark green on rest of thorax and first two abdominal segments, and rest of abdomen yellow green. Legs conspicuously wax-powdered and body studded with clumps of white wax. Alatae brown and green with much reduced hindwings. On *Indigofera spicata* in Malawi, Zimbabwe, Tanzania; not yet recorded from commercially grown indigo.

Anuraphis Del Guercio
Aphidinae: Myzini

A small palaearctic genus of medium-sized aphids, dull olive green to brown in life with rather short appendages. Typically they form 'pseudo-galls' by crumpling or rolling leaves of *Pyrus*, from which they migrate to the subterranean parts of Umbelliferae, or Compositae of the *Tussilago* group, where they are attended by ants. (The name *Anuraphis* has in the past been more widely applied to include species currently placed in *Brachycaudus* and *Dysaphis*.)

The ecology and life cycles of the four species of *Anuraphis* on pear in the Crimea were studied by Kolesova (1970, 1972).

Anuraphis farfarae (Koch)
Pear—Coltsfoot Aphid

In rolled or folded leaves of *Pyrus communis* in spring, each colony comprising a plump-bodied brown fundatrix and her green progeny, all of which develop wings and migrate to form colonies on the roots of *Tussilago farfara*. Anholocyclic overwintering on the roots of *Tussilago* may also occur (Kolesova, 1972). In Europe, and Central Asia east as far as Turkey. $2n = 12$.

Anuraphis pyrilaseri Shaposhnikov

Closely related to *A. farfarae* but migrating from pear to the roots of different secondary host plants (e.g. *Laser trilobum, Hippomarathrum crispum, Laserpitium hispidum*). In central and eastern Europe, Crimea and northern Caucasus, and redescribed from Sicily by Barbagallo and Stroyan (1982). $2n = 12$.

Anuraphis subterranea (Walker)
Pear—Hogweed Aphid   Plate 48

Rolling or folding the leaves of pear in a similar way to *A. farfarae*; but immature progeny of the fundatrix inside the gall are brown like their mother, and migrate when adult to the subterranean parts of several Umbelliferae, especially *Pastinaca* and *Heracleum*, where they may give rise to large, ant-attended colonies. Throughout Europe and Central Asia east as far as Iran. Closely related to *A. subterranea*, but much less common on pear, is another species, *A. catonii* Hille Ris Lambers, which migrates to the roots of *Pimpinella* species. Fundatrices of *A. catonii* on pear have dark marginal abdominal sclerites like *A. subterranea*, but these sclerites do not bear tubercles. Both species $2n = 26$.

Aphanostigma Börner
Phylloxeridae

This genus comprises two small species associated with *Pyrus* in the Mediterranean region and Japan, respectively, and a larger Japanese species galling *Ulmus*. *Aphanostigma* differs from *Phylloxera* in the absence of spiracles from abdominal segments 2–5.

Aphanostigma piri (Cholodkovsky)

Appearance in life: Very small, yellow phylloxerids laying eggs in bark crevices of pear trees, also attacking and sometimes destroying the buds, and in some pear varieties causing necrosis around the calyx of the fruit. Apterae 0.8–1.0 mm.

Host plant: Pyrus communis.

Distribution: France, Switzerland, Italy, Israel, Crimea and the Ukraine.

Biology: Holocyclic in France and Crimea, apparently anholocyclic in Israel. Leclant (1963) gives a general account of the life cycle and economic importance of *A. piri* in France, and Swirski (1954a) describes its population ecology in Israel. Swirski *et al.* (1969) studied differences in susceptibility of pear varieties to *A. piri* attack. $2n = 8, 9$. In Japan, pear trees are attacked by another, very similar species, *A. iaksviense* (Kishida), which has been the subject of extensive ecological studies (Tanabe and Mishima, 1929, 1930). $2n = 8$ and 9.

Aphidura Hille Ris Lambers Aphidinae: Myzini

A palaearctic genus of about 12 medium-sized yellowish to greenish aphids (the adults sometimes darker dorsally), mostly in the Mediterranean region and eastern Europe. They live mostly on Caryophyllaceae but at least one species may alternate from *Prunus*. The genus has been very little studied.

Aphidura pujoli (Gomez-Menor)

Small, brownish to green aphids with darker head, siphunculi, cauda, and anal plate, and some dark dorsal abdominal markings, living on *Dianthus caryophyllus* and certain other Caryophyllaceae (*D. commutulus, Silene parviflora*) in southern Europe (Spain, Portugal, France, Italy) east to the Ukraine. (Another *Aphidura* species, *A. picta* Hille Ris Lambers, with a blackish sclerotic pattern on the dorsal abdomen and black cylindrical siphunculi, has been described from cultivated *Dianthus* in Israel.)

Aphis L. Aphidinae: Aphidini

The largest aphid genus containing more than 400 species, most of which occur in the Northern Hemisphere, but a few are native to South America, New Zealand, and Australia. Mostly small to medium-sized aphids, often greenish but varying from pale to dark green, and sometimes yellow, brown, reddish, or black. Most species live on shrubs or herbs and a comparative few occur on trees. They usually live on the young foliage and often distort the leaves, but some species live on young twigs or at the base of the stem, or on the roots. Most species of *Aphis* are attended by ants which may construct earthen shelters for the species living just above soil level.

There are evident species groups associated with particular groups of plants, but no overall pattern of host association and aphid morphology, good enough to provide a useful subgeneric classification of the world fauna, has yet been detected.

At present there are no comprehensive accounts of the genus, even on a regional basis. Takahashi (1966) gave an account of the Japanese fauna; Hille Ris Lambers (1974) described a number of South American species and Stroyan (1972, pp. 51–62) revised those European species which can be regarded as a distinct subgenus, *Pergandeida*.

Aphis affinis Del Guercio

Small dark grey-green to almost black aphids on *Mentha* species, usually clustered on stem apices. Closely related to *A. gossypii* but with a relatively short terminal process to the antenna. Found in southern Europe, southern U.S.S.R., the Middle East, India, and Pakistan. Monoecious holocyclic on *Mentha*; Tuatay and Remaudière (1964) describe the sexual morphs in Turkey. Sagar and Singh (1981) compared control by various insecticides in India.

Aphis (Protaphis) anuraphoides Nevsky   Plate 25

Small aphids covered with greyish-white wax, with black siphunculi, usually ant-attended, on safflower and other *Carthamus* species in Central Asia, the Middle East, and north-east Africa. Some of the species of this group described from other Compositae in this region could be synonymous with *A. anuraphoides*, as the taxonomy of the group is little-worked and difficult. *A. (Protaphis) carthami* (Das) on wild safflower (*C. oxycantha*) in India and Pakistan, is a similar (or identical) species.

Aphis armoraciae Cowen   Plate 26

One of a group of closely related North American root-feeding *Aphis* (cf. *A. maidiradicis, A. menthaeradicis*). Pale green to olive green with a ore or less broken pattern of transverse dorsal dark bands which is better developed in the aptera than the alata. Mainly known from the western U.S.A. where it occurs commonly on the roots (and sometimes aerial parts) of plants in several families including Compositae, Cruciferae, Umbelliferae, and Gramineae (maize, wheat); but not so damaging to maize as *A. maidi-radicis*. Has ability to transmit mosaic viruses of beet, cauliflower, and celery. Aphids provisionally assigned to this species have been collected from roots of *Opuntia* species (Cactaceae) in Australia (Eastop, 1966) and South Africa, and from Compositae in Brazil, but there are no records from crop plants outside North America. Ant-attended, monoecious holocyclic, with alate males (males of *A. maidiradicis* are apterous). $2n = 8$.

Aphis citricola van der Goot   Spiraea Aphid, Green Citrus Aphid   Plate 34
(= *spiraecola* Patch)

Appearance in life: Small aphid, often curling and distorting leaves near stem apices of host plants, or on flower heads, and usually ant-attended. Body bright greenish yellow or yellowish green to apple green, with head brown. Legs and antennae mainly pale but siphunculi and cauda dark brown to black. Alatae have head and thorax dark brown, abdomen yellowish green with a dusky lateral patch on each segment. Apterae 1.2–2.2 mm, alatae 1.2–2.2 mm.

Host plants: A very polyphagous aphid with numerous host plants in over 20 families, but especially on Caprifoliaceae, Compositae, Rosaceae, Rubiaceae, and Rutaceae, and particularly on plants of shrubby habit. *Citrus* is probably its most important crop host.

Virus transmission: A vector of citrus tristeza virus, cucumber mosaic, pawpaw distortion virus, plum pox, potato virus Y, tobacco, etc., viburnum strain of lucerne mosaic.

Distribution: Probably of Far Eastern origin. It has been in North America at least since 1907, and was introduced more recently to the Mediterranean region (about 1939), Africa (1961), Australia (1926), and New Zealand (1931).

Biology: Holocyclic in North America, where *Spiraea* is the primary host, and also in Brazil (de Menezes, 1970). In Japan (Komazaki *et al.*, 1979) both *Spiraea* and *Citrus* serve as primary hosts. Anholocyclic in many parts of the world. There is an extensive literature, particularly in relation to the economic importance of this species on *Citrus*; most comprehensive accounts are by Barbagallo (1966a) in Italy, and Miller (1929) in Florida. For a more recent summary see Kranz *et al.* (1977, pp. 330–31). On other plants, especially Rosaceae, *A. citricola* is often confused in the literature with *A. pomi*, which is the name used by Cottier (1953) for his account of *A. citricola* in New Zealand. *A. citricola* has a shorter ultimate rostral segment than *A. pomi*, fewer caudal hairs, and no lateral tubercles on abdominal segments 2–4, whereas *A. pomi* usually has conspicuous tubercles on these segments. *A. citricola* has alate males, whereas those of *A. pomi* are apterous. $2n = 8$.

Aphis coronillae Ferrari (= *scaliai* del Guercio)

A shiny black *Aphis* of the *Pergandeida* group, closely related to *A. craccivora*, colonizing *Trifolium* species (*pratense, repens, dubium*) throughout Europe. Mixed populations of *A. coronillae* and *A. craccivora* possibly sometimes occur, and the separation of these two species is not always certain, hybridization being suspected. Monoecious holocyclic on *Trifolium*, with apterous males (those of *A. craccivora* are alate). Stroyan (1964) gives an account of this species (as *A. scaliai*).

Aphis craccae L.

Black aphid, both adults and immatures densely covered with greyish-white wax powder, usually living in dense clusters on the terminal growth of vetches, especially *Vicia cracca*. Occurs throughout Europe, also in Japan and China according to Tao (1958) and introduced into North America (Russell, 1966). Not recorded from *Vicia faba*, except by Tao (1962). Can transmit bean yellow mosaic and cucumber mosaic viruses.

Aphis craccivora Koch
Black Legume Aphid, Cowpea Aphid, Groundnut Aphid   Plate 28

Appearance in life: Small aphid, adults shiny black, immatures lightly dusted with wax. Young colonies are concentrated on the growing points of the host plant, and are regularly attended by ants. Apterae 1.4–2.0 mm, alatae 1.4–1.9 mm.

Host plants: Polyphagous, with a marked preference for Leguminosae, but also found in small colonies on plants of many other families.

Virus transmission: A vector of about 30 plant virus diseases including non-persistent viruses of beans, cardamon, peas, beet, cucurbits, and Cruciferae, and the persistent viruses subterranean clover stunt, groundnut mottle, and groundnut rosette.

Distribution: Originally probably palaearctic warm temperate, now virtually world wide; particularly well distributed through the tropics. C.I.E. Distribution Map **99** (revised 1983).

Biology: Anholocyclic almost everywhere, but sexual morphs have been recorded from Germany (Falk, 1960) and from India (R.C. Basu *et al.*, 1969). Males are alate. Principal ecological studies of *A. craccivora* have been by Réal (1955) on groundnuts in West Africa, and by Gutierrez *et al.* (1971, 1974a, 1974b) in relation to the transmission of subterranean clover stunt virus in south-east Australia. Müller (1977a) compares tropical and central European populations. For a concise account of the species see Kranz *et al.* (1977, pp. 324–25). $2n = 8$.

Aphis decepta Hottes & Frison

A yellow aphid with dark brown siphunculi, collected from the undersides of leaves of *Pastinaca sativa* (a European plant) in northern U.S.A. (Illinois, New York, Pennsylvania) and on *Heracleum lanatum* in Canada (Manitoba). Rarely collected, and not recorded as a pest of cultivated parsnips, but a vector of onion yellow dwarf virus.

Aphis fabae Scopoli   Black Bean Aphid   Plate 29

Appearance in life: Young colonies consist of matt black aphids on young shoots, older colonies spread over most of aerial parts of plant with many individuals developing white wax markings. Regularly ant-attended. Apterae 1.5–3.1 mm, alatae 1.3–2.6 mm.

Host plants: Very polyphagous on its secondary host plants, which include many crops. Particularly important for its direct feeding damage to *Vicia faba*, and as a virus vector in sugar beet.

Virus transmission: A vector of more than 30 plant pathogenic viruses, including non-persistent viruses of beans and peas, beets, crucifers, cucurbits, *Dahlia*, potato, tobacco, tomato, and tulip, and the persistent viruses beet yellow net and potato leaf roll.

Distribution: Widespread in temperate regions of the Northern Hemisphere, also in South America and Africa, but not common in the hotter parts of the tropics and the Middle East, where it tends to be replaced by *A. fabae solanella* (see below).

Biology: Over much of Europe *Aphis fabae* is heteroecious holocyclic, alternating between *Euonymus europaeus*, or sometimes *Viburnum opulus*, and a wide range of secondary host plants. In Europe, *A. fabae* is part of a complex of species utilizing *Euonymus* as primary host, the relationships of which are discussed by Müller (1982). The taxonomic status of forms which have spread to other parts of the world is still uncertain since morphological studies are inconclusive and the biology has been little studied outside Europe. In Europe *A. fabae* has been the subject of extensive ecological, behavioural, anatomical, and physiological studies. A recent general account is by Cammell (1981). $2n = 8$.

Aphis fabae solanella Theobald (= *euonymi* sensu Börner, 1952/53)

A member of the *A. fabae* group often regarded as a separate species, very similar to *A. fabae* s.s. but with slightly longer siphunculi, shorter cauda, and shorter lateral abdominal hairs. Heteroecious holocyclic with *Euonymus europaeus* as primary host in Europe, anholocyclic on secondary hosts in other parts of the world (Africa, the Middle East, India and Pakistan, South America). Polyphagous, but on fewer crops than *A. fabae* s.s.; found particularly on Solanaceae and especially *Solanum nigrum*, the leaves of which are crumpled and curled characteristically. $2n = 7$ and 8.

Aphis forbesi Weed   Strawberry Root Aphid   Plate 39

Appearance in life: A small dark, bluish-green aphid colonizing young aerial growth of strawberry in spring, and also feeding on the roots. Immature stages are yellow green and alatae almost black. Constantly attended by ants. Apterae 1.2–1.6 mm, alatae 1.3–1.8 mm.

Host plants: Only recorded from wild and cultivated *Fragaria vesca*.

Virus transmission: Not a vector of strawberry viruses but one of many known vectors of onion yellow dwarf.

Distribution: Native to North America. Introduced into Europe about 1928 (Balachowsky, 1933) and also recorded from Japan and South America.

Biology: Monoecious holocyclic on strawberry. Males apterous. Biology studied especially by Marcovitch (1925) who, while working with this species, discovered that aphid sexual morphs were induced by change in day length. Apparently not very injurious to strawberries, except perhaps in light sandy soils. $2n = 8^*$.

Aphis frangulae Kaltenbach

In central Europe, aphids closely related to, and virtually indistinguishable from, *A. gossypii* (q.v.) occur on potato and various other herbaceous plants (*Capsella, Lysimachia*, but not Cucurbitaceae). Unlike the anholocyclic *A. gossypii*, these other forms are holocyclic, with *Frangula alnus* as primary host. The *A. frangulae* complex is discussed by Thomas (1968) who recognizes a number of subspecies, only some of which colonize potato.

Aphis glycines Matsumura   Soybean Aphid   Plate 30

Appearance in life: Small yellow aphid with black siphunculi, colonizing the stem apices and young leaves of growing soybean plants, and later on undersides of leaves of mature plants.

Host plants: Heteroecious holocyclic in China with *Rhamnus dahurica* as primary host. Secondary hosts restricted to Leguminosae, mainly on soybean and wild *Glycine* species, but also recorded from *Pueraria javanica* and *Desmodium intortum*.

Virus transmission: Able to transmit abaca mosaic, beet mosaic, and millet red leaf viruses; also mungbean mosaic and bean yellow mosaic (Iwaki and Auzay, 1978).

Distribution: Far East including China, Japan, Korea, Thailand, North Borneo, Malaya, and the Philippines.

Biology: Wang *et al.* (1962) studied the life cycle and ecology of *A. glycines* in China.

Aphis gossypii Glover   Melon Aphid, Cotton Aphid   Plate 31

Appearance in life: Apterae are very variable in colour. Large specimens are dark green, almost black, but the adults produced in crowded colonies at high temperature may be less than 1 mm long and very pale yellow to almost white. Most commonly light green mottled with darker green, with dark siphunculi and a pale or dusky cauda. Often ant-attended. Apterae 0.9–1.8 mm, alatae 1.1–1.8 mm.

Host plants: Extremely polyphagous. Crop plants attacked include cotton, cucurbits, citrus, coffee, cocoa, eggplant, peppers, potato, okra, and many ornamental plants including *Hibiscus*. Regarded as a major pest of cotton and cucurbits on which it builds up large populations.

Virus transmission: Known to transmit over 50 plant viruses, including non-persistent viruses of beans and peas, crucifers, celery, cowpea, cucurbits, *Dahlia*, lettuce, onion, pawpaw, peppers, soybean, strawberry, sweet potato, tobacco, and tulips. A vector of the persistent viruses cotton anthocyanosis, lily rosette, lily symptomless, and pea enation mosaic.

Distribution: Virtually world wide, but in colder temperate regions confined to glasshouses (where it is a major pest). Particularly abundant and well distributed in the tropics, including many Pacific islands.

Biology: The taxonomic status of *A. gossypii* is problematic, and hence the interpretation of biological information is difficult. Although anholocyclic in Europe, *A. gossypii* is closely related to European *Aphis* species of the *frangulae* group utilizing *Frangula alnus* as primary host (Thomas, 1968). This indicates a palaearctic origin. However, Kring (1959) demonstrated holocyclic overwintering by *A. gossypii* in Connecticut, U.S.A., with *Catalpa bignonioides* and *Hibiscus syriacus* utilized as primary hosts. Either separate Nearctic and Palaearctic species are confused under the same name or, more probably, *A. gossypii* has re-acquired its holocycle in North America utilizing new primary host plants.

On a world-wide basis, *A. gossypii* comprises an indefinite number of anholocyclic lines, some of which may have particular host-plant associations. For example, *A. gossypii* occurs on chrysanthemums and cucumbers in British glasshouses, but aphids from chrysanthemums will not colonize cucumber, and vice versa—although both can be reared on cotton. The chrysanthemum-living form has acquired resistance to organophosphorus and carbamate insecticides, the cucumber-living form has not; this example shows that in certain respects it may be necessary to consider separate populations of *A. gossypii* as distinct taxonomic entities.

The literature on *A. gossypii* is extensive; see Barbagallo (1966b, pp. 217–23) and Kranz et al. (1977, pp. 328–29) for short introductory accounts. $2n = 8$.

Aphis grossulariae Kaltenbach European Gooseberry Aphid Plate 37

Appearance in life: A small greyish-green aphid forming dense colonies at tips of gooseberry shoots in spring, causing severe deformation and clumping of young leaves, and stunting growth. Apterae 1.5–2.2 mm, alatae 1.6–2.4 mm.

Host plants: European gooseberry (*Grossularia uva-crispa*) is the primary host, from which alatae migrate in late spring to *Epilobium* species (*adenocaulon, hirsutum, montanum, palustre, parviflorum, roseum*) and *Godetia hybrida*. Also recorded from *Ribes alpinum* and *R. aureum* but not known to overwinter in the egg stage on plants other than gooseberry.

Virus transmission: A vector of gooseberry veinbanding virus.

Distribution: Throughout Europe, east to the Ukraine.

Biology: Heteroecious holocyclic, the migration to *Epilobium* apparently being an obligatory part of the life cycle (R. Harrington, personal communication).

Aphis helianthi Monell Plate 40
  (= *heraclella* Davis; Addicott, 1981)

Adults of this species are usually deep olive green mottled with yellowish green, with black siphunculi. Colonies occur in late spring and summer on stems and leaves of wild carrot, celery, parsnip, other Umbelliferae, and wild plants in several other families in western and northern U.S.A. and Canada. Heteroecious holocyclic, with dogwood (*Cornus stolonifera*) as primary host. Apparently, cultivated umbellifers are rarely colonized, but ability to transmit a number of viruses, including celery mosaic, has been demonstrated. $2n = 8$.

Aphis (Cerosipha) humuli (Tseng & Tao)

Yellowish-green aphids on *Humulus japonica*, recorded from China and Taiwan. Biology not known.

Aphis idaei van der Goot Small European Raspberry Aphid Plate 21

Appearance in life: Spring colonies of small bright green aphids at tips of young canes causing severe leaf curl, later producing alatae and dispersing to other parts of plant. Summer apterae on undersides of leaves of fruiting canes are pale yellow. The rather long siphunculi, curved outward distally, are distinctive. Apterae 1.0–2.0 mm, alatae 1.2–1.8 mm.

Host plants: European raspberry, and some cultivated plants of raspberry parentage, including loganberry. Can also colonize *Rubus occidentalis*.

Virus transmission: A vector of raspberry leaf curl, raspberry vein chlorosis and rubus yellow net, but has failed to transmit raspberry leaf spot, raspberry ring spot, and raspberry yellow blotch.

Distribution: Throughout Europe, east to the Ukraine; introduced to New Zealand (Lowe, 1966).

Biology: Monoecious holocyclic on *Rubus idaeus*. Males are apterous. Dicker (1940) gives an account of the life cycle in England. Markkula and Rautapää (1963) record three Aphidiid species parasitizing *A. idaei* in southern Finland. $2n = 8$.

Aphis illinoisensis Shimer   Plate 41

Appearance in life: Small, rather shiny, deep reddish-brown to almost black aphids, with rather long and divergent dark siphunculi, sometimes occurring in large numbers on young shoots and leaves of grape vine, and later on the fruit clusters. Apterae 1.6–1.9 mm, alatae 1.3–2.0 mm.

Host plants: Vines (Vitaceae) are the secondary hosts, the autumn migrants in N. America going to *Viburnum prunifolium*, on which plant overwintering eggs are laid. Besides wild and cultivated *Vitis vinifera*, *V. tiliaefolia* and *Cissus sicyoides* are also recorded as host plants.

Distribution: East and central U.S.A., Central and South America as far south as Uruguay.

Biology: Heteroecious holocyclic in Virginia, U.S.A., where the life cycle was studied by Baker (1917).

Aphis intybi Koch

A small black aphid sometimes occurring in large numbers in spring on young growth of chicory (*Cichorium intybus*) and later at the stem bases in ant shelters. Monoecious holocyclic, only on *Cichorium* species, with apterous males (Tuatay and Remaudière, 1964). In Europe (Sweden, Germany, Hungary, Italy) and the Middle East (Iran, Israel, Turkey).

Aphis lambersi (Börner)   Permanent Carrot Aphid

Small dark green to almost black aphids feeding at stem bases and lower leaf sheaths of wild carrot, and occasionally on cultivated carrot. Monoecious holocyclic, with apterous males, and apparently specific to *Daucus* (very similar but distinct species occur on *Plantago* and *Taraxacum*). Stroyan (1955) and Pintera (1956) give brief accounts. Throughout Europe.

Aphis lupini Gillette & Palmer

Dusky olive to blackish-green medium-sized aphids on leaves and stems of *Lupinus* species. Monoecious holocyclic. In mid-western U.S.A. (Colorado, Utah, Idaho, Montana).

Aphis maidiradicis Forbes — Corn Root Aphid

Appearance in life: Apterae bluish green, lightly dusted with greyish-white wax; head dark, and dusky transverse bands on thoracic segments and abdominal segments 1, 7, and 8. Alata with head and thorax black, abdomen pale green with dark markings. Forms colonies on roots of host plants, invariably attended by ants.

Host plants: Principally known as a pest of maize (corn) in U.S.A., but aphids thought to be this species have been collected on the roots of many other plants including species of *Artemisia, Aster, Cosmos, Plantago, Polygonum, Portulaca,* and *Rumex.* It has also been reported attacking the roots of seedling cotton (Bondy and Rainwater, 1939). However, it is difficult to be certain about host-plant records in view of the problems of differentiating this aphid from *A. armoraciae* (q.v.).

Distribution: Throughout U.S.A., particularly in the corn belt. Aphids assigned to *A. maidiradicis* are also recorded from Jamaica and Brazil.

Biology: Monoecious holocyclic, with apterous males. Eggs are looked after by ants during the winter months. The life cycle on maize and relationships with ants were studied in detail by Davis (1909, 1917). $2n = 8$ and 9.

Aphis medicaginis Koch

Black aphids of the *Pergandeida* group, closely related to *A. craccivora*, on shoots and flower-heads of *Medicago* species, possibly including lucerne, in spring, later moving to stem bases. Closely related to, and possibly even synonymous with, *A. coronillae* on clover. In Northern and Central Europe; extra-European records in earlier literature probably all apply to *A. craccivora*. Biology unknown, but presumably monoecious holocyclic on *Medicago*.

Aphis menthaeradicis Cowen

Dull grey green with dark transverse bars, especially well marked in apterous adults. Ant-attended colonies on roots of mint, and perhaps on other plants (*Aster, Taraxacum, Artemisia*). Males apterous. Very closely related to *A. maidiradicis*, and there is doubt as to whether the two are distinct species.

Aphis nasturtii Kaltenbach Buckthorn-Potato Aphid Plate 32
(= *abbreviata* Patch)
(= *A. rhamni* of many authors)

Appearance in life: Apterous adults bright yellowish green, with siphunculi pale or only slightly dusky; the body colour is rather constant, in contrast to the considerable colour variation in the closely related *A. gossypii–A. frangulae* group (q.v.). Apterae 1.3–2.0 mm, alatae 1.2–2.0 mm.

Host plants: Primary host is buckthorn, *Rhamnus cathartica* and *R. alnifolia*. Polyphagous on secondary host plants in several families including Solanaceae (*Solanum nigrum*, potato), Cruciferae (*Nasturtium, Capsella*), and Polygonaceae (*Polygonum*).

Virus transmission: An efficient vector of the non-persistent potato viruses A, Y, and aucuba mosaic, but a poor vector of potato leaf roll. Populations of *A. nasturtii* do not usually seem to be a problem in seed potato production.

Distribution: Throughout Europe, the Middle East, North India and Pakistan, Far East (Japan), and North America.

Biology: Heteroecious holocyclic with *Rhamnus* as primary host (mainly *R. cathartica* in Europe, *R. alnifolia* in north-eastern U.S.A.), probably anholocyclic in warmer regions. The life-cycle and population dynamics in Maine, U.S.A., were studied by Shands and Simpson (1971) and in Poland by Galecka (1966). $2n = 8$.

Aphis neomexicana (Cockerell & Cockerell)
(= *ribigillettei* Allen & Knowlton) (= *sanborni* Patch)

Pale green aphids with pale siphunculi forming colonies at shoot tips of wild and cultivated gooseberries, and *Ribes* spp. (including *aureum, alpinum, nigrum*) in spring. A North American species, closely related to the European *A. grossulariae*. Heteroecious holocyclic, with a migration to *Epilobium* species (*lineare, coloratum, adenocaulon*: Patch, 1927). *A. ribigillettei* is considered to be the same species by Robinson and Rajanavongse (1976). $2n = 8$.

Aphis nerii Boyer de Fonscolombe Oleander Aphid Plate 36

Appearance in life: Apterae bright yellow with black siphunculi and cauda; antennae and legs also predominantly dark. Alata yellow and black with dark wing veins. Often forming dense colonies on younger stems of host plants. Apterae and alatae both 1.5–2.6 mm.

Host plants: Mainly on Asclepiadaceae (*Asclepias, Gomphocarpus,*

Calotropis) and Apocyanaceae (*Nerium oleander, Vinca*); also occasionally colonizing plants in other families (e.g. Euphorbiaceae, Compositae, Convolvulaceae).

Virus transmission: Ability to transmit sugar cane, pawpaw, and chili mosaic viruses has been demonstrated.

Distribution: Widely distributed through the Old and New World tropics and subtropics including many Pacific islands.

Biology: Apparently entirely anholocyclic; no sexual morphs have been described. The aposematic coloration of this species may be linked with the sequestration of cardiac glycosides from its host plants (Rothschild *et al.*, 1970). $2n = 8$.

Aphis newtoni Theobald

A small aphid with deep green coloration, forming colonies low down on leaf blades, and later on young flower stalks of *Iris* spp. (*foetidissima, pseudacorus, sibirica*). The lateral tubercles are large and conspicuous in both apterae and alatae. Recorded only from north-west Europe (Denmark, England, Germany, the Netherlands, Sweden). Monoecious holocyclic, with apterous males. $2n = 8$.

Aphis passeriniana (del Guercio)

Small, medium to dark green aphids with light wax dusting, forming colonies on the young shoots and in the flower heads of *Salvia* species, including *S. officinalis*. In Europe (England, Germany, Italy) and the Middle East (Israel).

Aphis pomi De Geer   Green Apple Aphid   Plate 35

Appearance in life: Small yellow-green aphids with rather large black siphunculi and dark cauda, colonizing young growth of host plants, causing only slight leaf curl. Apterae 1.3–2.2 mm, alatae 1.3–2.3 mm.

Host plants: Apple, pear and quince, and plants in several other genera of woody Rosaceae including *Chaenomeles, Cotoneaster, Crataegus, Mespilus,* and *Sorbus*.

Distribution: Europe, the Middle East (Iran, Israel, Turkey), and North America; but many North American records apply to *A. citricola* (q.v.). Oriental records of this aphid on a number of host plants (e.g. in Japan) all seem to apply to other species, although it is surprising that *A. pomi* has not spread world-wide.

Biology: Monoecious holocyclic, with apterous males. A detailed early account is given by Baker and Turner (1916). Numerous references to ecology and control of *A. pomi* in apple orchards will be found in *Review of Applied Entomology*. Potential biological control agents include hymenopterous parasitoids (Evenhuis, 1963), dipterous predators (Bouchard *et al.*, 1981; Marboutie, 1976), and entomophthorous fungi (Tsinovskii and Egina, 1977). $2n = 8$.

Aphis punicae Passerini
Plate 33

Small yellowish-green aphids typically colonizing the upper sides of mature leaves of pomegranate, concentrated along the mid-ribs and around the leaf margins. The siphunculi are pale on the basal half, darker apically. Colonies of this aphid may also occur on *Duranta plumieri, Plumbago capensis,* and *Lawsonia inermis*. Throughout the Mediterranean region, and in India and Pakistan. Monoecious holocyclic on *Punica*, with alate males, but in Israel there is also anholocyclic overwintering on *Duranta* (Swirski, 1954b). $2n = 8*$.

Aphis ribiensis Gillette & Palmer

Yellowish-green, with short, pale siphunculi, producing spring colonies curling young leaves of *Ribes* spp. (*aureum, nigrum*) in North America. Heteroecious, migrating for the summer to plants in marshy situations (*Rorippa* species, *Veronica americanum, Bidens cernua*). This species is discussed by Hille Ris Lambers (1974) who points out its similarity to *A. triglochinis* (q.v.) in Europe, also alternating between *Ribes* and water plants.

Aphis rubicola Oestlund (= *rubiphila* Patch)
Plate 27

Very small, pale yellowish-green aphids on new growth of wild and cultivated American red and black raspberries, and in summer on fruiting canes. In the U.S.A. and Canada. A similar species, *A. rubifolii* (q.v.), occurs on blackberry and may sometimes be found on red raspberry; however, *A. rubifolii* is short-haired and always has 5-segmented antennae, whereas *A. rubicola* has long, conspicuous hairs dorsally and laterally on the abdomen and on the femora, and the antennae are 6-segmented except in dwarf summer apterae. *A. rubicola* may also occur occasionally on blackberry. Monoecious holocyclic, with apterous males. A vector of raspberry leaf curl, but failed to transmit red raspberry mosaic. Rankin (1927) describes the life cycle in New York state, and Brodel and Schaefers (1980) have studied sexual morph production in the laboratory. $2n = 8$.

Aphis rubifolii (Thomas)

Very small, pale greenish-yellow aphids on wild and cultivated species of blackberry, and perhaps occasionally on red raspberry, in the U.S.A. Biology little known, but probably monoecious holocyclic, with apterous males. (It is likely that dwarf summer apterae of *A. rubicola* on raspberries have in the past been misidentified as this species, in which case it could transpire that both *A. rubicola* and *A. rubifolii* are more specific in their host-plant associations with blackberries and raspberries, respectively, than has formerly been recognized.)

Aphis ruborum (Börner)   Permanent Blackberry Aphid

Appearance in life: Spring forms are blue green with pale siphunculi, living in dense colonies on young shoots of host. Later generations live under leaves, in flowers, and on developing fruits, and the apterae are then very small and pale yellow-green in colour. Apterae 0.8–2.5 mm, alatae 1.2–2.0 mm.

Host plants: Wild and cultivated blackberries (*Rubus fruticosus* agg., *R. laciniatus*, etc.) and occasionally on strawberries in the autumn. Not known from raspberries in Europe, but recently collected by R.H. Converse from raspberries in Chile.

Distribution: Europe, North Africa, the Middle East, India, Pakistan, and (coll. 1981) Chile. Some Indian records may apply to *Aphis longisetosa* Basu, described from wild *Rubus*, which has longer hairs on the head and dorsal abdomen, and dark siphunculi and cauda.

Biology: Monoecious holocyclic on blackberries, with both apterous and alate males. Life cycle observations are recorded by Dicker (1940, p. 25, as an unnamed species), Hille Ris Lambers (1950b), and Stroyan (1955). $2n = 8$.

Aphis salviae Walker

Black *Aphis* of the *Pergandeida* group on Salvia species, in Europe and the Middle East. Possibly synonymous with *A. craccivora*, although Ilharco (1979) points out some differences.

Aphis sambuci L.   Elder Aphid   Plate 42
(= *A. sambucifoliae* Fitch)

Apterae of this aphid found in summer on the roots of Caryophyllaceae and certain other plants, are olive to dark green with a coating of mealy wax. They are assiduously attended by ants. The primary host is elder, *Sambucus* spp., on the young green stems of which dense colonies of non-waxy, dark green to almost black individuals build up in spring. The biology of *A. sambuci* in Europe has been studied by Jacob (1949) and Iglisch (1966). A very variable species; biological studies outside Europe, where secondary hosts are largely unknown, may show whether one or a group of species is involved. Takahashi (1923, p. 60) observed sexuparae and males developing on *Sambucus* spp. in

Japan. Aphids of the *A. sambuci* group occur in Europe, the Middle East, Japan, and North and South America. $2n = 8$.

Aphis schneideri (Börner)   European Permanent Currant Aphid   Plate 38

Appearance in life: Dark green aphids dusted with bluish-grey wax. Spring colonies on currant bushes stunt growth of the shoots and crumple the leaves into 'nests', usually visited by ants. Apterae 1.2–2.2 mm, alatae 1.6–2.2 mm.

Host plants: Redcurrant, blackcurrant, and in summer on certain ornamental *Ribes* species (*aureum, sanguineum*).

Virus transmission: A possible vector of gooseberry veinbanding virus.

Distribution: Europe.

Biology: Monoecious holocyclic on *Ribes* species. The effects of this aphid on the growth of currant bushes in spring can be severe. In the British literature there has been some past confusion of this species with *A. grossulariae*, which forms similar colonies on gooseberry but has much shorter antennal hairs. For a brief account see Stroyan (1955).

Aphis symphyti Schrank

Apterae dark green, dusted with bluish-grey wax, with dark, coarsely imbricated siphunculi and a pale cauda. Colonies on young shoots, and later on flower-heads and under leaves of comfrey (*Symphytum officinale*). Monoecious holocyclic.

Aphis triglochinis Theobald   Redcurrant–Arrowgrass Aphid

Small, brownish-green aphids which form spring colonies on young growth of *Ribes rubrum*, and sometimes *R. nigrum*, curling the leaves but not stunting growth so severely as *A. schneideri* (q.v.). Heteroecious holocyclic, migrating for the summer to plants in marshy situations (*Rorippa silvestris, Triglochin maritima, Barbarea* spp., *Myosotis palustris*). For a brief account of this aphid see Hille Ris Lambers and Dicker (1965). $2n = 8^*$.

Aphis vaccinii (Börner)

Bluish-black aphids with a slight powdering of bluish wax, on *Vaccinium* species (*myrtillus, uliginosum, vacillans*) and *Andromeda polifolia*. In Europe and as far east as Ukraine. Probably monoecious holocyclic. A short account is given by Fidler (1951).

Aphis varians Patch — Variable Currant Aphid

Dark blue-green to green aphids with pale siphunculi and dark cauda, colonizing spring growth of blackcurrant and gooseberry and sometimes on other wild or cultivated *Ribes* species in North America, crumpling young leaves and stunting growth. Closely related to *A. schneideri* in Europe, but this species is heteroecious holocyclic, migrating in late spring to Onagraceae, especially *Chaemonerium angustifolium*. For life cycle details see Patch (1927). $2n = 8$.

Aploneura Passerini — Pemphiginae: Fordini

A small palaearctic genus with alternation between galls on *Pistacia* and the roots of Gramineae and vines, but also a strong tendency to anholocycly on the roots of secondary host plants. The apterae have very short 4- or 5-segmented antennae; the alatae have 6-segmented but unusually short antennae which are distinctive because segments III and IV both bear a single, large secondary sensorium on the distal half, resembling the primary sensoria on V and VI. The apterae on roots produce copious white wax and are not attended by ants. The alatae of *Aploneura* hold their wings flat in repose, unlike *Baizongia* and *Slavum* (q.v.), in both of which the wings are held roof-like. In the case of *Slavum*, no satisfactory structural characters have been found to support its generic separation from *Aploneura*, which is therefore in doubt.

Aploneura ampelina Mokrzecky (= *Slavum lentiscoides* Mordvilko?)

Pale, spindle-shaped aphids in white wax on roots of vines which have sometimes been confused with the vine phylloxera, *Viteus vitifolii* (q.v.). Recorded mainly from southern Europe, Crimea, and Central Asia; but also in South Africa and probably more widespread. Roots of grasses may also be colonized. Alatae of *A. ampelina* may be distinguished from those of *A. lentisci* by the shape of the sensorium on antennal segment VI, which has its long axis across the segment in *ampelina*, along the segment in *lentisci*. Life cycle of *A. ampelina* is still unclear; it may be synonymous with *Slavum lentiscoides* which forms galls on *Pistacia vera* (Davatchi, 1958), but experimental work is needed to confirm this. See Mordvilko (1935, pp. 221–25) for further information.

Aploneura lentisci (Passerini) — Plate 148

Appearance in life: Apterae on grass roots pale yellow with darker head; body spindle-shaped with very short appendages, covered with fine white wax which is flocculent at posterior end. Apterae 1.1–3.0 mm, alatae 1.3–2.3 mm.

Host plants: Pistacia lentiscus (primary host) and numerous species of Gramineae; occasionally on roots of dicotyledons (e.g. *Ranunculus, Veronica*).

Distribution: West, central, and southern Europe, the Middle East, Central Asia, Africa (Morocco, Kenya, Nigeria, Zimbabwe), Australia, New Zealand, and Argentina. Not (yet) found in North America.

Biology: Holocyclic in the Mediterranean region, galling the leaves of *Pistacia lentiscus*. In other parts of its distribution area *A. lentisci* is anholocyclic on grass roots. However, the large flights in Britain during August consist almost entirely of sexuparae. Mordvilko (1935) gives an account of the life cycle, the root-feeding forms are described by Zwölfer (1958), and for a more detailed study see Zavattari (1921). Not visited by ants. $2n = 16$.

Asiphonaphis Wilson & Davis Aphidinae: Aphidini

A small North American genus with only two currently recognized species, related to *Aphis* but completely lacking siphunculi. Both alatae and apterae have prominent lateral tubercles on all abdominal segments.

Asiphonaphis pruni Wilson & Davis Plate 46

Apterae are pale yellow to whitish green, sometimes banded with darker green, with darker head and pronotum, in colonies on leaves and distorting the new growth of *Prunus* species (*armeniaca, serotina, virginiana*) in Canada and U.S.A. Robinson (1964) gives an account of this species on *P. virginiana* in Manitoba. Monoecious holocyclic, with apterous males.

Asiphonella Theobald Pemphiginae: Fordini

A genus of two described species with host alternation between *Pistacia*, on which they form large, cock's comb-like galls, and the roots of Gramineae, especially *Cynodon dactylon*. The apterae on grass roots have 3-faceted eyes which are noticeably protuberant, and well-developed segmental wax glands secreting a coating of white flocculent wax. Both apterae and alatae have 6-segmented antennae. The two species are difficult to distinguish on their secondary host plants, although the migrants from *Pistacia* are readily distinguishable. An account is given by Remaudière and Tao (1957).

Asiphonella cynodonti (Das)

Apterae on grass roots broadly oval, light green to dull green with a dark head, whole body covered with white wax powder. Primary host in Iran is *Pistacia khinjuk* (Davatchi, 1958) from which migration occurs to roots of *Cynodon dactylon* and probably certain other grasses. Recorded from Iran, Pakistan, and India. The original description (Das, 1918) includes biological observations.

Asiphonella dactylonii Theobald   Plate 147

Apterae on grass roots broadly oval, pale yellow, yellowish green to green, head darker, whole body covered in bluish-white wax. This species forms galls on *Pistacia chinensis* in China, migrating to roots of *Cynodon dactylon* and certain other grasses (e.g. *Oplismenus compositus, Pollinia ciliata,* and *Digitaria decumbens*). Also recorded from Egypt, Sudan, Zimbabwe, the Philippines, Bermuda, New Mexico, Guyana, Brazil, and Argentina. Presumably anholocyclic on grass roots outside the Far East. Blanchard (1944) gives an account of this aphid on *Cynodon* in Argentina, under the name *Paraprociphilus graminis*.

Astegopteryx Karsch   Hormaphidinae: Cerataphidini

An oriental genus of over 30 species with host alternation between *Styrax* and monocotyledonous plants, chiefly palms and bamboos. The morphs from galls on *Styrax* differ greatly in morphology from those on secondary host plants, which were described in different genera (*Trichoregma, Oregma*) until Hille Ris Lambers (1953b) showed that *Astegopteryx styracophila* on *Styrax benzoin* was the same species as *Oregma pallida* van der Goot on bamboos. Apterae of *Astegopteryx* on bamboos and palms are small aphids with well-developed frontal horns and segmentally arranged wax glands, forming large colonies on the leaves. Most species probably have the ability to maintain themselves continuously by parthenogenetic reproduction on their secondary hosts.

Astegopteryx species on bamboos

Small to very small yellow, green or brownish aphids with variable amounts of white wax, especially around sides of abdomen. Usually attended by ants. 15 species have been described from bamboos, 6 of them by van der Goot (1917/18) in Java. Further biological and taxonomic studies are needed to determine life cycles and clarify relationships within the group. Some of the best known or most distinctive species are listed below, but at present it is not possible to give a reliable key to their identification.

Species name	Body colour in life	Secondary host plant	Distribution	Reference
bambusae (Buckton)	Dark brownish green mottled with black	*Bambusa* spp.	Nepal, Pakistan, India, Ceylon, China, Thailand, Malaysia, Indonesia and Fiji	Doncaster (1966)
bambusifoliae (Takahashi)	Yellowish green with 2 dark green longitudinal stripes	*Bambusa* spp., *Phyllostachys lithopia, Dendrocalamus latiflorus*	Taiwan	Liao (1976)

flava (Takahashi)	Yellow with dark head and pronotum	'bamboo'	Malaysia	Takahashi (1950)
insularis (van der Goot)	pale green, with a large dark green patch	*Dendrocalamus latifrons*	Indonesia and Taiwan	Liao (1976)
malaccensis (Takahashi)	dark purplish brown	'bamboo'	Malaysia	Takahashi (1950)
minuta (van der Goot)	pale green with paired dark green patches	*Bambusa* spp., *Arundinaria* spp.	Nepal, India, Thailand, China, Japan, and Indonesia	Ghosh, Pal, and Raychaudhuri (1974)
styracophila Karsch (=*pallida* van der Goot)	greenish white with dark green patches	*Bambusa* spp., *Arundinaria* spp.	Thailand, Malaysia, and Indonesia	Hille Ris Lambers (1953)

Astegopteryx muiri (van der Goot)

According to the original description, small aphids of a darkish violet colour with marginal wax secretion, attended by ants. On undersides of leaves of certain Zingiberaceae (*Amomum, Languas*) in Malaysia; not recorded as a pest of cultivated ginger, but aphids provisionally assigned to this species have been collected on wild ginger in New Guinea and Sarawak.

Astegopteryx nipae (van der Goot)   Plate 133

Very small aphids; apterae with grey head and thorax, bright brownish-red abdomen, and a wax fringe, occurring in large numbers in clusters on the undersides of palm leaves (*Cocos nucifera, Elaeis guineensis, Musa sapientium, Zalacca edulis*), invariably attended by ants. Recorded from Malaysia, Indonesia, Solomon Islands, New Hebrides, Fiji, and Tonga.

Astegopteryx rappardi Hille Ris Lambers

Apterae small, with grey head and thorax and greyish-brown abdomen, in large numbers in clusters on undersides of palm leaves (*Cocos nucifera, Elaeis guineensis*), often collected with *A. nipae*. Besides the key characters, Hille Ris Lambers (1953) gives other features distinguishing this species from *A. nipae*. Alatae of *A. rappardi* are black, whereas those of *A. nipae* have a greenish abdomen. Infested leaves become spotted with yellow. Recorded from Malaysia, Indonesia, and Solomon Islands.

Astegopteryx rhapidis (van der Goot)

Apterae small, yellowish brown to dark brown, with a wax fringe, forming numerous colonies on the undersides of leaves of palms (*Cocos nucifera, Elaeis guineensis, ?Livistona altissima, Rhapis javanica, Kentia* sp., *Calamus* sp.). Always attended by ants. Recorded from Taiwan (Takahashi, 1931), Philippines, Palau Islands, Malaysia, Indonesia, and the Solomon Islands.

Atheroides Haliday   Chaitophorinae: Siphini

A small genus (5 species) of small to medium-sized markedly elongate aphids living monoeciously on grasses in Europe. Only one species occurs regularly on pasture grasses.

Atheroides serrulatus Haliday   Plate 7

Apterae with dorsal surface yellowish brown to brown, ventral surface pale and unsclerotized, with a clear line of demarcation between the two. Usually found living singly on the blades of various grasses (*Festuca, Poa, Agrostis, Dactylis, Deschampsia, Holcus, Hordeum*). Monoecious on Gramineae, with apterous males. Throughout Europe and as far east as Turkey and the Ukraine. $2n = 8$.

Aulacorthum Mordvilko   Aphidinae: Macrosiphini

Nearly 50 species of *Aulacorthum* are known of which 32 are from the eastern palaearctic and oriental region. The subgenus *Neomyzus* only occurs in this area except for the widely distributed glasshouse pest *A. (N.) circumflexum*. Most of the remaining species are western palaearctic; the few described from North America seem to belong to other genera, e.g. *Placoaphis*. The genus is defined by the parallel-sided inner faces of the spinulose antennal tubercles.

A. solani is one of the most polyphagous aphids and *A. magnoliae* in the East is also recorded from members of many different plant families. There is a similar lack of pattern in the hosts of the plant-specific species. Eight species of *Aulacorthum* are described from Compositae, 4 from Liliaceae, 3 each from Ericaceae and Labiatae, and 2 from Dipsacaceae, Lauraceae, Oleaceae, Rosaceae, and Rutaceae. Accounts are available from Europe (Hille Ris Lambers, 1947b, 1949), Japan (Miyazaki, 1971, in *Acyrthosiphon*), India (Raychaudhuri, 1980), Java (van der Goot, 1917/18), and Canada (Richards, 1972b).

Aulacorthum (Neomyzus) circumflexum (Buckton)
                                                    Mottled Arum Aphid   Plate 97

Appearance in life: Apterous adults shiny, nearly white or pale yellow to bright green, with distinctive sclerotic dorsal markings, consisting of transverse bands or paired patches on the thorax and a large, usually roughly horseshoe-shaped, patch on the abdomen. Antennae, legs, siphunculi, and cauda mainly pale. Alatae occur rarely. Apterae 1.2–2.6 mm, alatae 1.6–2.4 mm.

Host plants: Extremely polyphagous, feeding on numerous species of both monocots and dicots. It has even been found feeding on ferns and conifers. In temperate climates this aphid occurs particularly in glasshouses, and it is a common pest of house plants (e.g. *Calla, Cineraria, Cyclamen, Fuschia, Zantedeschia*).

Virus transmission: Shown to be able to transmit over 30 plant viruses, including the persistent viruses barley yellow dwarf and potato leaf roll, and nonpersistent viruses of beans, beet, cauliflower, celery, *Dahlia*, onion, potato, radish, soybean, tobacco, and tulip. However, in colder climates *A. circumflexum* is not important as a vector because it rarely occurs outdoors.

Distribution: Virtually world-wide, presumably due to transportation by man. Origin unknown, but South East Asia seems most likely.

Biology: Apparently entirely anholocyclic; sexual morphs have not been recorded. For additional information see Hille Ris Lambers (1949). $2n = 8$.

Aulacorthum (Neomyzus) dendrobii Basu

Small to medium-sized, adult apterae creamy yellow with brown to purplish dorsal segmental markings. Siphunculi distinctly swollen on distal half. Found colonizing young stems of *Dendrobium* sp. in West Bengal (A.N. Basu, 1969), where overwintering is anholocyclic. So far not encountered elsewhere. Sexual morphs are unknown.

Aulacorthum flavum Müller

Shiny yellow to yellowish green; on shoot tips and undersides of leaves of *Vaccinium* spp. (*uliginosum, myrtillus*). The apterae have yellowish-brown coloration around the bases of the siphunculi, and the alatae have a large dark brown dorsal abdominal patch. Recorded from Germany and Czechoslovakia. Monoecious holocyclic, with alate males. Müller (1958) describes all morphs and gives biological observations.

Aulacorthum magnoliae (Essig & Kuwana)   Plate 96

Appearance in life: Apterae medium sized, distinctively coloured with reddish head and prothorax, rest of body yellow green to green, tibiae and antennae usually wholly black, siphunculi pale basally with dark apices, cauda dark. Alatae have brownish transverse bands on dorsal abdomen. Mainly on leaves of host plants.

Host plants: Polyphagous, feeding on plants in over 20 different families, including *Citrus* and many ornamental shrubs and trees. Indian records are mainly from Cucurbitaceae (Raychaudhuri, 1980, p. 104).

Distribution: Japan, Korea, and India.

Biology: The biology is apparently little known. Takahashi (1923, p. 86) made life cycle observations in Japan, where alate females and a few alate males migrated to *Sambucus racemosa*, but no oviparae were produced. Males are also recorded from India (Raychaudhuri, 1980). $2n = 12$.

Aulacorthum rufum Hille Ris Lambers

Apterae varying from reddish brown to apple green, laterally and ventrally faintly powdered with wax; living rather well dispersed on shoot tips and under leaves of *Vaccinium* spp. (especially *V. myrtillus*). The alatae have dark brown segmental markings on the abdomen, especially laterally. In mounted preparations this species strongly resembles *A. solani*. Recorded from England, Scotland, Sweden, the Netherlands, and Czechoslovakia. Sexuales of this species have not been found and the life cycle is unknown (Hille Ris Lambers, 1947b).

Aulacorthum solani (Kaltenbach)
Glasshouse–Potato Aphid, Foxglove Aphid   Plate 98

Appearance in life: Colour of apterae variable; from a rather shiny, whitish green or yellow, in which case there is usually a conspicuous bright green or rust-coloured spot at the base of each siphunculus, to a uniformly dull green or greenish brown. Appendages mainly pale except for dark brown apices to the tibiae, siphunculi, and antennal segments. Alatae look quite different, with dark brown head and thorax, much darker antennae, legs, and siphunculi, and a variably developed pattern of transverse dark bars on the dorsal abdomen. Apterae and alatae 1.8–3.0 mm.

Host plants: Extremely polyphagous, colonizing plants in many different families of both dicots and monocots (but not Gramineae). Bulbs (especially tulips) often have large populations of *A. solani*, and it is a common pest in glasshouses and on pot plants. Common on potatoes.

Virus transmission: A vector of about 40 plant viruses including both persistent and non-persistent viruses of beet and potato. Müller *et al.* (1973) studied intraspecific variation in the ability to transmit pea enation mosaic virus.

Distribution: Probably of European origin, but now almost world-wide.

Biology: The biology of *A. solani* is complicated, like that of many of the most important aphid pest species, by the occurrence of numerous races or subspecies, including some with particular host-plant associations (Müller, 1970, 1976). Polyphagous forms of *A. solani* with both holocycly and holocycly occur. Holocyclic *A solani* have either apterous or (more rarely) alate males, and the unusual ability to overwinter as eggs on many different host plant species. Wave *et al.* (1965) give an account of the biology of *A. solani* in the north-eastern U.S.A. Robert and Rabasse (1977) discuss natural control by *Aphidius* spp. in western France. $2n = 10$.

Aulacorthum speyeri Börner

A distinctive aphid with most of the dorsal surface shiny black in adult apterae, except for pale yellow areas on the mesonotum and abdominal tergites 1 and 5. Legs and antennae are mainly pale, siphunculi dark at base and apex with a paler area in between, and cauda black. On *Convallaria majalis, Polygonatum multiflorum*, and certain other Liliaceae, and occasionally found in small numbers in concealed situations on other plants (e.g. *Lycopus, Potentilla*). Infested leaves of *Convallaria* develop large yellow spots and become desiccated. Monoecious holocyclic, with alate males. Recorded from England, Sweden, the Netherlands, Germany, Hungary, and Iran. Müller (1979) gives an account of this aphid. $2n = 10$. A similar species, but with a pale cauda, *A. watanabei* (Miyazaki), lives on *Convallaria keiskei* in Japan. $2n = 10$.

Aulacorthum vaccinii Hille Ris Lambers

Adult apterae with dorsal surface of body shiny black but antennae, legs, siphunculi, and cauda mainly pale with dark apices. Monoecious holocyclic on *Vaccinium* spp., with alate males. Originally described from Sweden on *V. uliginosum* (Hille Ris Lambers, 1952), but also recorded from Japan on *Leucothoë grayana* var. *oblongifolia* and *V. japonicum* (Miyazaki, 1971).

Baizongia Rondani   Pemphiginae: Fordini

Baizongia resembles *Aploneura* in having well-developed wax pore-plates, and resembles *Geoica* in the dorsal position of the anal aperture, the hairs on the anal plate and eighth abdominal tergite in both genera forming a 'trophobiotic organ' to retain droplets of honeydew until they are removed by ants. The apterae of *Baizongia* have 5-segmented antennae; the alatae have 6-segmented antennae, with a more normal distribution of sensoria than in *Aploneura*, and hold their wings roof-like in repose. Only one species is known.

Baizongia pistaceae (L.)   Plate 146

Appearance in life: Body of adult apterae on grass roots almost globular, whitish or pale yellow, powdered with wax. Head, prothorax, antennae, legs, and anal region light brown. Apterae 1.6–2.3 mm, alatae 1.8–2.0 mm.

Host plants: Pistacia terebinthus and *Pistacia palaestina* (primary hosts) and the roots of numerous grass species (*Agrostis, Dactylis, Festuca, Poa*), occasionally on wheat.

Distribution: Throughout Europe, the Middle East, North Africa, Kenya, India, and Pakistan.

Biology: Large, elongate horn-like galls are produced on *P. palaestina* in the Mediterranean region, about 10,000 alate migrants emerging from each gall. Wertheim (1954) describes the life cycle in Israel. In northern and central Europe, *B. pistaciae* is anholocyclic in grass roots. It is invariably attended by ants, especially *Lasius flavus*, and may overwinter in ants' nests. $2n = 24$.

Brachycaudus van der Goot — Aphidinae: Macrosiphini

A genus containing 44 palaearctic and 1 American species. Characterized by the rounded spiracular apertures, short cauda, and subapical annular incision below the siphuncular flange. Fourteen species belonging to 4 of the 5 or 6 recognized subgenera live on or alternate from *Prunus*. There are species groups associated with Ranunculaceae, Caryophyllaceae, Polygonaceae, Boraginaceae, and Scrophulariaceae, and two of the best known species (*helichrysi* and *cardui*) alternate from *Prunus* to Compositae. The single American species, *B. rocardiae*, resembles the palaearctic species *aconiti* and *napelli* and, like them, lives on Ranunculaceae. Remaudière (1952) revised the European species then known and the largest subgenus, *Acaudus*, has been revised by Burger (1975).

Brachycaudus (Thuleaphis) amygdalinus (Schouteden) — Plate 55

Appearance in life: Apterae green, squat-bodied with rather short legs and antennae, causing rolling of young almond or peach leaves and stunting new growth in spring. The leaves are rolled somewhat obliquely with respect to the mid-rib, rather than perpendicularly to the mid-rib as in *B. helichrysi*. Apterae 1.6–2.1 mm, alatae 1.1–1.7 mm.

Host plants: Primary host plant almond or peach, migrating for the summer to *Polygonum* spp. (*aviculare, equisetiforme*).

Distribution: England, France, Portugal, Mediterranean, the Middle East, the Ukraine, Pakistan, and South Africa.

Biology: Accounts are given by Swirski (1954a) for Israel and by Talhouk (1977) for Lebanon. The holocycle is facultative, some populations reproducing parthenogenetically throughout the year on *Polygonum*, even where primary hosts are available. Severe damage may be inflicted to almond trees in the Middle East by the spring generations. $2n = 12$.

Brachycaudus cardui (L.) — Thistle Aphid — Plate 58

Appearance in life: Adult apterae very variable in general appearance; body colour green, yellow, or reddish with an extensive sclerotic area on the dorsal abdomen which is usually shiny black in summer generations on secondary hosts, but often much paler in spring forms on the primary host. In rolled leaves of plum in spring, but much more commonly encountered in summer forming dense colonies on stems and flower heads of various Compositae, much attended by ants. Apterae 1.9–2.6 mm, alatae 1.6–2.3 mm.

Host plants: Primary host *Prunus domestica*, sometimes *P. spinosa, P. avium,* or *P. armeniaca*. Secondary hosts are various Compositae (notably *Carduus, Cirsium, Cynara, Arctium, Chrysanthemum, Tanacetum, Matricaria*) and Boraginaceae (e.g. *Cynoglossum, Symphytum, Echium, Borago*).

Virus transmission: A vector of several non-persistent viruses, including plum pox virus.

Distribution: Europe, Central Asia, the Middle East, North Africa, India (Kashmir), the U.S.A., and Canada.

Biology: Heteroecious holocyclic. Summer populations on thistles are often heavily parasitized by Aphidiinae. The relationship with ants is discussed by Blunck (1948). $2n = 12$ (Kuznetsova, 1968) or $2n = 10*$.

Brachycaudus helichrysi (Kaltenbach)   Leaf-curling Plum Aphid   Plate 56

Appearance in life: Very variable in colour; apterae in spring populations on damson and plum are green, brown, brownish green or brownish yellow, shiny with a slight wax bloom, causing leaves to roll up tightly perpendicular to the mid-ribs. Apterae in summer on herbaceous plants are small to very small aphids, in various shades of green, pale yellow, or almost white (sometimes pinkish), clustered on stems and flower heads. Apterae 0.9–2.0 mm, alatae 1.1–2.2 mm.

Host plants: Primary host plants are various *Prunus* species, notably *domestica, insititia, spinosa*. Secondary hosts are numerous species of Compositae (e.g. *Achillea, Chrysanthemum, Matricaria, Senecio, Erigeron, Ageratum*) and Boraginaceae (*Myosotis, Cynoglossum*), and sometimes plants in other families (e.g. *Cucurbita, Rumex, Alchemilla, Saxifraga, Veronica*). *B. helichrysi* on *Trifolium*, and perhaps on other Leguminosae, is usually regarded as a distinct form (*warei* Theobald).

Virus transmission: Able to transmit plum pox, cucumber mosaic, *Dahlia* mosaic, and *Cineraria* mosaic viruses. Sometimes flying in such large numbers that it is an important vector of non-persistent viruses in crops it does not colonize, e.g. potato virus Y in potatoes.

Distribution: Of palaearctic origin, now world wide.

Biology: Heteroecious holocyclic, but with widespread anholocycly in warm climates, and in glasshouses in temperate regions. The form *warei* is apparently anholocyclic on clover (Theobald, 1927, p. 255). Bennett (1955), Bell (1983), Stenseth (1970), and Karczewska (1975) studied the biology and life history of *B. helichrysi* in England, Northern Ireland, Norway, and Poland, respectively. Bennett found that eggs hatched on *Prunus* in November and December in England, the first instar fundatrices feeding on dormant flower buds through the winter; Bell also recorded mid-winter hatching of eggs in Northern Ireland, suggesting that there may be no egg diapause. In Norway and Poland, overwintering is in the egg stage. This species is the major aphid pest of plum, and is also of considerable importance on glasshouse chrysanthemums. $2n = 12$.

Brachycaudus persicae (Passerini)　　　　　　Black Peach Aphid　　Plate 59
　(= *persicaecola* Boisduval)
　(= *persicaeniger* Smith)

Appearance in life: Adult apterae shiny dark brown to black, with black siphunculi, most commonly found in large spring colonies on suckers and young stems of peach or nectarine. The leaves are only sparsely colonized and not rolled by this aphid. Apterae 1.5–2.2 mm, alatae 1.7–2.1 mm.

Host plants: Prunus species, especially *P. persica* but also *P. armeniaca, P. domestica, P. spinosus*. Has also been recorded from pear (*Pyrus communis*).

Virus transmission: Apparently not a vector of peach viruses.

Distribution: Europe (except Scandinavia?), Turkey and the Caucasus, southern Africa, Australia, New Zealand, and North and South America.

Biology: Monoecious, in part at least. In late summer and during the winter colonies occur on the roots of the host plant. However, very similar aphids also occur in summer on Scrophulariaceae, and could be a host-alternating form or subspecies of *B. persicae*. Anholocyclic overwintering on the roots of *Prunus* is common (Pelvat *et al.*, 1970), but winter eggs are also laid. Ant-attended. Life history observations of this species (as *Anuraphis persicae-niger*) were made by Malenotti (1923) and Pussard (1932). This aphid has frequently been confused in the literature with *B. prunicola* (Kaltenbach), which has much shorter siphunculi. $2n = 10^*$.

Brachycaudus (Appelia) prunicola (Kaltenbach)
　(= *prunifex* Theobald)

Appearance in life: Adult apterae shiny, black, dark green, or deep yellow green with black abdominal markings; immature stages grey-green. In spring colonies on new growth cause severe curling and decoloration of leaves. Apterae 1.4–2.4 mm, alatae 1.3–1.9 mm.

Host plants: Primary host usually *Prunus spinosus*, sometimes *P. domestica*, or *P. insititia*. When heteroecious (see below) the secondary host is *Tragopogon*.

Distribution: Not yet recorded outside Europe.

Biology: Heteroecious holocyclic with alternation between *Prunus* and *Tragopogon*, but part of the population—possibly a subspecies—may live without alternation on *P. spinosus*. For further information see Thomas (1962) and Mostafawy (1967). $2n = 12$.

Brachycaudus (Appelia) schwartzi (Börner)   Peach Aphid   Plate 57

Appearance in life: Adult apterae shiny yellow brown to dark brown with extensive black dorsal markings; immature stages yellow brown. Spring colonies cause severe curling and distortion of peach leaves. Apterae 1.4–2.1 mm, alatae 1.5–2.0 mm.

Host plants: Prunus persica, occasionally *P. serotina*.

Distribution: Europe, Iran, India, South America, and California.

Biology: Monoecious holocyclic on peach trees, with alate males. Brief accounts of biology and natural enemies are given by Pussard (1932) in France (as *Anuraphis amygdali*) and Zuniga (1967) in Chile (as *Appelia tragopogonis*). See also Mostafawy (1967) and Thomas (1962) for fuller information. $2n = 12$ (Gut, 1976).

Brachycaudus (Appelia) tragopogonis (Kaltenbach)

Shiny grey-brown to dark brown aphids on aerial parts of salsify and other *Tragopogon* species. Part of a complex of closely-related species including *B. prunicola* and *B. schwartzi* (q.v.); the taxonomic problems and host-plant relations of the group were considered by Thomas (1962), and detailed morphological and biological studies were made by Mostafawy (1967). Monoecious holocyclic on *Tragopogon* spp. in Europe; populations in western Europe which alternate between *Prunus* and *Tragopogon* are probably best regarded as a form of *B. prunicola*. Hille Ris Lambers (1948) described a subspecies with long dorsal hairs, *B. (Appelia) tragopogonis setosus*, on *T. longirostris* in Palestine, and collections from *Tragopogon* in the Middle East and Central Asia may be of this subspecies (Mostafawy, 1967). $2n = 12^*$ ($2n = 11$ for 1 sample of spp. *setosus**).

Brachycorynella Aizenberg   Aphidinae: Macrosiphini

A monotypic genus similar to *Brachycolus*.

Brachycorynella asparagi (Mordvilko)   Plate 64

Long-bodied green aphids covered with grey mealy wax, feeding only on *Asparagus* (*officinale, densiflorus*). The rather short appendages, small mammariform siphunculi, and elongate cauda are characteristic. Apterae and alatae 1.2–1.7 mm. Feeding causes 'rosetting' of asparagus ferns, the internodes being severely shortened and the leaves shortened and turned blue green in colour (Forbes, 1981). In eastern Europe (Bulgaria, Poland) and introduced into North America about 1969. $2n = 10^*$.

Brachyunguis Das   Aphidinae: Aphidini

A genus close to *Aphis*, with about 25 nominal species associated with xerophytes, mostly in Central Asia and the Middle East where they are common but inconspicuous on *Lycium, Tamarix*, and various Chenopodiaceae. A few species are described also from much the same group of plants in Arizona, California, and Utah. *Brachyunguis* have the terminal process clearly shorter than the base of antennal segment VI, and a pale cauda. Many of the nominal species are very similar to one another and their separate identity needs experimental confirmation.

Brachyunguis harmalae Das

Adult apterae are small (1.2–1.8 mm), deep green anteriorly but more yellowish in region of siphunculi, thickly dusted with white wax powder. Ant-attended. Recorded feeding on young leaves of *Citrus* in Israel (including one large attack—Swirski, 1963) and Sudan. This aphid is holocyclic on *Peganum harmalae*, with both apterous and alate males; Das (1918) gives a full account of its life history on this host plant. *Citrus* is apparently a reserve host, and other plants (*Loranthus acaciae, Suaeda* spp.) are also colonized.

Brevicoryne van der Goot   Aphidinae: Macrosiphini

A small genus associated with Cruciferae, with 5 species of palaearctic origin and one in the Canadian Arctic. The Central Asiatic *B. shaposhnikovi* was described from *Lonicera* and the North American *Aphthargelia symphoricarpi* which also lives on Caprifoliaceae is similar to *Brevicoryne*. The palaearctic Cruciferae-feeding genus *Lipaphis* is also closely related to *Brevicoryne*, as is the palaearctic *Hyadaphis* alternating between Caprifoliaceae and Umbelliferae. The short siphunculi and triangular cauda are characteristic of *Brevicoryne*. Eastop in Hodjat (1981) gives a key for the identification of alatae.

Brevicoryne brassicae (L.)   Cabbage Aphid, Mealy Cabbage Aphid   Plate 60

Appearance in life: Apterae medium sized, greyish green or dull mid-green, with dark head and dark dorsal thoracic and abdominal markings, the body covered with greyish-white mealy wax which is also secreted onto the surface of the plant and extends throughout the colony. Alatae with dark head and thorax and black transverse bars on dorsal abdomen. Apterae 1.6–2.6 mm, alatae 1.6–2.8 mm.

Host plants: Feeding virtually restricted to members of the Cruciferae, the mustard oil sinigrin providing the chemical stimulus necessary to elicit a feeding response. Many genera of Cruciferae are colonized. Field crops, especially cabbage, cauliflower, Brussels sprout, radish, swede, and mustard, are often attacked severely, but kale and rape are only lightly infested and turnips are virtually immune. Large colonies form under the leaves, and also on the flowerheads of seed crops, reducing the setting of seed.

Virus transmission: A vector of about 20 plant viruses including cabbage black ring spot, cabbage ring necrosis, cauliflower mosaic, and radish mosaic.

Distribution: Throughout all the temperate and warm temperate parts of the world.

Biology: Monoecious holocyclic in colder regions, anholocyclic where winter climate is mild. Males are alate. There is an extensive literature on this well-known species. Markkula (1953) provides a general account. Ecological studies have included computer modelling (Gilbert and Hughes, 1971), and the association with a hymenopterous parasitoid *Diaretiella rapae* has received particular attention (e.g. Paetzold and Vater, 1967). Early biological control attempts in Australia are reviewed by Wilson (1960). Physiological (e.g. van Emden and Bashford, 1969) and genetic (Dunn and Kempton, 1972) aspects of the relationship of *B. brassicae* with its host plants have also been studied in some depth. $2n = 16$.

Callaphis Walker　　　　　　　　　　Drepanosiphinae: Phyllaphidini
(= *Panaphis* Kirkaldy)

A genus of 2 described species, both living on *Juglans*.

Callaphis juglandis (Goetze)　　　　　Large Walnut Aphid　　Plate 10 (left)

Large, stout aphids (body length 3.5–4.3 mm) feeding in rows along the veins on the upper sides of leaves of *Juglans* spp. All parthenogenetic morphs are winged, with dark head and thorax, wings with brown-bordered veins, and abdomen yellow with broad, dark-brown, transverse bands. Throughout Europe and in Central Asia (Turkey, Iran, Iraq, Pakistan, north-west India), and (in 1928) introduced into the U.S.A. (California, Oregon) where populations seem to have increased since the successful biological control of

Chromaphis juglandicola. It seems that these two species cannot readily coexist, since colonies of *Callaphis juglandis* on upper surfaces of walnut leaves would be adversely affected by a rain of honeydew from colonies of *Chromaphis* on the undersides of the leaves above (Olson, 1974). Holocyclic on *Juglans*, males alate.

Callaphis nepalensis Quednau

A similar aphid to *C. juglandis*, differing in the characters given in the key; also, *C. nepalensis* has most of the forewing veins without brown borders, and a relatively longer last rostral segment. Recorded from *Juglans regia* in Nepal and Tibet. Colour and habitat in life not known.

Capitophorus van der Goot   Aphidinae: Macrosiphini

About 30 species of pale, sometimes almost translucent, slender aphids with elongate appendages. The apterae bear long capitate hairs at least on the head and posterior abdominal segments; the alatae have only much shorter hairs, a dark dorsal abdominal patch, and numerous sensoria on antennal segments III, IV, and usually V. The genus is widely distributed in the palaearctic and oriental regions, and there are also 5 North American species. The heteroecious species have primary hosts in the Elaeagnaceae, the majority migrating to Compositae and the rest to Polygonaceae. Other species are monoecious on Compositae and Polygonaceae.

Capitophorus was once used in a much broader sense to include most Macrosiphini with capitate hairs, including species currently placed in *Chaetosiphon, Cryptomyzus, Myzaphis,* and *Pleotrichophorus*. Regional taxonomic revisions of *Capitophorus* include Hille Ris Lambers (1953—western Europe); Corpuz-Raros and Cook (1974—North America), Raychaudhuri *et al.* (1980a—India), and Miyazaki (1971—Japan).

Capitophorus elaeagni (del Guercio)   Plate 68
(= *braggii* Gillette)

Small to medium sized, pale greenish white to yellowish green, almost translucent aphids with pale appendages, except that the apices of the very long siphunculi are distinctly dusky. Alatae have an almost rectangular solid black dorsal abdominal patch. Heteroecious, with *Elaeagnus* species as primary hosts, migrating to thistles (*Cirsium, Carduus*) and also to *Cynara scolymus* where it can do economic damage. Widely distributed through the temperate and warm temperate regions of the world. For a fuller account see Hille Ris Lambers (1953a) or Cottier (1953). A very similar species, **C. carduinus** (Walker) is monoecious holocyclic on thistles in Europe and has also been recorded from *Cynara scolymus* in south-east France (Robert, 1969a). Apterae of *C. carduinus* tend to have much smaller capitate spinal hairs on the anterior abdominal tergites than *C. elaeagni*, and some collections of *Capitophorus* from *Cynara* in Brazil and Uruguay have been assigned to *C. carduinus* on that basis, but the morphological discrimination between these two species is not at all certain. $2n = 16$.

Capitophorus horni (Börner)

Small to medium sized, very pale green aphids often with two rather faint, brighter green dorsal longitudinal stripes; appendages almost colourless, including the tips of the siphunculi. In alatae the sclerotic patch on the dorsum is paler than in *C. elaeagni*, and often segmentally divided. Monoecious holocyclic on *Cirsium* spp. in Europe, and also sometimes attacking *Cynara scolymus*. Robert (1969a) has studied the biology of *C. horni* in western France. $2n = 16$.

Cavariella del Guercio   Aphidinae: Macrosiphini

A northern hemisphere genus of about 30 species, half of them from Asia. Characteristic of this genus are the 2 hairs close together on the eighth abdominal tergite, which are placed on a prominent tubercle or supracaudal process. In alatae this tubercle is usually smaller than in apterae, but its presence is indicated by the close proximity of the two hairs. Species of *Cavariella* mostly alternate between *Salix* and Umbelliferae. Taxonomic revisions are available for Europe (Hille Ris Lambers, 1947b), Siberia (Ivanovskaya, 1980), Japan (Miyazaki, 1971), India (Raychaudhuri, 1980), and North America (Stroyan, 1969a).

Cavariella aegopodii (Scopoli)   Willow-Carrot Aphid   Plate 70

Appearance in life: Apterae small to medium sized, green or yellowish green, rather elongate oval and dorso-ventrally flattened, the dorsal surface of the body roughened by numerous small depressions. Alatae have a black dorsal abdominal patch. Apterae 1.0–2.6 mm, alatae 1.4–2.7 mm.

Host plants: Primary hosts are *Salix* spp., chiefly *S. fragilis* and *S. alba*, migration occurring to carrots, celery, and many other wild and cultivated Umbelliferae. Colonies form on the leaves and umbels.

Virus transmission: A vector of the persistent viruses carrot motley dwarf and carrot red leaf, the semi-persistent parsnip yellow fleck and *Anthriscus* yellows (Murant, 1978), and several non-persistent viruses including celery mosaic.

Distribution: Widespread throughout the temperate and warm temperate regions.

Biology: Heteroecious holocyclic in most parts of the world, but often anholocyclic on Umbelliferae in warmer climates (Müller and Schöll, 1958). Dunn (1965), and Dunn and Kirkley (1966) studied the life cycle and host-plant relations in Britain. Chapman *et al.* (1981) investigated attraction and repulsion by plant odours. $2n = 10$.

Cavariella archangelicae (Scopoli)

Adult apterae straw-coloured, sometimes greenish but always very pale; siphuncular wax (discharged by prodding abdomen) colourless to slightly greenish (see *C. konoi* below). Heteroecious, migrating from *Salix* spp. to *Angelica* spp. (including *Archangelica*) in Europe and Iceland. Recorded from *Myrrhis odorata* in Iceland. $2n = 6$.

Cavariella konoi Takahashi

Adult apterae green; if yellow green, then with faint darker green longitudinal stripes; siphuncular wax distinctly yellow (see *C. archangelicae* above). Preserved or mounted specimens are difficult to separate from *C. archangelicae*; a full discussion of the differences is given by Stroyan (1964). Heteroecious, migrating from *Salix* spp. to *Angelica* spp., and also recorded from *Apium, Cicuta, Myrrhis*, and *Sium*. Now apparently of holarctic distribution (Europe, Iceland, North America, Mexico, Japan). Stroyan (1969) discusses the North American records and the geographical variation of this species. $2n = 8^*$.

Cavariella theobaldi (Gillette & Bragg)

Adult apterae rather bright green, distinguishable from other *Cavariella* on cultivated Umbelliferae by their rather dusky, slightly tapering (not swollen) siphunculi. Heteroecious, migrating from *Salix* spp. to *Heracleum sphondylium* and *Pastinaca sativa*, and sometimes found on certain other wild Umbelliferae (*Aegopodium, Angelica, Chaerophyllum*). Bem and Murant (1979) report on the transmission of parsnip yellow fleck virus and five other viruses from *Heracleum* by this aphid. In Europe, Turkey, and north-eastern North America. $2n = 8$.

Cerataphis Lichtenstein   Hormaphidinae: Cerataphidini

About 6 South East Asian species, some probably alternating from galls on *Styrax* to monocotyledons including bamboo, palms, screw pines, and orchids in their native regions, but several species more widely distributed by commerce on their palm and orchid secondary hosts. The apterae appear rather scale-like insects being round, dorso-ventrally flattened, and with a marginal ring of wax. Alatae are normal-looking aphids but have the media of the forewing once-forked and the antennae only 5-segmented. Hille Ris Lambers (1953b) distinguishes 3 economically important species.

Cerataphis orchidearum (Westwood) Orchid Aphid, Scale Aphid

Adult apterae sedentary; small, flattened, broadly oval, dorsal surface dark reddish brown to black dusted with wax, with a fringe of radiating plates of white wax. Apterae 1.0–1.6 mm long × 0.7–1.2 mm broad, alatae 1.0–2.0 mm. On various Orchidaceae, including vanilla, in the tropics, and in European and North American glasshouses. Earlier records of *C. lataniae* from orchids probably all refer to this species. Anholocyclic; sexual phase not known. See Heie (1980) or Zimmerman (1948—under *C. lataniae*) for further information. $2n = 18^*$.

Cerataphis palmae (Ghesquière) Plate 135
(= *variabilis* Hille Ris Lambers)

Small, sedentary aphids with a dark brown, flattened, almost circular body fringed with white wax; legs short and hidden under body. Apterae 1.0–1.8 mm long × 0.8–1.3 mm broad, alatae 1.3–2.0 mm. On leaves of palms; a common pest of coconut and oil-palm throughout the tropics. Anholocyclic; sexual phase unknown.

Two other species of *Cerataphis* have been recorded from cultivated Palmaceae. **C. lataniae** (Boisduval) resembles *C. palmae*, and the two have been confused in the past, but it lacks the one or more pairs of thick dagger-like hairs on the underside of the head of *C. palmae*. Eastop (1966) gives other distinguishing characters. It appears to be widespread on palms, especially *Latania*, through the tropics and in glasshouses. **C. formosana** Takahashi was described from *Cocos* sp. in Taiwan (Takahashi, 1924), and differs from the two previous species in the shape and position of the frontal horns.

Ceratoglyphina van der Goot Hormaphidinae: Cerataphidini

A monotypic genus distinguished from other Hormaphidinae with frontal horns by the rounded cauda and anal plate.

Ceratoglyphina bambusae van der Goot

Small (1.4–2.1 mm), oval-bodied aphids, dark brown dorsally with yellow siphuncular region and legs, and with white wax powder especially on lateral parts of body, presumably arising from the narrow lateral segmentally divided glandular strips. Dense colonies form on the basal parts of the leaves of bamboos (*Dendrocalamus, Bambusa*), often attended by ants. In China, Taiwan, Indonesia, Malaysia. A subspecies, *C. bambusae bengalensis* A.K. Ghosh, occurs on bamboos in Assam and West Bengal, India, and is discussed by M.R. Ghosh et al. (1974).

Ceratovacuna Zehntner   Hormaphidinae: Cerataphidini

About 10 species in South East Asia, some alternating from galls on *Styrax* to bamboo, or living only on bamboo and other large Gramineae such as sugar cane. Body colour is brownish or greenish yellow or yellow, densely covered with wax wool in life. Taxonomic accounts are available for Japan (Takahashi, 1958), Korea (Paik, 1965), India (M.R. Ghosh et al., 1974), and Taiwan (Tao, 1966; Liao, 1976—bamboo-feeding species). Earlier authors placed the species in *Oregma*.

Ceratovacuna lanigera Zehntner   Plate 134

Appearance in life: Apterae small to medium-sized, pale green or brownish covered with white woolly wax, forming dense colonies on lower sides of leaves of host plant, often attended by ants. Alatae with brown-black head and thorax, and dusky transverse bands on dorsal abdomen. Apterae 1.4–2.3 mm, alatae 2.0–2.5 mm.

Host plants: Saccharum officinale and *Miscanthus* spp.

Distribution: India, Nepal, Bangladesh; throughout East and South East Asia; Fiji and Solomon Islands.

Biology: Anholocyclic on Gramineae; the sexual phase is not known. A serious pest of sugar cane in South East Asia, but in Japan and Fiji it has only been found on *Miscanthus*, suggesting geographical variation in its host plant preferences. Takano (1934) describes population fluctuations, natural enemies, and attempts at biological control in Taiwan, and Rueda and Calilung (1975) describe developmental stages, natural enemies, and population changes in the Philippines. $2n = 12*$.

Ceratovacuna species on bamboos

At least 6 species occur on bamboos. Life cycles are largely unknown, but a sexual phase on *Styrax* is suspected for some species in South East Asia. Some of the available information and references are tabulated below:

Species name	Body colour in life	Host plant	Distribution	Reference(s)
hoffmanni Takahashi	Not recorded	bamboos, *Schizostachyum lima*	China, Taiwan, Sumatra	Takahashi (1936) Tao (1966)

Species name	Body colour in life	Host plant	Distribution	Reference(s)
indica Ghosh, Pal, & Raychaudhuri	Yellow	bamboo	West Bengal	M.R. Ghosh, Pal, and Raychaudhuri (1974)
japonica (Takahashi) (= *brevicornis* Takahashi)	Brown or pinkish brown	*Bambusa, Sasamorpha, Phyllostachys*	Taiwan, Japan, Korea	Takahashi (1958) Liao (1976)
longifila (Takahashi)	Dark brown (with long wax filaments)	*Bambusa*	Taiwan	Liao (1976)
silvestrii (Takahashi) (= *subglandulosa* Hille Ris Lambers & Basu?)	Yellow to brown (or black*)	*Arundinaria, Bambusa*	Taiwan, India*	Liao (1976) M.R. Ghosh et al. (1974)
sp. C of Takahashi 1958	Not recorded (no wax?)	*Sasa, Phyllostachys*	Taiwan, Japan	Paik (1965)

* Indian specimens are recorded as reddish brown to black in life, which may call into question the synonymy of *C. silvestrii* and *C. subglandulosa*.

Cervaphis van der Goot   Greenideinae: Cervaphidini

A small genus in India and South East Asia distinguished by the remarkably long and numerous hair-bearing processes on the body of the apterae, the marginal processes being branched, with each branch ending in a hair. In alatae these processes are reduced (except on the head) to low, flat, hair-bearing tubercles. The siphunculi are long, cylindrical, slightly curved outwards and slightly swollen subapically where there is a ring of hairs. Hille Ris Lambers (1956a) and Ghosh (1982b) revised the genus.

Cervaphis rappardi Hille Ris Lambers   Plate 130

Apterae are yellow in life, sometimes greenish yellow, with colourless appendages, on flowers, flowerstalks, and sometimes on leaves, young shoots, or young fruit of certain varieties of cocoa, attended by ants (*Dolichoderus*). Feeding may cause flower heads to shrivel and fall off. Colonies also occur on *Nephelium lappaceum* which may be the native host plant. Alatae are mainly black, the abdomen green with black dorsal markings. Recorded from Indonesia, Malaysia, New Guinea, Philippines. It is possible that a second species, *C. schouteniae* van der Goot, living on *Actinophora* in South East Asia, may also utilize cocoa as a 'reserve' host plant; Hille Ris Lambers (1956a) distinguishes between these two species. A.N. Basu (1961) described a subspecies, *C. rappardi indica*, from pigeon pea in Assam and West Bengal.

Chaetomyzus Ghosh & Raychaudhuri   Aphidinae: Macrosiphini

A monotypic genus, only known from India.

Chaetomyzus rhododendri Ghosh & Raychaudhuri

Medium-sized, rather elongate oval aphids on leaves (?) of *Rhododendron arboreum*, found in Assam and Himachal Pradesh, India. Easily recognized by the distinctly clavate siphunculi and pointed dorsal tubercles of both apterae and alatae. The alata has a dark dorsal abdominal patch. Colour in life is not recorded. See Chakrabarti, Ghosh and Chowdhuri (1971) for further description.

Chaetosiphella Hille Ris Lambers   Chaitophorinae: Siphini

A small genus of aphids on grasses closely related to *Atheroides* but characterized by the very long and acute last segment of the rostrum.

Chaetosiphella berlesei (Del Guercio)

Dark grey, very bristly, elongate-oval aphids living on leaf blades of *Festuca, Deschampsia,* and *Corynephorus* in Europe (Belgium, the Netherlands, Poland, Sweden, and Yugoslavia). See Stroyan (1955) for additional information.

Chaetosiphon Mordvilko   Aphidinae: Macrosiphini

A genus of about 20 species widely distributed in the Northern Hemisphere and mostly associated with Rosaceae of the *Rosa−Fragaria−Potentilla* group, characterized by having 5 hairs on all first tarsal segments, apterae with conspicuous capitate hairs, cylindrical siphunculi and a short cauda, and alatae with inconspicuous hairs and dorsal abdominal pigmentation. The economically important species, here placed in the subgenus *Pentatrichopus*, were formerly included in *Capitophorus*. They are not attended by ants.

Accounts of *Chaetosiphon* are available for Europe (Hille Ris Lambers, 1953a), Japan (Miyazaki, 1971), India (David *et al.*, 1971), and North America (Schaefers, 1960; Richards, 1963).

Chaetosiphon chaetosiphon (Nevsky)

Medium-sized, shiny, mid- to light green aphids, holocyclic monoecious on *Rosa* spp., especially *R. alpina*. The alata has a large black dorsal abdominal patch, but green siphunculi and cauda. Not ant-attended. Recorded from French Pyrenees, Swiss Alps, Central Asia, and Kashmir. Hille Ris Lambers (1953a) gives the fullest account of this species.

Chaetosiphon gracilicorne David, Rajasingh, & Narayanan

Small, yellow-green to dark green aphids recorded from *Rosa macrophylla* and *R. moschata* in Himachal Pradesh, India (Chakrabarti, 1976).

Chaetosiphon (Pentatrichopus) coreanum (Paik)

Small, pale yellow to almost white aphids on undersides of young leaves of *Rosa* spp. (*rugosa, polyantha* var. *genuina*), recorded from Japan and Korea. Monoecious holocyclic on roses, with a dwarf summer form (Miyazaki, 1968a). $2n = 8*$.

Chaetosiphon (Pentatrichopus) fragaefolii (Cockerell)
Strawberry aphid   Plate 73

Appearance in life: Apterae small, elongate oval, translucent yellowish white to pale greenish yellow. Alatae with head and thoracic lobes black, abdomen very pale greenish white with a brown-black dorsal patch. Apterae 0.9–1.8 mm, alatae 1.3–1.8 mm.

Host plants: Mainly cultivated strawberry varieties; less frequently found on certain wild *Fragaria* spp. (especially *F. chiloensis* in North America, but rarely on *F. vesca* in Europe), and *Potentilla* spp. (including *P. anserina*). Occasionally found on cultivated roses. See also Schaefers and Allen (1962).

Virus transmission: A vector of the persistent viruses strawberry crinkle, strawberry lesion, strawberry mild yellow edge, and strawberry vein chlorosis, and the non-persistent strawberry mottle and strawberry veinbanding.

Distribution: Presumably of North American origin but now occurs everywhere in the world that strawberries are cultivated.

Biology: Parthenogenetic forms occur throughout the year on strawberry. Oviparae and both alate and apterous males have been recorded, but although produced readily by some clones in greenhouse or laboratory cultures, they are very rare in the field, and successful overwintering as an egg stage is unknown. The biology and population ecology of *C. fragaefolii* is relatively well known (Dicker, 1952; Hille Ris Lambers, 1953a; Schaefers and Allen, 1962). Earlier American authors, however, may have worked with mixtures of *C. fragaefolii* and *C. thomasi*, and Richards' (1963) account includes *C. thomasi*. A detailed study has been made on the influence of various factors on the production of alatae (e.g. Schaefers and Judge, 1972). $2n = 14*$.

Chaetosiphon (Pentatrichopus) glaber David, Rajasingh, & Narayanan

The apterae are small, pinkish-brown or green aphids with transverse dark bands on the dorsal abdomen, recorded from *Rosa macrophylla* in Himachal Pradesh, India (David, Rajasingh and Narayanan, 1971).

Chaetosiphon (Pentatrichopus) jacobi Hille Ris Lambers

Apterae similar to those of *C. thomasi* in life but darker, the sclerotic cuticle of the dorsal surface quite brown in colour. In western U.S.A. on thin-leaved wild *Fragaria* spp., especially *F. californica*, only colonizing cultivated strawberries in greenhouse or laboratory conditions. Life cycle unknown; no males have been found. Strawberry crinkle virus has been shown to multiply in this aphid (Sylvester *et al.*, 1974). Other strawberry viruses transmitted by *C. jacobi* are mild yellow edge, lesion, mottle, and veinbanding.

Chaetosiphon (Pentatrichopus) minor (Forbes) (= brevipilosus Baerg)

Very small (1.0–1.4 mm) yellow-green aphids on stems and undersides of leaves of *Fragaria* spp. in eastern North America, Venezuela and Japan (Miyazaki, 1971). In the U.S.A. this species is recorded mainly from cultivated strawberries. However, Richards (1963) reports it mainly from wild *Fragaria* spp. in Ontario, Canada, and Frazier (1975) found it vectoring strawberry mild yellow edge virus in *F. virginiana* in North Carolina. Holocyclic on strawberries, with alate males. See Schaefers (1960) for further information.

Chaetosiphon (Pentatrichopus) tetrarhodum (Walker)   Plate 74

Appearance in life: Apterae are spindle-shaped, small to medium sized, pale green to yellow-green, rarely reddish, on underside of younger leaves of *Rosa*. Alatae have head and thorax blackish, abdomen with a black central patch. Apterae 1.0–2.6 mm, alatae 1.2–2.4 mm.

Host plants: Rosa spp., especially 'wild' species (*R. rugosa, R. rubiginosa, R. carina*), but also on cultivated varieties.

Virus transmission: Able to transmit the non-persistent strawberry viruses mottle and veinbanding.

Distribution: Occurs on roses almost everywhere in the world, except South East Asia.

Biology: Monoecious holocyclic on roses in Europe, but probably anholocyclic in warmer parts of the world. The sexuales in autumn are much darker in colour than the parthenogenetic morphs; the oviparae are olive green, the males are apterous, very small and narrow bodied, dark green to black. For a fuller account see Hille Ris Lambers (1953a). $2n = 14$.

Chaetosiphon (Pentatrichopus) thomasi Hille Ris Lambers

Appearance in life: Very similar to *C. fragaefolii* (q.v.) but apterae with body on average somewhat broader and the dorsal cuticle more heavily sclerotized. Apterae 1.0–2.6 mm, alatae 1.2–2.4 mm.

Host plants: Mainly on cultivated strawberries and certain wild *Fragaria* spp., especially *F. chiloensis*. This species also occurs on both wild and cultivated *Rosa*, much more frequently than *C. fragaefolii*. Certain *Potentilla* spp. may also be colonized.

Virus transmission: A vector of the persistent virus strawberry mild yellow edge, and the non-persistent strawberry mottle and veinbanding viruses.

Distribution: Widespread in North America, and also found in Brazil and Chile (Carrillo, 1974).

Biology: The life cycle of *C. thomasi* in North America is probably variable with anholocycly in warmer regions but the sexual phase is apparently more functional than in *C. fragaefolii*, and overwintering in the egg stage may predominate in the north-eastern U.S.A. (Schaefers and Allen, 1962). Both apterous and alate males occur.

Chaitoregma Hille Ris Lambers & Basu

Hormaphidinae: Cerataphidini

A small, perhaps monotypic, genus characterized by the shape of the frontal horns and the heavily sclerotized dorsum.

Chaitoregma tattakana (Takahashi)

Apterae are small, yellowish brown to deep violet-black, covered in life with fine wax dust. They form large colonies on the undersides of leaves, the infested parts showing signs of withering. Not ant-attended. Recorded from *Arundinaria* sp. in the southern Himalayas (Hille Ris Lambers and Basu, 1966), and from *Bambusa* sp. and *Yoshania nittakayamensis* in Taiwan (Liao, 1976).

Chromaphis Walker

Drepanosiphinae: Phyllaphidini

A genus of 2 species associated with *Juglans*. Small pale yellow aphids, closely related to *Callaphis*. Adult viviparae all alate.

Chromaphis hirsutustibis Kumar & Lavigne

Adults all alate, pale yellow, often with black patches on abdominal tergites 4 and 5. Long hairs on frons and hind tibia, and dark spots at ends of wing veins,

Colopha

are characteristic. On *Juglans regia* in India (Himachal Pradesh, Uttar Pradesh) and Nepal. Chakrabarti (1978) described the sexual morphs.

Chromaphis juglandicola (Kaltenbach)   Walnut Aphid   Plate 10 (right)

Appearance in life: Adults all alate, pale lemon yellow, sometimes yellowish brown or pinkish, with pale brown thoracic lobes and (in autumn generation) paired brown spots on abdominal tergites 4 and 5. The antennal segments have dark apices and there is a black spot near the apex of each hind femur. Nymphs often have small paired dark spots on several abdominal tergites. Body length 1.2–2.3 mm.

Host plants: Specific to *Juglans regia*; occasionally recorded on other *Juglans* spp. in California.

Distribution: Europe, Central Asia, India, Pakistan, and introduced into North America.

Biology: Davidson (1914) studied the life history and biology of *C. juglandicola* very comprehensively in California, describing all morphs and development stages. Intensive ecological studies have been carried out in California (e.g. Sluss, 1967), and successful biological control of *C. juglandicola* has been obtained with *Trioxys pallidus* (Aphidiinae) from Iran (van den Bosch *et al.*, 1979). Natural control of *C. juglandicola* (and *Callaphis juglandis*) in walnut orchards in Italy was studied by de Biase and Calambuca (1979). $2n = 8*$.

Colopha Monell   Pemphiginae: Eriosomatini

A genus of about 6 Northern Hemisphere species, the better-known members of which form closed cock's comb-shaped galls on the upper surfaces of *Ulmus* leaves, from which they migrate for the summer to the basal parts of Gramineae or Cyperaceae.

Colopha ulmicola (Fitch)

Small aphids covered in white wax wool on the roots of Gramineae (*Aira, Agrostis, Deschampsia, Leersia, Triticum, Zea*) in North America. In spring this aphid forms cock's comb-like galls on *Ulmus* (*americana, fulva, racemosa*). For biological information see Patch (1910), who discusses this aphid under the names *C. ulmicola* and *Tetraneura graminis*, which are now regarded as synonyms.

Coloradoa Wilson   Aphidinae: Macrosiphini

About 30 Old World species associated with Anthemidae (Compositae).

Characteristic features are the spatulate body hairs shaped like partly opened fans, the stiletto-shaped last rostral segment, and the reduced or absent ocular tubercles. Heinze (1960) revised the European species.

Coloradoa rufomaculata (Wilson)   Green Chrysanthemum Aphid   Plate 67

Very small (1.0–1.7 mm) green aphids on stems and undersides of leaves of cultivated chrysanthemums. The siphunculi vary in pigmentation from dusky just at the apex to almost wholly dark, and are symmetrically swollen just before the apex. Distributed throughout the world on cultivated chrysanthemums, mainly occurring in glasshouses in cold temperate regions. Cottier (1953) gives a full description. Probably anholocyclic almost everywhere, although oviparous females are known from Australia and New Zealand. It is a vector of chrysanthemum virus B, and can also transmit tomato dwarfing virus.

Corylobium Mordvilko   Aphidinae: Macrosiphini

A monotypic genus resembling *Macrosiphum* but with apterae bearing long, thick, capitate hairs placed on tubercles. Hille Ris Lambers (1947a) gives an account of the genus. The other *Corylus*- and *Carpinus*-feeding species subsequently placed in *Corylobium* are now in *Macrosiphum* (*Neocorylobium*).

Corylobium avellanae (Schrank)   Plate 108

Medium-sized (1.7–2.7 mm) green, spindle-shaped aphids living on shoot apices, petioles, and undersides of youngest leaves of *Corylus* species, especially *C. avellana*, in Europe eastwards as far as the Ukraine and Turkey, and recently found in Canada (British Columbia). Often dense colonies occur on suckers growing up through the bush. Monoecious holocyclic, with alate males. Natural enemies, and especially the impact of a pathogen, are discussed by Viggiani in (avallaro (1983, p. 109–113). $2n = 10^*$.

Cryptaphis Hille Ris Lambers   Aphidinae: Macrosiphini

This genus contains one Old World and one New World species living concealed on the basal parts of Gramineae, and resembling *Metopolophium* except for their long, capitate hairs. It seems likely that 4 Asian species described from Labiatae and Geraniaceae, which are generally similar to the grass feeders but have spinulose heads, are closer to *Perillaphis* and/or *Eomyzus*.

Cryptaphis poae (Hardy)   Plate 103

Rather small (1.3–1.8 mm), very shiny, reddish-green to dark brown aphids forming small colonies on bases of grasses just below soil level, or under stones; usually it occurs in the shade or near water, and is unlikely to occur in open pasture. On various grasses, especially *Festuca, Holcus*, throughout Europe.

Monoecious holocyclic, with apterous males. Hille Ris Lambers (1947a) gives a full description of this species under the name *C. setiger*. $2n = 20^*$.

Cryptomyzus Oestlund   Aphidinae: Macrosiphini

A genus of about 10 Old World species associated with *Ribes* and/or Labiatae. The apterous parthenogenetic females have long antennae, swollen siphunculi, secondary sensoria on the third antennal segment (except fundatrices), and long capitate hairs on the body and often on the legs and antennae. The first antennal segment usually has an angular projection at its apex. Two species have been introduced into North America, presumably as eggs on *Ribes*. Accounts of *Cryptomyzus* are available for Europe (Hille Ris Lambers, 1953a) and Japan (Miyazaki, 1971). The species were often included in earlier accounts of *Capitophorus*.

Cryptomyzus galeopsidis (Kaltenbach)
European Blackcurrant Aphid   Plate 92

Appearance in life: Apterae on *Ribes* are usually pale greenish white, often with a faint green dorsal longitudinal stripe, but sometimes distinctly yellow, living rather dispersed on undersides of leaves without causing any host-plant reaction. Alatae have dark transverse bands on the dorsal abdomen, sometimes coalescing into a single patch. Apterae and alatae 1.3–2.6 mm.

Host plants: Primary hosts are *Ribes* species, mainly *R. nigrum* but also *R. - rubrum* (see below) and *R. grossularia*. Migration occurs to several species of *Lamium* (*purpureum, amplexicaule,* but not *album*) and *Galeopsis*.

Distribution: Europe, Iceland, and introduced into North America (Alberta, New Brunswick, New Hampshire).

Biology: Heteroecious holocyclic between *Ribes* and its labiate hosts. However, there are non-migrating populations of *C. galeopsidis* on *R. rubrum* and *R. nigrum*, which have each been given the status of subspecies by Hille Ris Lambers (1953a). These life-cycle variants are further discussed by Heie (1962). $2n = 12$.

Cryptomyzus ribis (L.)
Redcurrant Blister Aphid   Plate 93

Appearance in life: Apterae on *Ribes* are medium-sized, broadly oval aphids, very pale green to pale yellow, often almost white, in colour. Feeding by this species on the undersides of the young leaves causes a characteristic host-plant reaction, in which purplish or brownish-red blisters are formed on the upper surface of the leaves, the aphids living in the concavities thus formed underneath. Alatae usually have 3 broad transverse bars on the dorsal abdomen coalescing to form a trapezoid-shaped patch. Apterae 1.2–2.5 mm, alatae 1.5–2.2 mm.

Host plants: Primary host is almost always *Ribes rubrum*, but occasionally it is found on other *Ribes* spp. Keep and Briggs (1971) surveyed the host range in England. Secondary hosts are usually *Stachys* species, although there are records from *Galeopsis, Lamium,* and *Leonurus.*

Virus transmission: Able to transmit several non-persistent viruses including mosaic viruses of beet, cauliflower, cucumber, Dahlia, and watermelon, also cabbage black ring spot and onion yellow dwarf, but not known to be a vector of *Ribes* viruses.

Distribution: Europe, Japan, and North America.

Biology: Heteroecious holocyclic, migrating from *R. rubrum* to *Stachys*. Populations, perhaps of a separate subspecies, can remain on redcurrants all summer, and apterous as well as alate males are known. Hille Ris Lambers (1953a) discusses the life cycle and biology. Sokolova and Ponomareva (1968) investigated the nature of the reaction of *Ribes* to *C. ribis.* $2n = 12$. A very similar species, *C. korschelti* Börner, migrates from *Ribes alpinum* to *Stachys*.

Diuraphis Aizenburg   Aphidinae: Macrosiphini

About 10 species of these rather elongate, wax-covered aphids with inconspicuous siphunculi, mostly living in the rolled leaves of Gramineae, have been described in Europe (7) and North America (3). The group was little known until *D. (H.) tritici* was described from Colorado, U.S.A., in 1911, and there were serious outbreaks of *D. noxia* on barley in Russia in 1912. The subgenus *Holcaphis* differs from *Diuraphis* s.str. by lacking a process on the eighth abdominal tergite.

Diuraphis (Holcaphis) frequens (Walker)

Rather small, very elongate (1.3–2.1 mm) green aphids with darker green heads, the body covered with grey wax powder, usually colonizing *Agropyron repens*, bunching the still unfurled leaves and stopping growth. Also recorded from *Lolium*. Widespread in Europe, eastwards to Turkey. Monoecious holocyclic on *Agropyron* with apterous males (Hille Ris Lambers, 1939b). $2n = 14$ (Gut, 1976).

Diuraphis noxia (Mordvilko)   Plate 65

Appearance in life: Apterae rather small, spindle-shaped, pale yellow green or grey-green dusted with white wax powder. Alatae have a pale green abdomen.

Infested leaves are rolled into tubes and desiccated, and infested ears become bent. A vector of barley yellow dwarf virus (Orlob, personal communication). Apterae 1.4–2.3 mm, alatae 1.5–2.0 mm.

Host plants: Mainly on barley and wheat, rarely on other cereals or grasses. Particularly injurious to late-sown barley in continental climates.

Distribution: Of Palaearctic origin, now widespread; southern Europe, Central Asia, Middle East, North Africa, South Africa, and Argentina. Not yet in North America or Australasia.

Biology: Monoecious holocyclic on *Hordeum* and *Triticum* in cold temperate climates, probably anholocyclic elsewhere. There is an extensive Russian literature, one of the most comprehensive studies being by Grossheim (1914; or see *Review of Applied Entomology* 3: 307–8). Berest (1980) has studied the parasite and predator complex of *D. noxia* in the Odessa and Nikolaev regions of the U.S.S.R. Dürr (1983) describes this aphid in South Africa. In northern, western, and central Europe a very similar species with a shorter process on the eighth abdominal tergite, **D. muehlei** Börner, feeds specifically on *Phleum pratense*, turning the leaves yellow (Stacherska, 1975). $2n = 10$.

Diuraphis (Holcaphis) tritici (Gillette)   Western Wheat Aphid

Pale yellowish-green, small (1.1–1.9 mm), elongate aphids with a powdery white coating of wax, on leaves of wheat in North America (Colorado, Montana, Illinois). Blue knot-grass, *Agropyron occidentale*, is also regularly colonized and may be the native host. Monecious holocyclic with apterous males. See Parker (1916) for an early account of the biology.

Dysaphis   Aphidinae: Macrosiphini

An Old World genus of about 100 known species, characterized by a short cauda, often short siphunculi, and the frequent presence of spinal tubercles on the head and abdominal segments 7 and 8. The species distort and often discolour the leaves of Pyroidea (*Crataegus, Malus, Pyrus, Cotoneaster, Sorbus*), and/or live concealed at the bases of the stems of Umbelliferae and (more rarely) Compositae, Valerianaceae, Plantaginaceae, Polygonaceae, or Liliaceae. A few species live on members of other plant families. They are ant-attended on both primary and secondary hosts, and on secondary hosts the ants usually build earth 'shelters' around them. Shaposhnikov (e.g. 1956) and Stroyan (1957, 1963) have published extensively on this genus.

Dysaphis angelicae (Koch)

Yellowish-green to greenish-grey aphids at leaf bases of *Angelica sylvestris* and

A. archangelica in northern Europe, attended by ants. Heteroecious holocyclic, with *Crataegus* as primary host. Not recorded from regions where *Angelica* is grown commercially.

Dysaphis anthrisci Börner   Apple–*Anthriscus* Aphid

One of several sibling species of *Dysaphis* (the others being *devecta, chaerophylli, radicola*) causing similar galls on apple in spring, and only really separable by their life cycles and/or secondary host-plant relationships. Heteroecious holocyclic with migration to leaf bases of *Anthriscus sylvestris*. The fundatrices in the galls on apple are blue grey with a wax bloom; the alatae from the galls have a dull pinkish-grey abdomen with a large black dorsal patch. See Stroyan (1963). Distributed throughout Europe. $2n = 12$.

Dysaphis apiifolia (Theobald)   Hawthorn–Parsley Aphid   Plate 52
(= *ferruginea-striata* Essig)

Appearance in life: Yellowish-grey to greenish-grey aphids, lightly dusted with wax, forming dense colonies at the leaf bases of celery, parsley, or fennel, at or slightly below soil level, attended by ants which usually build up earth 'shelters' around them. Alatae produced by these colonies have a dull greenish abdomen with a black dorsal patch. Apterae 1.4–2.6 mm, alatae 1.5–2.4 mm.

Host plants: Primary host of the subspecies *petroselini* in Europe (see below) is *Crataegus*. Summer forms and anholocyclic populations live on *Apium graveolens, Foeniculum vulgare, Petroselinum sativum*, and certain other Umbelliferae (*Anethum, Conium, Levisticum, Smyrnium*). Records of this species from cultivated carrots are mostly referable to *D. foeniculus*.

Virus transmission: A vector of celery mosaic and celery crinkle leaf mosaic viruses, but apparently not of the persistent virus celery leaf spot.

Distribution: Europe, the Middle East, Central Asia, north Africa, South Africa, Australia, and North and South America.

Biology: The holocyclic form in northern and central Europe, migrating from *Crataegus* to Umbelliferae including parsley, is currently regarded as a separate subspecies, **petroselini** (Börner). The biology of this form is fully described by Stroyan (1963). In other parts of the world, *D. apiifolia* (*s. lat.*) is anholocyclic on secondary host plants. $2n = 12$.

Dysaphis bonomii (Hille Ris Lambers)   Permanent Parsnip Aphid

Apterae are pale to dull greyish green with dark dorsal bands or patches, lightly dusted with wax, at the stem bases of *Pastinaca sativa*, but not recorded from cultivated varieties. Monoecious holocyclic, with apterous males (Stroyan, 1963). Only in Europe.

Dysaphis brancoi (Börner) — Apple–Valerian Aphid

Apple leaf galls of this aphid when fully mature are a deep purplish red, darker than those of *D. devecta*. The fundatrix is blue grey evenly covered with white wax; alatae migrating from the galls have a dull pinkish-grey abdomen with a very extensive dark dorsal patch. Heteroecious holocyclic, migrating from apple to *Valeriana*. Throughout Europe, but the form in Britain is regarded as a separate subspecies (*rogersoni*), and in continental Europe there are also closely related species or subspecies which are monoecious on *Valeriana* (Stroyan, 1963).

Dysaphis chaerophylli (Börner) — Apple–Chervil Aphid

Leaf galls on apple in spring and the morphs within them are difficult to distinguish from those of other members of the species complex (*devecta, anthrisci, radicola*). Heteroecious holocyclic, with migration to leaf bases of *Chaerophyllum* (*bulbosum, hirsutum, temulum*). Only in Europe.

Dysaphis crataegi (Kaltenbach) — Hawthorn–Carrot Aphid

Yellowish-grey to greenish-grey aphids, lightly wax-dusted, forming dense ant-attended colonies on carrots at or a little below ground level. Heteroecious holocyclic, with *Crataegus* as primary host. *Daucus* is the preferred secondary host of *D. crataegi s. str.*, but certain other Umbelliferae, e.g. *Myrrhis* and *Anthriscus*, may also be colonized, and wild or cultivated *Pastinaca sativa* is the secondary host of a separate subspecies in Europe, **D. crataegi kunzei** (Börner). See Stroyan (1963) for a full account. In Europe, Central Asia, Middle East, and introduced into the U.S.A. $2n = 12$.

Dysaphis cynarae (Theobald)

Pinkish-grey to dirty olive-coloured aphids dusted with white wax powder, in ant-attended colonies on *Cynara* at the base of the stem and just below soil surface. Recorded from the Mediterranean region (Egypt, Sicily), and from Brazil. In Sicily this species is apparently anholocyclic on *Cynara scolymus*, and occurs also on *C. cardunculus* (on which there might be a holocycle), *Silybum marianum, Lupsia galactites,* and *Cirsium syriacum* (Barbagallo, 1974).

Dysaphis devecta (Walker) — Rosy Leaf-curling Aphid — Plate 51

Appearance in life: The aphids roll the edges of apple leaves and turn them red. The leaf-galls in spring contain a mixture of normal, bluish-grey wax-powdered apterae and dark green or reddish alatiform apterae with varying degrees of dorsal sclerotization and pigmentation. Apterae 1.8–2.4 mm, alatae 1.6–2.5 mm.

Host plants: Specific to apple, including some ornamental varieties.

Distribution: Only in Europe.

Biology: Monoecious holocyclic on apple, with apterous males. This aphid has a peculiar life cycle; sexual morphs are produced before mid-summer, after only three parthenogenetic generations (Hille Ris Lambers, 1945; Forrest, 1970). Alston and Briggs (1977) investigated the genetic basis of resistance in apple to *D. devecta* in south-east England. Forrest and Dixon (1975) studied gall formation.

Dysaphis foeniculus (Theobald) Plate 49

Appearance in life: Apterous adults grey green with a light dusting of greyish-white wax, forming dense colonies on the basal parts of the host plant at or a little below ground level, usually attended by ants. Alatae have a dark green abdomen with a large black dorsal patch. Apterae 1.6–2.3 mm, alatae 1.6–2.5 mm.

Host plants: On several species of Umbelliferae including *Daucus, Foeniculum, Apium, Anethum, Ferula,* and also recorded occasionally from *Rumex*.

Virus transmission: Apparently *not* a vector of carrot motley dwarf disease.

Distribution: Mediterranean area, Iraq, Turkey, India, Pakistan, Africa, Australia, New Zealand, and North and South America.

Biology: Little studied; apparently anholocyclic everywhere, no sexual morphs have been described. Morphologically close to *D. tulipae*. $2n = 12^*$.

Dysaphis ossiannilssoni Stroyan

Greyish-green aphids in ant-attended colonies at leaf bases or on leaf collar of *Angelica sylvestris* and *A. archangelicae*. Probably monoecious holocyclic on *Angelica*. Only so far recorded from Sweden.

Dysaphis (Pomaphis) plantaginea (Passerini) Rosy Apple Aphid Plate 54
(= *Anuraphis roseus, Sappaphis mali*)

Appearance in life: Apterae in crumpled leaf galls on apple are dull pinkish to slate grey or purplish grey, with a greyish-white wax bloom. Alatae from galls have a reddish-grey abdomen with an extensive dark dorsal patch. Apterae 2.1–2.6 mm, alatae 1.8–2.4 mm.

Host plants: Primary host is apple (sometimes on pear in warmer climates); secondary hosts are *Plantago* species.

Distribution: Europe, Central Asia, Middle East, North Africa, and North America. Also recorded from Japan and Taiwan, based on supposed synonymy of *Myzus plantagifolii* Shinji and *M. plantagicola* Takahashi with *D. plantaginea*. Despite the similarity of Japanese specimens from *Plantago* to the exules of *plantaginea*, the absence of records from the Far East of the more conspicuous spring form on apple and the different feeding site on *Plantago* in Japan (at the bases of the leaf petioles at or near ground level) suggest that *Dysaphis plantagifolii* (= *plantagicola*) is really a distinct taxon not colonizing apple.

Biology: Heteroecious holocyclic, in Europe at least, migrating from leaf galls on apple to plantains where they feed along the veins under the leaves. *Plantago lanceolata* seems to be a preferred host. Populations may remain on apple into the summer, and in some areas anholocyclic overwintering on plantains is suspected. There is an extensive literature on this well-known pest aphid, including detailed laboratory studies of the factors controlling polymorphism (Bonnemaison, 1971). Gall formation was studied by Forrest and Dixon (1975). Ecological aspects studied include winter egg survival in Italy (Baronio, 1971), parasites in Czechoslovakia (Starý, 1975), and feasibility of chemical (Memmi et al., 1979), biological (Marboutie, 1976), and integrated (Holdsworth, 1970) control. $2n = 12$.

Dysaphis (Pomaphis) pyri (Boyer de Fonscolombe)

Pear–Bedstraw Aphid   Plate 53

Appearance in life: Medium to rather large, oval, brownish-red to dark brown aphids, covered with wax meal, distorting and yellowing the leaves of pear in spring. The siphunculi are black. Alatae have a brownish-red abdomen and a black dorsal patch. Apterae 1.7–3.2 mm, alatae 1.8–2.5 mm in spring, 1.3–2.5 mm in autumn.

Host plants: Cultivated *Pyrus communis* is the main primary host; sometimes on wild pears and there are also records from *P. nivalis* and *P. ussuriensis*. Secondary hosts are species of *Galium*, especially *G. mollugo* and *G. aparine*.

Distribution: Europe, Central Asia, the Middle East, North Africa, northern India, Nepal, and Pakistan.

Biology: Heteroecious holocyclic, migrating to *Galium* after about three generations on pear, which in some years can result in heavy primary host infestations. Life history and population biology have been studied in some detail by Savary (1953) in Switzerland and Kolesova (1974) in Crimea. See also Stroyan (1957). $2n = 12$.

Dysaphis radicola (Mordvilko)   Apple–Dock Aphid

Leaf galls of this species on apple in spring, and the aphids within them, are difficult to distinguish from other members of the species complex (*anthrisci, chaerophylli, devecta*). Heteroecious holocyclic, with migration to roots of *Rumex* species, and also recorded from *Rheum*. Anholocyclic overwintering on *Rumex* roots occurs in Britain, where the apple generations are not known (Stroyan, 1963). Throughout Europe; alate aphids provisionally assigned to this species have been trapped in the U.S.A. (Louisiana, New York), but there are no American records from host plants. $2n = 12$.

Dysaphis (Pomaphis) reaumuri (Mordvilko)   Leaf-rolling Pear Aphid

Appearance in life: Colonies on pear distort and yellow the leaves and cause spiral coiling of the petioles and flower stalks. Apterae are rather small to medium-sized, broadly oval aphids, light green (in early spring) to dark green, sometimes with a brownish or reddish tinge in summer, covered with white wax powder which is densest in early spring. The siphunculi are pale. Alatae have a dirty yellow-green abdomen with a black dorsal patch. Apterae 1.0–2.3 mm, alatae 1.5–2.0 mm.

Host plants: Main primary host is *Pyrus communis*, wild or cultivated; also recorded from *P. salicifolia, P. syriaca,* and *P. korshinskyi.* Secondary hosts are *Galium* spp. (*mollugo, aparine, taurica*).

Distribution: Southern Europe and Central Asia.

Biology: Heteroecious holocyclic, migrating from pear to *Galium*, but small populations can stay on pear throughout the summer and produce oviparae in autumn. Kolesova (1974) gives a detailed account of the life history of *D. reaumuri* in Crimea, and Grigorov (1977) discusses its biology in Bulgaria. $2n = 12$.

Dysaphis tulipae (Boyer de Fonscolombe)   Tulip Bulb Aphid   Plate 50

Appearance in life: Apterae are very pale yellow, grey or pink, appearing almost white due to a covering of white wax powder, with dark siphunculi. Alatae have a pale yellowish-grey, sometimes pinkish, abdomen, with a central black dorsal patch and black siphunculi. Apterae 1.5–2.5 mm, alatae 1.5–2.3 mm.

Host plants: A large number of monocotyledonous species in Liliaceae, Iridaceae, Araceae, and Musaceae, including members of the genera *Arum, Chionodoxia, Crocus, Freesia, Gladiolus, Iris, Lilium, Moraea, Musa, Scilla,* and *Strelitzia. Iris* spp. are perhaps the commonest hosts.

Virus transmission: A vector of the persistent lily symptomless virus, and the non-persistent tulip breaking virus.

Distribution: Almost cosmopolitan, except that it has apparently not been noted in South America.

Biology: Apparently entirely anholocyclic; sexual morphs have not been found. The aphids spend the dormant period of their host plants under the scales or in crevices of the bulbs, corms, or rhizomes. Sometimes attended by ants. For additional information see Stroyan (1963). $2n = 11$ and 12.

Eomacrosiphum Hille Ris Lambers — Aphidinae: Macrosiphini

A monotypic genus distinguished from *Macrosiphum* by the first tarsal segments with 5 hairs, although the siphunculi and other features of the type species are also very distinctive.

Eomacrosiphum nigromaculosum (MacDougall) — Black-and-Red Rose Aphid

Apterae in life are bright red with black bands or patches on the dorsal abdomen, black siphunculi held at right angles to the body, and black and yellow legs. Feeding mainly on stems of wild and cultivated *Rosa*, in western North America. Monoecious holocyclic with alate males (Palmer, 1952).

Ericolophium Tao — Aphidinae: Macrosiphini

A monotypic genus related to *Neoacyrthosiphon*; the main point of difference is that the second tarsal segments of *Ericolophium* are not spinulose.

Ericolophium itoe (Takahashi)

Rather small, elongate oval aphids, body reddish brown with a greenish tint, and black appendages. Immatures are yellowish green with darker green longitudinal stripes. On *Rhododendron* species in Taiwan and Japan. Life cycle is apparently unknown.

Eriosoma Leach — Pemphiginae: Eriosomatini

A genus of about 20 species. These include 6 species of *Eriosoma s. str.* in North America which migrate from 'rosette' leaf galls on *Ulmus* to form woolly wax masses on the twigs of Pyroidea or are monoecious on *Ulmus* (*E. rileyi*); and about 13 species of the subgenus *Schizoneura* from the palaearctic region, alternating between leaf galls on *Ulmus* and the roots of Pyroidea, *Ribes*, or Compositae. Danielsson (1982) gives an account of the species found on *Ulmus* in Europe and of those found on *Ribes* roots. Akimoto (1983) provides a revision of the Japanese species. In North America systematic accounts are available for Illinois (Hottes and Frison, 1931) and the Rocky Mountain region (Palmer, 1952).

Eriosoma (Schizoneura) grossulariae (Schüle)

Small to medium-sized light reddish to yellowish-white aphids living in white wax wool on roots of gooseberry, and also on *Ribes sanguineum*, producing bluish-grey to black sexuparae in autumn which migrate to *Ulmus* spp., on which open, curled-leaf galls are produced in spring. Throughout Europe, Japan, and introduced into North America. This species has previously been confused with *E. (S.) ulmi*; see Danielsson (1982). $2n = 10^*$.

Eriosoma lanigerum (Hausmann)   Woolly Apple Aphid, American Blight
Plate 137

Appearance in life: Apterae are small to medium-sized, purple, red, or brown, covered with a thick white flocculent wax; on roots, trunk, or branches of host plant, often causing deformation and cancer-like swellings of the bark. Alatae have a reddish-brown abdomen with woolly white wax posteriorly. Apterae 1.2–2.6 mm, alatae 1.8–2.3 mm.

Host plants: Mainly on apple, on which it can be a severe pest; also found on *Crataegus, Cotoneaster*, and more rarely on *Cydonia, Pyrus*, and *Sorbus*, but not so injurious to these plants.

Virus transmission: Not a vector of apple viruses.

Distribution: Co-extensive with apple cultivation throughout the world. Origin is unknown, possibly North American (see below).

Biology: Anholocyclic on apple throughout the world. However, despite extensive studies by workers in many countries, many questions remain about the biology of *E. lanigerum*. Colonies persist through the winter on the roots of the host plant, but authors differ about whether there is a regular seasonal migration from aerial parts to roots and back. Hoyt and Madsen (1960) studied dispersal by first instar larvae from crowded populations. An apparently abortive sexual phase occurs in many parts of the world, sexuparae being commonly observed, and sexuales and winter eggs may sometimes be produced on apple, but the eggs usually do not hatch. In North America, *Ulmus americana* was regarded as the primary host following work by Patch (1912a), but Patch's aphid (and the species studied by later workers in eastern U.S.A., e.g. Schoene and Underhill, 1935) is now thought to be another species, *E. herioti* Börner, which migrates from *U. americana* to the roots of *Crataegus, Malus*, and *Sorbus americana* (Börner, 1952/53). Thus, the holocycle and primary host, if any, of *E. lanigerum*, and whether it was originally a North American species, remain in doubt.

The chalcidoid parasite, *Aphelinus mali*, has been introduced into many countries in attempts to control *E. lanigerum*, with varying success (see, for example, Wilson, 1960; Hafez, 1978). Thakur and Dogra (1980) discuss biological control and other work on *E. lanigerum* in India. $2n = 12$.

Eriosoma (Schizoneura) lanuginosum (Hartig)   Elm Balloon-Gall Aphid
Plate 138
Eriosoma (Schizoneura) pyricola Baker & Davidson

Two closely related species alternating between *Ulmus* spp. and the roots of *Pyrus communis*. The apterae on pear roots are small to medium sized, rather elongate oval, pale pink or yellow to red, living in white wax wool chiefly on the smaller fibrous rootlets, sometimes on underground parts of suckers. The alatae (sexuparae) produced in autumn have a shiny dark green to almost black abdomen with a tuft of white wax on the posterior segments. The spring forms on *Ulmus* live in large, closed, green or brown bloated leaf galls of variable shape near the ends of twigs.

E. (S.) lanuginosum occurs in Europe, the Middle East, and Central Asia; de Fluiter (1933) gives an account of its biology in the Netherlands. *E. (S.) pyricola*, which was for a long time synonymized with *lanuginosum* until the work of de Fluiter, is apparently native to southern Europe and introduced into North America (where it can utilize *U. americanus* as a primary host), Australia, and New Zealand. One of the two species, probably *pyricola*, is also recorded from South Africa. Baker and Davidson (1916, 1917) gave a full description of the life cycle of *pyricola*, and the influence of the host plant on sexupara production has been investigated (Swenson, 1971). *E. (S.) lanuginosum* has $2n = 10^*$.

Eriosoma (Schizoneura) ulmi (L.)   Elm–Currant Aphid

Small, oval, light red to brownish-red apterae living in white wax wool on roots of currant bushes (*Ribes nigrum, R. rubrum, R. aureum*), producing bluish-grey to black alate sexuparae in autumn which migrate to *Ulmus* species, on which open, curled-leaf galls are produced in spring. In Iceland, Europe, Central Asia. Janiszewska-Cichocka (1969, 1971) has made a detailed study of this aphid in Poland. However, Danielsson (1982) has shown that the aphid normally occurring on roots of *Grossularia* is a similar but distinct species, *E. (S.) grossulariae* (q.v.). $2n = 10^*$.

Eucallipterus Schouteden   Drepanosiphinae: Phyllaphidini

A genus of 2 species associated with Tiliaceae.

Eucallipterus tiliae (L.) Lime Aphid Plate 11 (left)

Small to medium-sized black and yellow aphids living on the undersides of leaves of *Tilia* spp. All viviparae are alate, with dark-bordered wing veins. Monoecious holocyclic on *Tilia* in Europe, Central Asia, North America, and New Zealand. The biology and population dynamics of this aphid have been studied intensively in England (Barlow and Dixon, 1980). $2n = 10$.

Eucarazzia del Guercio Aphidinae: Macrosiphini

A genus of two species associated with Labiatae. The very swollen siphunculi, and the wing pigmentation of the alatae, are very distinctive features.

Eucarazzia elegans (Ferrari) Plate 87

Rather small, broadly oval, green aphids feeding on the undersides of leaves, shoots, or flowers of *Mentha* spp. and various other Labiatae (*Salvia, Coleus, Lavandula, Nepeta*, etc.), in the Mediterranean region and Middle East, and also recorded from Kenya and South Africa. The alatae have the swollen part of the siphunculi dark and the cylindrical basal part pale, and their wings have dark triangular spots at the ends of all the veins. The life history is unrecorded; sexual morphs are unknown (Hille Ris Lambers, 1953a). $2n = 12*$.

Fimbriaphis Richards Aphidinae: Macrosiphini

A genus of 8 American aphids associated with Rosaceae, Ericaceae, and Liliaceae, and of 2 European species from Ericaceae which are similar to the European genus *Ericaphis*. Prior to 1959 the species were mostly placed in *Macrosiphum* or *Myzus*. Prior (1971) gave a key to the British species then known.

Fimbriaphis fimbriata Richards

Rather small to moderate-sized, broadly spindle-shaped light-coloured (yellow-green?) aphids on strawberries in western North America (Oregon, British Columbia). Alatae have a dark dorsal abdominal patch. No observations of the living aphids have apparently been recorded, and the biology is not known. Closely related to, and possibly synonymous with, *F. scammelli*.

Fimbriaphis gentneri (Mason)

Rather small, broadly spindle-shaped, light-coloured (green?) aphids on *Pyrus communis* in western North America (California, Idaho, Oregon, British Columbia). Also recorded from *Crataegus* spp. (Hille Ris Lambers, 1966c) and *Amelanchier* sp., one or more of which are presumably the native host(s). The alata has a broken dark dorsal patch. Monoecious holocyclic, with alate males.

All morphs have been described (Mason, 1947), but no observations of the living aphids have apparently been recorded.

Fimbriaphis latifrons (Börner)

Greenish-yellow to yellowish-green, slightly shiny aphids on young shoots and undersides of leaves and *Vaccinium uliginosum*; it also lives on *Empetrum* spp. In Iceland, northern, and western Europe. Remaudière (1952) gives an account of this species in the Pyrenees (as *Myzodium lagarrigueri*). Monoecious holocyclic with apterous males.

Fimbriaphis lilii (Mason)

One of two closely related species of small to medium-sized pale-coloured aphids on lilies in the western United States (see also *F. scoliopi*). *F. lilii* was described from *Lilium candidum* in Oregon, and is also recorded from lilies in Washington. The alata has a large black dorsal abdominal patch. Biology is unrecorded.

Fimbriaphis scammelli (Mason)   Plate 107

Small to medium-sized, broadly spindle-shaped pale yellow-green aphids on young shoots of various species of Ericaceae. The alata has dark brown dorsal abdominal markings only partly fused into a central patch, with a conspicuous pale spot between the siphunculi. *F. scammelli* was described from cranberry (*Vaccinium oxycoccus*) in New Jersey, U.S.A. *F. fimbriata* subspecies *pernettyae*, described by Prior from various Ericaceae (*Andromeda, Pernettya, Vaccinium* spp.) in England, is probably a synonym. Aphids of the *scammelli/pernettyae* group have been collected on cultivated blueberries in Oregon and Washington, and there are also records from Quebec and the Netherlands. Monoecious holocyclic, with alatae males. $2n = 14^*$.

Fimbriaphis scoliopi (Essig)

Small to medium-sized pale yellow to almost white aphids on the leaves of lilies (*Lilium* spp.) in the western United States (California, Oregon). Also recorded from *Scoliopus bigelovii*. The alata has a whitish to pale yellow abdomen with a large black dorsal patch. Biology is unknown.

Fimbriaphis wakibae (Hottes)

Small to medium-sized, pale brown or greenish-yellow aphids with the dorsal abdomen sclerotic and variably tanned, often with darker brown spinal and lateral regions. The siphunculi are brown, with pale areas around their bases. Alatae have a pale reddish-brown to olive-green abdomen with a dark dorsal abdominal patch, which has a clear window in it posteriorly, between the siphunculi. This species was originally described from *Pedicularis* (Scrophulariaceae) in Colorado, U.S.A., but has since been recorded mainly from Rosaceae. On wild *Fragaria vesca* in England, and colonizing young growth of cultivated strawberries in culture (Prior, 1971). Also collected from *Rosa canina* in England, and from *Rubus pubescens* in Canada. In North America (Colorado, Pennsylvania, North Carolina, Manitoba), and England. Monoecious holocyclic on *Fragaria* in England (Prior, 1971); life cycle in North America not recorded. $2n = 12^*$.

Forda von Heyden   Pemphiginae: Fordini

An Old World genus of about 8 species alternating in the Mediterranean region between galls on *Pistacia* and the roots of Gramineae. Several species show a marked tendency to anholocycly on grass roots. Both *F. formicaria* and *F. marginata* have been introduced into North America. Accounts are available of the morphs on *Pistacia* by Davatchi (1958); of the root-living morphs by Mordvilko (1935); from northern, central, and southern Europe (Heie, 1980; Zwölfer, 1958; Roberti, 1939b), and India (Chakrabarti, Maity, and Bhattacharya, 1982).

Forda formicaria von Heyden   Plate 142

Appearance in life: Apterae are medium-sized, oval aphids with dorsal surface of body highly domed, and short appendages, varying in colour from off-white to dull yellow (often with a darker medio-dorsal stripe), to various shades of dark green or bluish green; on roots of various grasses and cereals, attended by ants, or in ants' nests. Alatae have a dull green abdomen with dark dorsal transverse bands. Apterae (from roots) 2.0–3.0 mm, alatae 2.0–3.3 mm.

Host plants: Primary hosts are *Pistacia terebinthus* and *P. palaestina*. Secondary hosts are numerous species of Gramineae, including *Agropyron, Agrostis, Bromus, Cynodon, Dactylis, Deschampsia, Festuca, Hordeum, Lolium, Poa, Secale,* and *Triticum.*

Distribution: Europe, the Mediterranean region, the Middle East, Central Asia, Siberia, and North America.

Biology: Heteroecious holocyclic in the Mediterranean region, forming crescent-shaped galls by rolling leaflets of *Pistacia*. For an account of the life history in Italy see Roberti (1939). Probably anholocyclic overwintering occurs even where the primary hosts are available, but in northern Europe and North America populations are exclusively anholocyclic on roots of Gramineae, often overwintering in ants' nests. $2n = 18, 20, 21,$ or 22.

Forda hirsuta Mordvilko

This is the only *Forda* species known to gall *Pistacia vera*. The galls are bright red, and are formed by rolling downward and back of the leaf edges to form pockets, sometimes with several pockets occurring side by side along the edge of one leaf (Davatchi, 1958). Alatae from the galls usually have 7–13 secondary sensoria on antennal segment III, many of them almost ringing the segment. Heteroecious holocyclic, giving rise to colonies on roots of Gramineae (e.g. *Agropyron, Cynodon, Hordeum*). The apterae on grass roots are straw-coloured to orange, rather large (over 2.5 mm), very broadly oval, the dorsal surface highly domed. In Crimea, Caucasus, Iran, and Iraq. Probably some populations are anholocyclic on grass roots. $2n = 18$.

Forda marginata Koch   Plate 143
(= *follicularia* Passerini)

Appearance in life: Apterae on grass roots are small to medium sized with short legs and antennae; body highly domed dorsally, brownish yellow or greenish yellow in colour. Alatae have a green abdomen with brown transverse bands particularly on the posterior tergites. Apterae (from roots) 1.7–3.1 mm, alatae 2.1–3.1 mm.

Host plants: Primary hosts are *Pistacia mutica, P. palaestina* and *P. terebinthus*. Secondary hosts are numerous species of Gramineae including *Agropyron, Agrostis, Avena, Bromus, Dactylis, Festuca, Hordeum, Poa, Secale,* and *Triticum*.

Distribution: Europe, the Mediterranean region, the Middle East, India, Siberia, and North America.

Biology: Heteroecious holocyclic in the Mediterranean region and Middle East, forming galls on *Pistacia* by folding and swelling the edges of the leaflets (Roberti, 1939). Exclusively anholocyclic on roots of secondary hosts in north and central Europe and in North America, or overwintering in ants' nests. The morphological variation shown by these anholocyclic populations suggests that they may correspond to a group of closely related species, but the taxonomy will only be clarified by further studies of primary host-plant relationships. $2n = 17, 18, 20,$ and 22.

Forda orientalis George

According to the original description, apterae are yellowish white, rather large (about 3.5 mm), oval, very much arched dorsally, and alatae have a yellowish-brown abdomen. On roots of Gramineae (*Sorghum, Saccharum, Pennisetum, Botriochloa, Triticum*) in India (David, 1969), Pakistan, and Iran. The sexual phase is unknown.

Geoica Hart
Pemphiginae: Fordini

An Old World genus of about 10 species alternating between galls on *Pistacia* and the roots of Gramineae, and sometimes Cyperaceae. The taxonomy is difficult, particularly around *G. utricularia*. Accounts are available for the morphs on *Pistacia* by Davatchi (1958); for nothern, central, and southern Europe (Heie, 1980; Zwölfer, 1958; Roberti, 1939b, respectively); and for north-east India (Raychaudhuri, Pal, and Ghosh, 1980).

Geioca lucifuga (Zehntner)

Appearance in life: Medium-sized, pale yellowish or yellowish-brown almost globular aphids with a light covering of white wax, living on roots of Gramineae, usually attended by ants. Alatae have a yellowish abdomen with a broad transverse bar on each tergite, largest on the posterior segments. Apterae 1.2–3.3 mm, alatae 1.9–3.0 mm (root forms).

Host plants: Many species of Gramineae, including rice, sugar cane, *Cynodon dactylon, Hordeum, Triticum, Zea*. Recorded also from the roots of grape vine in Central Asia (Nevskii, 1929). Varietal differences in rice susceptibility to *G. lucifuga* were noted by Dani and Majumdar (1978).

Distribution: Throughout the world except in cold temperate northern regions. Yugoslavia, Central Asia, the Middle East, throughout Africa, Mauritius, India, Pakistan, Sri Lanka, China, Taiwan, Japan, Java, New Guinea, the Philippines, the Solomon Islands, Australia, New Zealand, Argentina, and Brazil.

Biology: Anholocyclic on roots of Gramineae. The holocycle has not been described, and the primary host is not known; Mordvilko (1935) suggests that it may be *Pistacia chinensis*. Tanaka (1961) discusses the biology of this aphid on rice roots in Japan.

Geoica setulosa (Passerini) (= herculana Mordvilko)

Off-white or pale greenish-grey, almost globular aphids on roots of grasses (e.g. *Agrostis, Briza, Festuca, Holcus*), attended by ants. In north-west and central Europe, Italy, Iran, and Turkey. In Iran there is a holocycle with *Pistacia khinjuk* as the primary host (Davatchi, 1958). In Europe *G. setulosa* is exclusively anholocyclic on grass roots, often overwintering in the nests of *Lasius flavus*. $2n = 20$ and 24.

Geoica utricularia (Passerini)   Plate 145
(= *eragrostidis* Passerini, = *squamosa* Hart)

Appearance in life: Apterae are off white, cream or yellowish white, lightly dusted with wax, broadly oval to almost globular, on roots of host plants, invariably ant-attended. Alatae have a pale yellowish-green abdomen with dark transverse bars, largest on the more posterior tergites. Apterae 1.6–3.0 mm, alata (one specimen) 2.0 mm (from roots).

Host plants: Primary hosts are *Pistacia terebinthus, P. atlantica, P. mutica, P. palaestina,* and *P. vera* (but see below). Secondary hosts are numerous species of grasses and cereals (e.g. *Agrostis, Avena, Bromus, Deschampsia, Festuca, Hordeum, Lolium, Phleum, Poa, Triticum, Zea*).

Distribution: Throughout Europe and in North Africa (Morocco), the Middle East, Central Asia, and North America.

Biology: The name *G. utricularia* is probably being applied to a complex of species and anholocyclic races. Davatchi (1958) described subspecies from Iran associated with different *Pistacia* species as primary host plants, and Wool and Koach (1977) found electrophoretic differences between inhabitants of galls on two *Pistacia* spp. in Israel. Gall formation by *G. utricularia* on *Pistacia* is described by Wertheim (1954); the roughly spherical galls differ in size and colour according to the species of host plant. A full account of the life history in Italy is given by Roberti (1939). In central and northern Europe and North America *G. utricularia* is anholocyclic on roots of Gramineae. Populations from grass roots may differ from each other in chaetotaxy, particularly of the last rostral segment, eighth abdominal tergite, and cauda. $2n = 16, 17,$ and 18.

Geopemphigus Hille Ris Lambers Pemphiginae: Fordini

A genus for 1 (or 2) neotropical species related to *Baizongia* and *Asiphoniella*.

Geopemphigus flocculosus (Moreira) Plate 149

Apterae (according to original description) pale dirty greenish yellow, small to medium sized, with tufts of white wax on abdomen. Alatae have transverse brown-black bars on the dorsal abdomen, and long flocks of filamentous white wax. On roots of *Ipomaea* spp., including *I. batatas.* Also recorded from *Parthenium hysterophorus* and *Coreopsis* sp. (Compositae); but specimens from these hosts have relatively long antennal segment III and last rostral segment, differences which could be nutritionally induced or possibly indicative of a distinct, undescribed species. In southern U.S.A. (Florida, Georgia), Central America, Caribbean, South America as far south as Sao Paulo, and Brazil. Life cycle is unknown, probably mainly anholocyclic; if a holocycle exists, then it will probably involve gall formation on an American *Pistacia* (Hille Ris Lambers, 1961). See also Smith *et al.* (1964).

Glabromyzus Richards Aphidinae: Macrosiphini

A small genus of about 4 nearctic species associated with *Rhus*. At least one of them is heteroecious, migrating to Gramineae, but life cycles and host relations of the group are in need of further study.

Glabromyzus howardii (Wilson)

Small to medium-sized, light brown to greenish-yellow aphids with black clavate siphunculi, in flower-heads of wild and cultivated Gramineae (barley, wheat, oats, *Dactylis glomerata, Elymus canadensis, Phleum pratense, Setaria, Stipa*). Alatae have a brownish-yellow or greenish-yellow abdomen with dusky markings. Widely distributed in North America, and specimens from *Setaria geniculata* in Brazil in the BM(NH) collection may also be assigned to this species. Heteroecious holocyclic with *Rhus* (subgenus *Toxicodendron*) species as primary hosts. This aphid is frequently synonymized with *G. rhois* by American authors (e.g. Palmer, 1952). Although the evidence seems to favour retention of the name *G. howardii* for grass-feeding *Glabromyzus*, it is possible that other species have a partial or complete migration from *Rhus* to Gramineae, and are being confused with *G. howardii* on secondary host plants (see Hille Ris Lambers, 1966c).

Glabromyzus rhois (Monell) Plate 122

Medium-sized yellow-brown aphids with black clavate siphunculi, living on undersides of leaves of sumac (*Rhus glabra, R. typhina*). Alatae have a brownish-yellow or greenish-yellow abdomen with dusky markings. Monoecious holocyclic, with alate males, on *Rhus*, or possibly with a partial migration to Gramineae. Other *Glabromyzus* species overwinter on *Rhus* in North America, but it is not clear whether any other species utilize sumac; the host-plant relationships of the group require further clarification (see Hille Ris Lambers, 1966c).

Glyphinaphis van der Goot   Hormaphidinae: Cerataphidini

A monotypic genus characterized by the absence of frontal horns and the spine-like body hairs of the apterous morph.

Glyphinaphis bambusae van der Goot

Small, pale olive-green to reddish-brown aphids on young shoots or at bases of leaves of bamboos (*Bambusa* spp., *Pleioblastus latiflorus*); also recorded from *Thysanolaena maxima*. Alatae, according to the original description, have a bluish-black abdomen. Recorded from India, China, Taiwan, Japan, and Singapore. Little has been written about the biology since van der Goot (1917/18, p. 232).

Greenidea Schouteden
Greenideinae: Greenideini

A genus of about 45 eastern Asiatic species extending from Japan to eastern Australia and from India to the Philippines. They live mainly on the shoots and young foliage of trees, particularly Fagaceae, but also Moraceae, Betulaceae, Juglandaceae, and more rarely on members of evolutionarily more recent families such as Myrtaceae, Rosaceae, and Rubiaceae. In the few species for which life-cycle information is available the sexuales occur in spring or summer, both sexes are winged and the eggs are depressed and flattened; but the sexual stages of many of the more common and widespread species are unknown and they are presumed to be mainly anholocyclic. The genus was reviewed by Raychaudhuri (1956) and more recent accounts are available for Japan(Takahashi, 1962), Indian fruit trees (A.K. Ghosh, 1975) China (Zhang and Zhong, 1979), and Taiwan (Liao, 1978; Tao 1962b).

Greenidea (Trichosiphon) anonae (Pergande)

Small aphids with a dark brown abdomen and yellow-brown head and thorax, living on undersides of young leaves of *Anona* species (also recorded from *Artabotrys, Antidesma, Baccauria, Polyalthia, Symplocos*). The siphunculi are black, curved outwards apically, about one-fifth of body length in apterae and almost one-half of body length in alatae. Sometimes ant-attended. In India, Japan, Indonesia, and Malaysia.

Greenidea artocarpi (Westwood)

Medium-sized, rather long-bodied pale brownish-green aphids feeding on undersides of young leaves, usually near the main trunk, of *Artocarpus* spp. (*heterophylla, incisa*). The siphunculi are brown black and very long and slender; almost half as long as body in apterae, and nearly as long as body in alatae. In southern India and Sri Lanka. Records of *G. artocarpi* from South East Asia probably apply to *G. ficicola*. David (1956) provides some biological notes.

Greenidea decaspermi Takahashi

Apterae are rather small, pear-shaped aphids, yellowish brown with dorsal abdomen darker brown to black. Alatae have a more elongate body. Siphunculi are black except at their apices, about two-fifths of body length in apterae and about three-quarters of body length in alatae. Described from *Decaspermum fruticosum* in Taiwan, but David et al. (1969) record it from *Psidium guajava* in India. Biology is not known.

Greenidea ficicola Takahashi
Plate 131

Apterae are small to medium sized, yellowish brown to dark brown, with pear-shaped body. The dark brown, hairy siphunculi are very long (at least one-third of body length) and curved outwards apically. Alatae have an elongate body with dark brown abdomen and siphunculi about two-thirds of body length. The aphids feed on *Ficus* spp, on the undersides of young leaves and on shoots, but sometimes concentrate on the fruits. In India, Bangladesh, Nepal, eastern Russia, China, Taiwan, Japan, Philippines, Indonesia, Malaysia, and Australia. Biology is little known; presumably mainly anholocyclic, since no sexual morphs have been recorded. Occasionally visited by ants. $2n = 22$.

Greenidea (Trichosiphon) formosana Maki

Dark brown aphids with pear-shaped body and long yellowish-brown siphunculi, about one-third of body length and curved outwards apically. Alatae are more narrow-bodied with longer, darker siphunculi about three-quarters of body length. Feeding on young shoots and undersides of young leaves of *Psidium guajava* and other Myrtaceae (*Rhodomyrtus, Eugenia, Melaleuca*). In India, Bangladesh, Nepal, China, Taiwan, Japan, Loochoo, Java, Sumatra, and the Philippines. Biology is little known and sexual morphs are unrecorded. Takahashi (1923, 1924) provides some biological notes and descriptions of immature stages. $2n = 18$ (Kulkarni and Kacker 1979).

Greenidea mangiferae Takahashi

Apterae are yellowish brown to dark brown with pear-shaped body and very long black siphunculi about one-half of body length. Alatae are quite narrow-bodied with siphunculi almost three-quarters of body length. Only known from Taiwan, feeding on *Mangifera indica* and *Euphoria longana*. Biology is unknown.

Hyadaphis Kirkaldy   Aphidinae: Macrosiphini

A palaearctic genus of about 15 described species associated with Caprifoliaceae and/or Umbelliferae. The best-known species alternate between *Lonicera* and the aerial parts of Umbelliferae. Morphologically, *Hyadaphis* are similar to *Lipaphis*, and both these genera were placed in *Rhopalosiphum* in some earlier accounts. There is no satisfactory account of the world fauna.

Hyadaphis coriandri (Das)   Plate 62

Appearance in life: Apterae are rather small, broadly oval, rather short-legged aphids, mainly dirty greenish in colour with dark green dorsal mottling and rust-red patches around the bases of the short, dark brown siphunculi; in life the body is variably dusted with white mealy wax. Alatae have a pale green abdomen with black dorsal markings, and reddish-brown patches around the siphuncular bases. Apterae and alatae 1.3–2.1 mm.

Host plants: Apparently capable of colonizing numerous species of Umbelliferae (*Anethum, Carum, Coriandrum, Cuminum, Daucus, Foeniculum, Pimpinella*); *Coriandrum* is particularly susceptible to attack. Bhargava *et al.* (1971) screened varieties of fennel for resistance to *H. coriandri*. It is also occasionally found colonizing plants outside the Umbelliferae (*Mentha longifolia, Amaranthus spinosus, Glycine max*).

Distribution: Portugal, Spain, the Mediterranean region, the Middle East, Central Asia, India, Pakistan, and Africa.

Biology: Probably anholocyclic over much of its range. A holocycle involving an unknown alternative host is suspected in northern India, where alate males occur in January (Das, 1918). The aphids live mainly in the umbels of the host plant. $2n = 13$ and $14*$.

Hyadaphis foeniculi (Passerini) Honeysuckle Aphid Plate 61

Appearance in life: Apterae are small to medium-sized, predominantly greyish green with dark antennae, legs, siphunculi, and cauda; the body variably dusted with fine white wax. Alatae have the abdomen green mottled with darker green and with a darker patch around the base of each siphunculus. Apterae and alatae 1.3–2.3 mm.

Host plants: Primary hosts are *Lonicera* spp. (*caprifolium, ciliosa, etrusca, japonica, periclymenum, xylosteum*). Secondary hosts are numerous species of Umbelliferae (*Aethusa, Angelica, Anthriscus, Apium, Carum, Cicuta, Conium, Daucus, Foeniculum, Levisticum, Pastinaca, Peucedanum, Pimpinella, Sium, Torilis*).

Virus transmission: A vector of about 12 plant viruses including celery crinkle leaf mosaic, celery mosaic, and celery yellow spot. Apparently not a vector of carrot motley dwarf.

Distribution: Of palaearctic origin, now almost cosmopolitan; Europe, Middle East, southern Africa, India, Australia, New Zealand, and North and South America.

Biology: Heteroecious holocyclic, migrating from *Lonicera* to Umbelliferae, where it forms colonies in the flower heads. It is likely, as suggested by Börner (1952), that the name as used here covers a complex of morphologically similar species in Europe, perhaps with different *Lonicera* species as primary hosts. More than one of these taxa may have become distributed around the world. For accounts of this aphid see Cottier (1953), Heinze (1960), or Tremblay (1963). $2n = 12$ and 14 (probably representing two different taxa).

Hyalomyzus Richards   Aphidinae: Macrosiphini

A genus of about 6 species of *Myzus*-like aphids, the alatae not having the black dorsal abdominal patch which is typical of *Myzus*.

Hyalomyzus eriobotryae (Tissot)

Rather small to medium-sized brown aphids overwintering and forming spring colonies on loquat (*Eriobotrya japonica*) in south-eastern U.S.A. It occurs in eastern North America from Florida to Ontario and is also recorded from Washington and North Dakota. Heteroecious holocyclic, migrating from *Eriobotrya* and *Crataegus* (which is presumably the native primary host) to *Lycopus virginicus* (Smith, 1960b).

Hyalopterus Koch   Aphidinae: Aphidini

A genus now restricted to 2 or more species related to *Rhopalosiphum* but with smaller siphunculi, alternating between *Prunus* spp. and *Phragmites*. The many only distantly related species with small siphunculi formerly placed in *Hyalopterus* are now distributed in other genera. The precise number of species of *Hyalopterus* is uncertain. Specimens from plums and close relatives tend to have more slender siphuncili and smaller lateral abdominal tubercles than those from apricot, peach, and almond, and these two forms are generally regarded as distinct species, although the differences are not always consistent on a world-wide scale. Here they are most conveniently treated together:

Hyalopterus amygdali (Blanchard)   Mealy Peach Aphid

Hyalopterus pruni (Geoffroy)   Mealy Plum Aphid   Plate 14

Appearance in life: Apterae are small to medium-sized rather elongate aphids, pale green mottled with darker green, covered with white wax meal, on the undersides of leaves of *Prunus* in spring. Infested leaves do not curl. Alatae have a green abdomen with white wax patches on each segment. Apterae and alatae 1.5–2.6 mm.

Host plants: Primary hosts of *H. pruni* are *Prunus* spp. including *armeniaca, amygdalus, domestica, domestica* x *insititia, persica,* and *spinosa*. Primary hosts of *H. amygdali* are *Prunus amygdalus* and *P. persica* (including possibly var. *nectarina*). Secondary host of both species is usually *Phragmites communis,* sometimes *Arundo donax*.

Virus transmission: *H. pruni* is a 'weak' vector of plum pox virus (Miniou, 1973), and is also able to transmit the persistent virus millet red leaf.

Distribution: *H. pruni* is cosmopolitan, but may have geographical races or subspecies, e.g. in South East Asia (Eastop, 1966). *H. amygdali* occurs in Europe, the Mediterranean region, the Middle East, and Central Asia to Pakistan.

Biology: The biologies of both species are similar, with alternation from *Prunus* to *Phragmites*. There have been detailed studies of the life history and population biology of *H. pruni* in California (L.M. Smith, 1936), Japan (Shiga, 1975), Norway (Stenseth, 1970), Egypt (El-Kady *et al.*, 1977), and Iraq (Starý, 1970; including influence of parasitism on populations). Populations of *Hyalopterus* on *Phragmites* along the rivers of tropical Africa, e.g. the Gambia, are presumably entirely parthenogenetic. $2n = 10$.

Hyperomyzus Börner   Aphidinae: Macrosiphini

A genus of about 17 species associated with *Ribes* as primary hosts, most of them having liguliflorous Compositae as secondary hosts, but some (subgenus *Hyperomyzella*) alternate to Scrophulariaceae. Except for the swollen siphunculi they resemble *Nasonovia*, which have a similar biology. *Hyperomyzus s.str.* (4 species) and the subgenus *Hyperomyzella* are of palaearctic origin while 8 of the 12 known species of the subgenus *Neonasonovia* are American. Hille Ris Lambers (1949), Heinze (1961), and Müller (1969) give accounts of the European species, and Hille Ris Lambers (1974) gives further information on some species of the subgenus *Neonasonovia*. The American species are mostly found in accounts of *Amphorophora*, in which genus many of the species were originally described.

Hyperomyzus lactucae (L.)   Blackcurrant – Sow Thistle Aphid   Plate 88

Appearance in life: Apterae are medium-sized, broadly spindle-shaped, opaque green aphids with pale legs and siphunculi, feeding on the undersides of young leaves of *Ribes* spp., which curl slightly and acquire yellow spots as a result of the aphids' feeding. Alatae have a rather broken central dark patch on the dorsal abdomen. Apterae and alatae 2.0–2.7 mm.

Host plants: Primary hosts are *Ribes* species, primarily *R. nigrum*, occasionally other species including *R. rubrum*. Keep and Briggs (1971) did extensive surveys to establish the host range in England. Secondary hosts are *Sonchus* spp., especially *S. oleraceus*. *Lactuca sativa* is not colonized.

Virus transmission: A vector of about 12 non-persistent viruses (but not of currant viruses) including lettuce mosaic, and the persistent viruses lettuce necrotic yellows (Boakye and Randles, 1974) and sowthistle yellow vein (Sylvester and McClain, 1978) which have both been shown to be transmitted transovarially.

Distribution: Palaearctic in origin, now widespread in Europe, Mediterranean, the Middle East, Central Asia, India, Pakistan, Japan, Australia, New Zealand, and North and South America.

Biology: Heteroecious holocyclic in Europe, migrating from *Ribes* to *Sonchus*, apparently anholocyclic on *Sonchus* in warmer parts of the world. Hille Ris Lambers (1949) gives a general account. Singer and Smith (1976) studied spring populations on *R. nigrum* and the effects on them of plant growth regulators. Randles and Crowley (1970) studied migratory activity from *Sonchus* in relation to seasonal incidence of lettuce necrotic yellows. $2n = 12$. In the Old World tropics, *H. lactucae* is replaced on ligulatiflorous Compositae by another species, *H. carduellinus* (Theobald), which has more slender siphunculi, shorter hairs on the extremities, and a wider host-plant range.

Hyperomyzus pallidus Hille Ris Lambers   Gooseberry – Sow Thistle Aphid

Appearance in life: Apterae are opaque greenish or yellowish white with pale appendages, forming colonies at shoot tips and on undersides of young, curled leaves of gooseberry in spring. Alatae are rather pale with a large solid black trapezoid patch on the dorsal abdomen. Apterae 2.1–3.0 mm, alatae 2.3–2.7 mm.

Host plants: Primary hosts are *Ribes* species, primarily *grossularia*, occasionally *alpinum* but not *nigrum* or *rubrum*. Keep and Briggs (1971) surveyed the *Ribes* host range in England. Secondary host is apparently always *Sonchus arvensis*.

Virus transmission: Has not been tested.

Distribution: Europe eastwards to Moldavia, and introduced into the U.S.A.

Biology: Heteroecious holocyclic between *Ribes* and *Sonchus* in Europe and North America. Karczewska and Stasiak (1973) have studied the life cycle and population ecology in Poland.

Hyperomyzus (Neonasonovia) picridis (Börner & Blunck)

Shiny green to yellow-green aphids with dusky, markedly swollen siphunculi, forming spring colonies on *Ribes alpinum*, rarely on gooseberries, migrating for the summer to *Picris* spp. (*echioides, hieracioides*). Hille Ris Lambers (1949) gives a general account. $2n = 12^*$.

Hyperomyzus (Hyperomyzella) rhinanthi (Schouteden)
                             Currant – Yellow Rattle Aphid   Plate 89

Appearance in life: Apterae are yellowish to yellowish-green aphids with a large black dorsal abdominal patch and black, strongly swollen siphunculi, on young shoots and suckers of redcurrant in spring, curling and distorting the leaves into more-or-less distinct 'leaf-nests'. Alatae have black transverse bars on the dorsal abdomen. Apterae 2.4–2.9 mm, alatae 2.1–2.6 mm.

Host plants: Primary host is *Ribes rubrum*, occasionally recorded in autumn on other *Ribes* species. Secondary hosts are *Rhinanthus* spp. (especially *major*) and *Euphrasia* spp. Recorded from *Fragaria vesca* in Bulgaria.

Distribution: Throughout Europe and Iceland, mostly in cool, moist situations.

Biology: Heteroecious holocyclic between *Ribes* and Scrophulariaceae in Europe. On the summer host the aphids live in the flowers (Hille Ris Lambers, 1949). Rather uncommon and therefore of little economic importance in continental Europe, but common in Iceland where it can persist throughout the short summer on redcurrant (Heie, 1964). $2n = 12^*$.

Hyperomyzus (Neonasonovia) ribiellus (Davis)   Ornamental Currant Aphid

Pale yellowish-green aphids with pale appendages and slightly swollen siphunculi colonizing shoot apices and young leaves of ornamental *Ribes* (*aureum, odoratum*), rarely on other *Ribes* species. Monoecious holocyclic with alate males. Only in North America. $2n = 12$.

Hysteroneura Davis

Aphidinae: Aphidini

A monotypic genus characterized by the hind wing with a single oblique vein, the long terminal process of the antenna and the long pale cauda.

Hysteroneura setariae (Thomas)   Rusty Plum Aphid   Plate 20

Appearance in life: Small, brown aphids with dark siphunculi and a pale cauda, usually forming colonies at the bases of the spikelets of Gramineae, sometimes on leaves or unripe seeds, often attended by ants. Alatae have a greenish-grey abdomen. Apterae and alatae 1.3–2.1 mm.

Host plants: Primary host in North America is usually *Prunus domestica*. Secondary hosts are numerous species of Gramineae including *Cynodon, Eragrostis, Eleusine, Hordeum, Oryza, Panicum, Pennisetum, Saccharum, Setaria, Sorghum,* and *Triticum*. Sometimes on Cyperaceae and on seedlings of oil palms and coconuts.

Virus transmission: A vector of sugar cane mosaic virus. Also able to transmit soybean mosaic, onion yellow dwarf, cucumber mosaic, and water-melon mosaic.

Distribution: Of North American origin, now widely distributed throughout the warmer parts of the world; its rapid spread is shown by the fact that in 1967 it was recorded for the first time from Australia, South Africa, and southern India.

Biology: In North America *H. setariae* is heteroecious between *Prunus* and Gramineae. Elsewhere it is mainly anholocyclic on Gramineae. Carver (1976) found sexuales and eggs on *Prunus persica* in South Australia, but it is not known whether the holocycle was completed. The population biology of *H. setariae* on rice in India was studied by Garg and Sethi (1978). $2n = 12$.

Illinoia Wilson   Aphidinae: Macrosiphini

A genus for about 45 North American species and one from the Caucasus (U.S.S.R.). *Illinoia* are mostly characterized by weakly clavate siphunculi bearing distally a few rows of polygonal reticulation. The species from cultivated *Rhododendrum* mostly bear 5 hairs on the first tarsal segments. The subgenus *Masonaphis* contains about 8 species, 6 of them on Ericaceae in North America. *Illinoia s.str.* contains about 30 species, 9 of them (including the Russian one) associated with Ericaceae, with 6 of the species known from eastern North America. Seven species of *Illinoia s.str.* and one *Masonaphis* live on Compositae. The 3–4 species of the subgenus *Oestlundia* are associated with *Rubus* and the 2 subspecies of *Amphorinophora* live on *Lonicera*. MacGillivray (1958) revised *Illinoia* (= *Ericobium*) using the name *Masonaphis* for the full genus.

Illinoia azaleae (Mason)   Plate 118

Medium-sized, spindle-shaped, rather shiny, deep green; on young stems and leaves of rhododendrons, especially *R. indicum* (azaleas). Siphunculi are long, slightly but distinctly clavate, pigmented light to mid-brown distally but with pale bases. Alatae have dark marginal abdominal sclerites and rather dark wing veins. In North America, and introduced into Europe where it is a common pest on potted azaleas, often causing defoliation. Also in New Zealand and South Africa. Besides *Rhododendron* spp., *I. azaleae* occurs occasionally on other Ericaceae (*Vaccinium* spp., *Andromeda palifolia*, *Leucothoe catesbaei*), and it has also been collected from *Viola tricolor* and *Tulipa* sp. Able to transmit cucumber mosaic virus. Life cycle not known; oviparae occur occasionally but no males have been recorded. $2n = 10*$.

Ilinoia borealis (Mason)

Medium-sized, spindle-shaped; colour in life not recorded. Siphuncili long, slightly clavate and usually evenly mid- to dark-brown in colour, not pale at base. On new growth of *Vaccinium* spp., *Gaultheria procumbens* (Ericaceae), and also recorded from *Myrica asplenifolia* (Myricaceae) and *Clintonia borealis* (Liliaceae). In north-western U.S.A. (Pennsylvania, Maine). Biology not known.

Illinoia davidsoni (Mason)

Rather large, pale-coloured, spindle-shaped; siphunculi dark brown, moderately clavate. On *Rubus parviflorus* in western North America, and once recorded from loganberry. Able to transmit thimbleberry ringspot virus. Migration between *Rubus* and *Arnica* (Compositae) is suspected (MacGillivray, 1958); males are alate. $2n = 12$.

Illinoia lambersi (MacGillivray)

Medium-sized to rather large, spindle-shaped, green, pink or yellow; siphunculi paler at base, darker distally, moderately clavate. Easily confused with *I. azaleae*, but *I. lambersi* is on average larger and its different colour forms often occur together, whereas *I. azaleae* is always green. On *Rhododendron* spp. (*indicum, japonicum, molle, ponticum,* etc.), often injurious to young growth of hybrid cultivars. In western North America, and introduced into Europe (England, Denmark, the Netherlands) and South America (Chile). Monoecious holocyclic with alate males in California (Hille Ris Lambers, 1966), method of overwintering in Europe unknown (Hille Ris Lambers, 1973). $2n = 10$ (Gut, 1976).

Illinoia pepperi (MacGillivray)

Medium-sized, green, spindle-shaped; siphunculi dark (sometimes pale at base), distinctly clavate, legs and antennae also mainly dark. On *Vaccinium* spp. (*corymbosum, pennsylvanicum, stramineum*) in north-eastern North America. Monoecious holocyclic, with alate males. Recorded as a vector of blueberry shoestring virus.

Illinoia rhododendri (Wilson)

Medium-sized, green or pink, sometimes with the dorsal cuticle sclerotic and rather smoky; siphunculi rather pale and only very slightly swollen. On *Rhododendron* spp. in western U.S.A. Life cycle uncertain but may be abnormal, with sexual morphs recorded early in the season (MacGillivray, 1958).

Illinoia rhokalaza (Tissot and Pepper)

Medium-sized, yellowish brown, spindle-shaped, with appendages mainly dark brown to shiny black. Siphunculi brown-black, sometimes paler at extreme bases, rather thick, only slightly clavate. Immatures have a waxy bloom. On

Rhododendron spp. (*canescens*, *maximum*) and *Kalmia latifolia* in eastern U.S.A. (Maine to Florida). Life cycle is apparently unknown.

Illinoia rubicola (Oestlund)   Spotted-winged Raspberry Aphid

Large, greenish yellow; apterae often with a broad dark green longitudinal mid-dorsal stripe. Siphunculi conspicuously clavate, often dusky to dark. Alatae have a dark pterostigma and a dark spot at the apex of each forewing. On *Rubus* spp., especially *R. idaeus strigosus*, widely distributed in North America. A vector of raspberry (black) necrosis, but not of raspberry leaf curl. Monoecious holocyclic, with alate males. $2n = 12$.

Illinoia wahnaga (Hottes)

Medium-sized, pale whitish green; siphunculi pale except at apices. Described originally from *Smilacina* in Colorado, U.S.A.; found by Robinson (1965) in large numbers on *Convallaria majalis* in Manitoba. Monoecious holocyclic on *Convallaria*, with apterous males. $2n = 10$.

Indiaphis Basu   Aphidinae: Macrosiphini

A genus of 3 described species in India, 2 of which live on *Rhododendron* spp. They are small to medium-sized aphids with frontal tubercles weakly developed or absent, no distinct ocular tubercles (triommatidia) and stout, almost flangeless siphunculi. Apterae of *I. crassicornis* Basu are very pale, glassy white, dusted with wax, with 6-segmented antennae and pale brown siphunculi, curved outwards apically. Apterae of *A. rostrata* are pale yellow and have 5-segmented antennae (Ghosh and Raychaudhuri, 1972). Both are recorded from unnamed *Rhododendron* species in north-east India.

Indomasonaphis Verma   Aphidinae: Macrosiphini

A monotypic oriental genus near *Masonaphis* but without secondary sensoria on the antennae of apterae, with dorsal abdominal hairs arising from tubercles, and with a very hairy cauda.

Indomasonaphis indica Verma

Rather large, elongate oval, lemon yellow, with pale brown clavate siphunculi, on undersides of leaves of *Rhododendron* sp. in north-east India (Verma, 1972).

Indomegoura Hille Ris Lambers

A small oriental genus associated with *Staphylea* as primary host.

Indomegoura indica (van der Goot)

Large, orange yellow, covered in white wax, with stout, dark siphunculi, on leaves of *Hemerocallis* (*aurantiaca, exaltata, middendorffii*), which is the secondary host plant. Migration (in Japan) is to *Staphylea bumalda* and *Euscaphis japonica* (Miyazaki, 1971). However, there is some uncertainty about the life cycle as Takahashi (1923), in describing the biology, recorded apterous sexuparae and both apterous and alate males on the primary host plant. In Japan, Korea, China, Taiwan, and India. $2n = 10*$.

Jacksonia Theobald   Aphidinae: Macrosiphini

A monotypic genus related to *Myzus*, distinguished by the distinctive form of the siphunculi and the large, protruberant secondary sensoria on antennal segments III–V of alatae.

Jacksonia papillata Theobald

Small, brownish or olive green, sometimes dull greenish yellow or reddish; living concealed on colourless basal parts of grass stems. Recorded from *Poa* spp., *Dactylis glomerata* and *Festuca rubra*; perhaps feeding on concealed parts of plants outside the Gramineae, especialy mosses (Müller, 1973). Apparently entirely anholocyclic. Stroyan (1950) gives an account of this species, which is very widely distributed (Iceland, Europe, New Zealand, Antipodes, Auckland Islands, Macquarie Isles, the U.S.A., Argentina, South Georgia Island).

Juncomyzus Hille Ris Lambers   Aphidinae: Macrosiphini

A small oriental genus of about 5 species distinguished from other *Myzus*-like genera by such characters as the low diverging antennal tubercles, absence of a dorsal black patch in the alatae, and frequent presence of sensoria on antennal segment III of apterae. The species are associated with either *Rhus* or *Juncus* (and *Carex*), but the suspected alternation between these two host-plant groups has not yet been demonstrated. Miyazaki (1971) gave an account of the Japanese species.

Juncomyzus rhois (Takahashi)

Small, shiny, yellowish brown with mainly black appendages, on *Rhus trichocarpa* and *R. vernicifera* in Japan. Biology not known. Miyazaki (1971) distinguishes a second blackish-brown species on *R. trichocarpa* under the name *J. niger*.

Kaltenbachiella Schouteden   Pemphiginae: Eriosomatini

A genus of about 4 species associated with *Ulmus* as primary hosts and

Labiatae as secondary hosts, and a fifth species recently described from galls on *Carpinus*. A.K. Ghosh (1981) reviewed and keyed the species.

Kaltenbachiella pallida Haliday

Very small (0.9–1.3 mm), yellowish-white aphids secreting flocculent wax, living in summer on subterranean parts of Labiatae (*Mentha, Galeopsis, Origanum, Thymus*). Immature stages are pale orange yellow. In autumn sexuparae migrate to *Ulmus* spp. (*campestris, glabra*), where in spring the fundatrices produce pale rounded galls clothed densely with short hairs, projecting from both sides of the leaf at the base of the mid-rib. When the gall is mature alatae leave it through a star-like aperture which forms at the top. In Europe, North Africa, Central Asia, and the Middle East. $2n = 28$.

Lachnus Burmeister   Lachninae: Lachnini

A genus of about 14 Old World and 1 New World species with characteristically pigmented wings, usually associated with Fagaceae and attended by ants. It is a taxonomically difficult genus as indicated under *L. roboris*. Accounts of the European fauna are available by Heinze (1962) and Szelegiewicz (1962), of the Middle Eastern fauna by Bodenheimer and Swirski (1957), and of the Japanese fauna by Sorin (1980). Until about 1930 the species of *Lachnus* were often placed in *Pterochlorus* and the name *Lachnus* applied to the large conifer-feeding genus *Cinara*.

Lachnus roboris (L.)   Plate 3 (left)

Large, dark blackish-brown, rather shiny aphids with very long legs giving them an almost spider-like appearance, on twigs of *Quercus* spp., and sometimes in large numbers on *Castanea vesca* in south-central Europe. Alatae have maculate forewings. Invariably ant-attended. Monoecious holocyclic with alate males; Michel (1942) gave a detailed account of the biology on oaks. Hille Ris Lambers (1956b) synonymized several *Lachnus* described from various *Quercus* species under the name *roboris*, considering variations in length and shape of body hairs and the small-bodied, long-legged forms which occur in summer in southern Europe to be environmentally induced. It nevertheless seems likely that *L. roboris* is a complex of species with different host-plant associations and karyotypes (in which case the name proposed for the southern European lachnid on *Castanea* is *L. castaneae*; Hille Ris Lambers, (1967a). *L. roboris s. lat.* occurs in Europe, eastwards to the Ukraine, and in Morocco. Chromosome numbers of $2n = 7, 8, 10, 11, 13, 16$ and 17 have been recorded; a sample from *Castanea* in Portugal had $2n = 10$ (Blackman)*.

Lachnus tropicalis (van der Goot)   Plate 3 (right)

Large, shiny brown-black long-legged aphids on twigs of *Quercus* spp. and

Castanea spp. in India and South East Asia. Alatae have maculate forewings with a greater area pigmented than in *L. roboris*. Ant-attended. Monoecious holocyclic with alate males. $2n = 12$, 13, or 16*.

Lipaphis Mordvilko   Aphidinae: Macrosiphini

A genus of about 10 western palaearctic species associated with Cruciferae, characterized by the weakly developed antennal tubercles, absence of secondary sensoria from the antennae (of apterae) which are about half as long as the body, and the weakly swollen siphunculi. Closely related to *Brevicoryne*; both these genera differ from *Hyadaphis* in usually having only 2 hairs on the first segment of the hind tarsus. Only one species occurs outside the western palaearctic and Middle East. Accounts are available by Doncaster (1954), Heinze (1960), and Prior (1971).

Lipaphis erysimi (Kaltenbach)   Turnip Aphid, Mustard Aphid   Plate 63
(= *Rhopalosiphum pseudobrassicae*)

Appearance in life: Apterae are small to medium-sized, yellowish green, grey green, or olive green, with a white wax bloom; in humid conditions often more densely coated with wax. Alatae have a dusky green abdomen with conspicuous dark lateral sclerites, and dusky wing veins. Sometimes in large numbers on the undersides of leaves, which may curl and turn yellow, or in inflorescences of host plants. Apterae 1.4–2.4 mm, alatae 1.4–2.2 mm.

Host plants: Many genera and species of Cruciferae (*Barbarea, Brassica, Capsella, Erysimum, Iberis, Lepidium, Matthiola, Nasturtium, Raphanus, Rorippa, Sinapis, Sisymbrium, Thlaspi*, etc.).

Virus transmission: A vector of about 10 non-persistent plant viruses, including cabbage black ring spot and mosaic diseases of cauliflower, radish, and turnip.

Distribution: Virtually world wide: C.I.E. Distribution Map A203 (1965).

Biology: A holocycle has been observed on radish in Japan, with apterous males (Kawada and Murai, 1979). Sexuales have also been reported from Europe, India, and New Zealand, but anholocycly predominates, at least in warmer climates. There is a large literature. Paddock (1915) gives a detailed account of biology in Texas, U.S.A. (as *Aphis pseudobrassicae*). Lal (1969) searched for varietal resistance to *L. erysimi* in rape and mustard in India, and Prasad and Phadke (1980) studied population dynamics on different *Brassica* species and varieties. The European form of *L. erysimi* is not normally a pest of cabbages. See Cottier (1953) for a short general account. $2n = 6$, 8, 9, and 10 (reasons for the variation in chromosome number are uncertain; possibly more than one taxonomic entity is involved).

Longicaudus van der Goot   Aphidinae: Macrosiphini

A genus for two species from Europe and one from the Himalayas, with short siphunculi, long hairy cauda, and alatae with a long third antennal segment, equal in length to IV + V + VI and bearing numerous (60–110) tuberculate sensoria. Hille Ris Lambers (1965) discussed the species of *Longicaudus* and their taxonomic affinities.

Longicaudus trirhodus (Walker)   Plate 72

Apterae on rose are rather elongate oval, medium-sized (2.0–2.7 mm), yellowish green to pale apple green, lightly wax-dusted, easily identified by the combination of very short siphunculi and a long hairy cauda. Alatae have an irregular-shaped black patch on the dorsal abdomen. In small colonies on leaves and blossom buds of roses in spring, migrating for the summer to *Aquilegia* and *Thalictrum* species (Ranunculaceae). Throughout Europe, Mongolia, Japan, and North America. $2n = 12$ (Gut, 1976).

Longistigma Wilson   Lachninae: Lachnini

A genus of 2 or 3 species, one in North America and the other(s) in the Himalayas and Taiwan. Very large aphids characterized in alatae by the elongate pterostigma extending around the tip of the forewing. Bark feeders on many species of trees.

Longistigma caryae (Harris)   Giant Bark Feeder   Plate 1

Appearance in life: Apterae pale brownish grey with a slight bluish-white wax bloom; with conspicuous rows of dark dorsal spots and dark siphunculi. On bark of host plants. Apterae 5.1–6.6 mm, alatae 4.7–7.2 mm.

Host plants: Bissell (1978) tabulates 292 host plant records, from tree species in 24 genera and 16 families (but especially *Acer, Carya, Castanea, Juglans, Platanus, Pyrus, Quercus, Salix, Tilia,* and *Ulmus*).

Distribution: Throughout the U.S.A., except the north-west; also recorded from Manitoba and Ontario, Canada. A fossil specimen identified as this species has been found in Iceland (Heie and Friedrich, 1971).

Biology: Monoecious holocyclic in northern U.S.A.; Wilson (1909) describes the life cycle on *Carya* in Washington D.C. Anholocyclic in more southerly states (Bissell, 1978). Marked fluctuations in population size occur from year to year; in some years the aphid is very abundant.

Macrosiphoniella del Guercio

A genus of about 115 Old World and 5 North American species, often having cauda and siphunculi about the same length, with the siphunculi reticulated over the distal half. Most species feed on Anthemidae and have a stilletto-shaped last rostral segment. At least half the species feed on *Artemisia* and 6–8 species are known from each of *Achillea, Aster, Helichrysum, Centaurea*, and other Cynareae, and fewer from *Chrysanthemum* and other Anthemidae. Host alternation does not occur in *Macrosiphoniella*. The species are found throughout the Northern Hemisphere but have a distinctly 'continental' distribution, about one-quarter of them being described from Central Asia, western Siberia, and Mongolia. Accounts are available for western Europe (Hille Ris Lambers, 1938), European U.S.S.R. (Shaposhnikov, 1964), Central Asia (Holman and Szelegiewicz, 1974, 1978; Ivanovskaya, 1977; Szelegiewicz, 1980), Korea (Szelegiewicz, 1980), Japan (Miyazaki, 1971), China (Tao, 1963), and India (Basu and Raychaudhuri, 1976b). The North American species are mostly found in earlier accounts of *Macrosiphum* (e.g. Hottes and Frison, 1931; Palmer, 1952).

Macrosiphoniella oblonga (Mordvilko)

Large (3–5 mm), narrow bodied, mid-green to pale green with a darker green mid-dorsal stripe, with relatively short siphunculi similar in length to the long cauda, and very long, thin legs. Monoecious holocyclic on *Artemisia* spp., with very narrow-bodied, apterous males. Also on cultivated *Chrysanthemum* spp. (*indicum, morifolium, frutescens*) in glasshouses. Lives singly on undersides of older leaves, sometimes on flower stems. Throughout Europe and in Japan (Miyazaki, 1971). Bochen (1978) studied its biology in Poland. Indian records probably apply to *M. paraoblonga* Basu and Raychaudhuri (Raychaudhuri, 1980, p. 173). $2n = 12$ (Gut, 1976).

Macrosiphoniella sanborni (Gillette)   Chrysanthemum Aphid   Plate 117

Appearance in life: Apterae are shiny, dark red-brown to blackish brown, broadly spindle-shaped, with black, relatively short and thick siphunculi, shorter than the black cauda. Alatae are similarly pigmented. Apterae 1.0–2.3 mm, alatae 1.8–2.6 mm.

Host plants: Usually cultivated *Chrysanthemum* spp. (*indicum, morifolium, frutescens*), on undersides of leaves. In South East Asia it sometimes occurs on other Compositae (*Artemisia, Aster*).

Virus transmission: Able to transmit chrysanthemum vein mottle and chrysanthemum virus B.

Distribution: Of east Asian origin, not distributed world-wide.

Biology: Anholocyclic on *Chrysanthemum*; no sexual morphs are known. It overwinters in glasshouses in colder regions. Tamaki and Allen (1969) studied population dynamics and competition with *Aphis gossypii* on glasshouse chrysanthemums. $2n = 12$.

Macrosiphoniella tanacetaria (Kaltenbach)

Pale grey green, dusted with fine wax powder, with black antennae, legs, siphunculi, and cauda; medum sized to large, broadly spindle-shaped. Principal host plant is *Tanacetum vulgare*, on which it is monoecious holocyclic with alate males, but the range of reserve hosts seems greater than in most *Macrosiphoniella*, including records from other *Tanacetum* spp., species of *Achillea, Aster, Chrysanthemum,* and *Matricaria*, and also *Salvia officinalis* (Labiatae). On *Tanacetum* in summer, the flowers and flower stems are colonized. Throughout Europe, Morocco, Israel, and North and South America. The form in the Mediterranean region has siphunculi longer relative to the cauda and may be a subspecies; this appears to be the form (*bonariensis* Blanchard) that has been introduced into South America. $2n = 12$.

Macrosiphoniella yomogifoliae (Shinji)

Deep mid-green, dusted lightly with wax powder, with mainly dark antennae and legs, black siphunculi and brown cauda. On *Artemisia* spp. and *Chrysanthemum* spp. (*morifolium, nipponicum*) in South East Asia (Japan, Korea, China, Taiwan, Malaya). This species is closely related to, and may be regarded as a subspecies of, the European *M. artemisiae* (Boyer de Fonscolombe).

Macrosiphum Oestlund   Aphidinae: Macrosiphini

A genus of about 120 species with elongate legs and antennae, with long siphunculi having polygonal reticulation over the distal 5–30% (mostly 8–18% in apterae and 15–25% in alatae), with hairs of medium length, and with little dorsal abdominal pigmentation. Several of the well-known species (*rosae, euphorbiae, pallidum*) alternate from *Rosa* to herbaceous secondary hosts but most species are monoecious on a wide variety of herbs and shrubs. As far as known most species have alate males. About half the species are described from North America, although some of these have seldom or never been recognized again and may really be synonyms of better known species. About 36 species are known from Europe and the remainder are from Central Asia, India, and the Far East. Three South American species may also belong in *Macrosiphum*. *Sitobion*, often treated as a subgenus of *Macrosiphum*, is here treated as a separate genus. Accounts of *Macrosiphum* are available for Europe (Hille Ris Lambers, 1939), Germany (Müller, 1969), Switzerland (Meier, 1961), Central Asia (Nevsky, 1929; Narzikulov and Umarov, 1969), Japan (Miyazaki, 1971),

Korea (Paik, 1965), China (Tao, 1963), and India (David, 1976). North American *Macrosiphum* are reviewed by Palmer (1952—Rocky Mountain region), Hottes and Frison (1931—Illinois), Patch (1919—eastern U.S.A.), and Soliman (1927—California). MacGillivray (1968) gives more recent information on some North American species.

Macrosiphum albifrons Essig   Lupin Aphid

Large (3.2–4.5 mm), pale bluish-grey-green aphids, dusted with white wax, on stems and leaves of *Lupinus* spp. Adult apterae have pale brownish siphunculi, whereas the siphunculi of immatures are uniformly dark. Widespread in North America except in south-eastern U.S.A., and introduced into England (Stroyan, 1981). Monoecious holocyclic, with alate males, in North America, but the population in England seems to be anholocyclic. Frazer and Gill (1981) have studied developmental parameters of *M. albifrons* in Canada. $2n = 10$.

Macrosiphum centranthi Theobald

Apterae are medium-sized to rather large (2.0–3.6 mm), broadly spindle-shaped, whitish green, yellowish green to mid-green with darker spinal stripe; on leaves and stems of Valerianaceae (*Centranthus, Valeriana*), but also sporadically on various other plants in summer. Recorded from England, Switzerland, Mozambique, South Africa, and India. Indian populations seem more polyphagous, with records from *Chrysanthemum, Cineraria, Mangifera, Rosa*, etc. (David, 1976). Monoecious holocyclic, with alate males, on Valerianaceae in Switzerland (Meier, 1961). Life cycle elsewhere is unknown, possibly mainly anholocyclic. It may be distinguished from *M. rosae* by the pale head, and from *M. euphorbiae* by the dark brown (as opposed to reddish) eyes, longer and more conspicuous body hairs, and longer antennae (usually more than 1.5 times body length). Müller and Schöll (1958) give an account of *M. centranthi* in South Africa. $2n = 10*$.

Macrosiphum cockerelli Hottes   Hottes' Green Goldenglow Aphid

Pale green, medium-sized to large aphids; monoecious holocyclic, with alate males, on *Rudbeckia montana* in western U.S.A. (Colorado, Idaho, Utah). Hottes' (1949) account under the name *M. rudbeckiarum* applies to this species.

Macrosiphum (Neocorylobium) coryli Davis

Small to medium-sized, pear-shaped, pale green with reddish or brownish head and prothorax; antennae and siphunculi mainly dark, cauda very pale. On undersides of leaves and at shoot apices of *Corylus americana* and *C. cornuta*, especially in shaded situations. Widespread in North America. Life cycle unknown.

Macrosiphum (Neocorylobium) corylicola Shinji

Medium sized, spindle-shaped, dull yellow to yellowish green, often dark brown dorsally; antennae and siphunculi mainly black, cauda dusky. On *Corylus* spp. (*heterophylla, sieboldiana*) and *Carpinus laxiflora* in Japan. For further details see Miyazaki (1971). Life cycle unknown.

Macrosiphum euphorbiae (Thomas)   Potato Aphid   Plate 109
(= *solanifoli* Ashmead)

Appearance in life: Adult apterae medium-sized to rather large, spindle-shaped or pear-shaped, usually some shade of green but sometimes yellowish, pink, or magenta, often rather shiny. Eyes are distinctly reddish. Legs, siphunculi, and cauda mainly same colour as body, but siphunculi often darker towards apices; antennae usually only dark apically, but sometimes almost entirely dark. Immatures rather long bodied, paler than adults with a dark spinal stripe and a light dusting of whitish-grey wax. Alatae have pale greenish to yellow-brown thoracic lobes, and usually only the antennae and siphunculi noticeably darker than in the aptera. Apterae 1.7–3.6 mm, alatae 1.7–3.4 mm.

Host plants: Primary host *Rosa* spp.; highly polyphagous on secondary hosts, feeding on over 200 plant species in more than 20 different families. Solanaceae, especially *Solanum tuberosum*, are particularly favoured secondary hosts.

Virus transmission: A vector of over 40 non-persistent viruses and 5 persistent viruses including beet yellow net, pea enation mosaic, pea leaf roll, and potato leaf roll. However, it appears to be unimportant as a vector of potato leaf roll under field conditions in comparison with *Myzus persicae* (Robert, 1971).

Distribution: Apparently of North American origin, now almost world wide, although it has only recently spread through Central Asia and the Middle East, and has not yet been recorded from the Indian subcontinent (except Ceylon).

Biology: Heteroecious holocyclic in north-eastern U.S.A. with wild or cultivated *Rosa* spp. as primary host plants. Shands *et al.* (1972) studied the ecology of populations on *Rosa palustris* in Maine. In Europe and probably elsewhere *M. euphorbiae* is mainly anholocyclic, although sexual morphs are sometimes produced in small numbers and the holocycle may sometimes occur (Möller, 1971b). The literature on *M. euphorbiae* is very large. Meier (1961) provides a general account of the aphid in Europe. Barlow (1962) studied its development on potato, MacGillivray and Anderson (1964) studied the factors controlling sexual morph production. Parasites and hyperparasites were studied in North America by Shands *et al.* (1965) and Sullivan and van den Bosch (1971). $2n = 10$.

Macrosiphum funestum (Macchiati) — Scarce Blackberry Aphid

Apterae are medium-sized to rather large (1.9–4.0 mm), spindle-shaped; rather dull mid- to dark green, or magenta to reddish brown; siphunculi are dark brown to black, sometimes with a paler base, and the antennae are also often quite dark. On young shoots and leaves of *Rubus* spp., especially *R. fruticosus*, rarely on *R. idaeus*. There are also records from *Galium* sp. and *Geranium robertianum*. Monoecious holocyclic on *Rubus*, with alate males; probably partly anholocyclic in areas with mild winters. Throughout Europe, and eastwards to Moldavia and Turkey. *M. funestum* can be distinguished from *M. euphorbiae* and similar species by its hairy last rostral segment (15–23 accessory hairs). Dicker (1940) provides biological observations (under the name *rubifolium* Theobald). $2n = 10$.

Macrosiphum gei (Koch)

Usually rather large (1.9–4.5 mm), elongate spindle-shaped, dull mid- to blue-green or purple to wine red. Monoecious holocyclic with alate males, on *Geum* spp. (especially *urbanum*), but in summer also colonizing several species of Umbelliferae (*Anthriscus, Chaerophyllum, Myrrhis, Torilis*). Occurs throughout Europe, and possibly in North America, although most North American records (e.g. Hottes and Frison, 1931) refer to *M. euphorbiae*. *M. gei* can be distinguished from *M. euphorbiae* by its long body hairs (e.g. those on head 55–98 μm long, as opposed to 28–48 μm long in *M. euphorbiae*). $2n = 10$ (Gut, 1976).

Macrosiphum hamiltoni Robinson

Medium-sized, dull pale green aphids on undersides of leaves of wild and cultivated (ornamental) *Humulus lupulus*, so far recorded only from Manitoba, Canada. Apparently monoecious holocyclic on *Humulus* as many oviparae were found, but males are still not described. Very similar to *M. euphorbiae*. See Robinson (1968) for further information.

Macrosiphum lilii (Monell)

Medium-sized to rather large, spindle-shaped, yellow-and-red aphids with dark siphunculi and a very long terminal process to the antenna (more than 6 times longer than base of segment VI). On *Lilium* spp. (and possibly other Liliaceae) in eastern U.S.A. A vector of lily rosette, *Ornithogalum* mosaic and *Tigridia* mosaic viruses. The aphids were described from bulbs imported from Japan (Comstock, 1879), but there have been no subsequent Japanese records. See also Patch (1923, p. 308). Life cycle unknown.

Macrosiphum martorelli Smith

Medium-sized, spindle-shaped, bright yellow-green aphids with siphunculi dark except at base, attacking flowers, young leaves, and pods of *Theobroma cacao* in the Caribbean (Cuba, Dominican Republic, Jamaica). Attended by the ant *Solenopsis geminata* (Smith, 1960a). Life cycle unknown.

Macrosiphum mordvilkoi Miyazaki

Apterae are medium-sized to rather large, green or yellowish green with head and prothorax black, shiny, or slightly waxy. Antennae and siphunculi black. Immatures are paler green, dusted with wax. On undersides of young leaves of *Rosa rugosa* in Japan, Korea, and eastern U.S.S.R. (Primorskaya); not recorded from cultivated roses. Monoecious holocyclic (Miyazaki, 1968a). Closely related to *M. rosae*.

Macrosiphum pachysiphon Hille Ris Lambers

Apterae are rather large, very pale pink with black, rather stout siphunculi. Dorsal body hairs are numerous and conspicuous. Alatae have a dusky sclerotic pattern on the dorsal abdomen. On *Rubus* spp. (especially *lasiocarpus*) and sometimes on other Rosaceae (*Rosa, Potentilla*) in India and Pakistan. Life cycle unknown. $2n = 18$ (Kurl, 1980b).

Macrosiphum pallidum (Oestlund)   Plate 110
(= *pseudorosae* Patch)

Medium-sized to large (2.1–4.6 mm), spindle-shaped, green or pink; usually with antennae, femoral apices, tibiae, tarsi, and most of siphunculus dark, but cauda pale. Head dusky but not black (cf. *M. rosae*). Apterae may be distinguished from those of *M. euphorbiae*, which sometimes have dusky antennae and siphunculi, by the presence of evident lateral tubercles on some abdominal segments. In *M. euphorbiae*, any such tubercles are small and inconspicuous. On wild *Rosa* and various other Rosaceae (*Agrimonia, Fragaria, Potentilla*), also recorded from *Eupatorium, Senecio* (Compositae), *Cicuta* (Umbelliferae), and Chenopodium (Chenopodiaceae). Widely distributed in North America. Life cycle seems to be still unknown. MacGillivray (1968) reviews this species.

Macrosiphum pechumani MacGillivray

Medium-sized, broadly spindle-shaped, milky white with dark brown to black head, antennae, legs, siphunculi, and cauda. Monoecious holocyclic, with alate males, on *Convallaria majalis* and *Smilacina racemosa* (the latter probably being the native host), so far only recorded from New York State, U.S.A. For further details see MacGillivray (1966).

Macrosiphum rosae (L.)   Rose Aphid   Plate 111

Appearance in life: Apterae are medium-sized to rather large, broadly spindle-shaped, shiny, mid- to dark green or deep pink to red-brown or magenta; with shiny black head and siphunculi. Antennae and legs bi-coloured yellow and black, cauda pale yellow. Dorsal abdomen sometimes with small black sclerites. Alatae have very distinct black sclerites along sides of abdomen. Apterae 1.7–3.6 mm, alatae 2.2–3.4 mm.

Host plants: Primary hosts are wild and cultivated *Rosa* spp. Secondary hosts are Dipsacaceae (*Dipsacus, Succisa*) and Valerianaceae (*Centranthus, Valeriana*); occasionally in summer on other Rosaceae (*Fragaria, Geum, Pyrus, Malus, Rubus*) and Onagraceae (*Chamaenerion, Epilobium*).

Virus transmission: Able to transmit at least 12 plant viruses including the persistent virus strawberry mild yellow edge. Not a vector of rose mosaic or rose streak.

Distribution: Distributed throughout most of the world on cultivated roses, but apparently not in eastern Asia. Records from Japan apply to *M. mordvilkoi*.

Biology: Heteroecious holocyclic between *Rosa* and Dipsacaceae or Valerianaceae in temperate regions. However, some populations remain on rose throughout the summer; when the autumn return migration from secondary hosts is occurring, these non-migrating populations produce few sexuales, and many overwinter anholocyclically in regions of mild winter climate. In warmer climates *M. rosae* may be completely anholocyclic on roses. Tomiuk and Wöhrmann (1980) studied its population ecology in Germany. $2n = 10$.

Macrosiphum rudbeckiarum (Cockerell)   Cockerell's Green Goldenglow Aphid

Pale green, medium-sized, with mainly pale appendages, on *Rudbeckia laciniata* and *R. montana* in New Mexico, Colorado, Idaho, and Utah, U.S.A. Life cycle unknown. Hottes' (1949) account under this name refers to *M. cockerelli*.

Macrosiphum stellariae Theobald

Rather narrowly spindle-shaped, medium-sized to large (1.8–3.9 mm), yellowish green to mid-green or rose red. Apices of femora and distal, reticulated parts of siphunculi are conspicuously pigmented, dark grey to black; in *M. euphorbiae* these parts may be dark brown, but not black. Monoecious holocyclic, with alate males, on Caryophyllaceae (*Dianthus, Gypsophila, Silene, Stellaria*) and sometimes on other plants (e.g. Ranunculus), usually in small, loose colonies on growing shoots. Able to transmit beet yellows virus under experimental conditions. In Europe

(Czechoslovakia, Germany, U.K.); there are also specimens in the B.M. (N.H.) collection from New Zealand and British Columbia. Möller (1971a) provides an account of this aphid. $2n = 10$.

Macrosiphum tiliae (Monell)

Medium-sized, pink or green with antennae and siphunculi black except at bases. Monoecious holocyclic, with alate males, on *Tilia americana*, especially on suckers around base of tree. Widespread in the U.S.A. Davis (1914, p. 83) gives an account of this species.

Macrosiphum (Neocorylobium) vandenboschi Hille Ris Lambers

Medium-sized, rather elongate oval, light green; eyes black, siphunculi mainly black, paler at base. On *Corylus cornuta* var. *californica* in California (Hille Ris Lambers, 1966c).

Macrosiphum zionensis Knowlton

Large, green, with long mainly black antennae and black siphunculi. On *Lupinus* sp. in Idaho and Utah. Life cycle not known.

Maculolachnus Gaumont   Lachninae: Lachnini
   (= *Longiunguis* Van der Goot)

A genus of 5 species related to *Lachnus* but without any clear pattern of pigmentation on the forewings, and with dorsal hairs often placed on small dark sclerites. Mainly associated with Rosaceae, with two species on *Rosa*.

Maculolachnus sijpkensi Hille Ris Lambers

Medium-sized to rather large (2.4–3.8 mm), yellowish brown to dark brown or blackish, on wild *Rosa* spp. (*acicularis, fendleri*); in colonies on stems near ground, probably ant-attended. Monoecious holocyclic, with apterous males. In North America and Mongolia. North American records of *Lachnus rosae* Cholodkovsky probably all apply to this species. $2n = 10$.

Maculolachnus submacula (Walker)   Rose Root Aphid   Plate 4

Medium-sized to rather large (2.7–3.8 mm), yellowish brown to dark chestnut brown, feeding on stems near ground or (in summer) on surface roots of wild and cultivated *Rosa*, always attended by ants. Monoecious holocyclic, with apterous males. In Europe, east to Ukraine, and in India. Böhmel and Jancke (1942) studied the embryology of the overwintering eggs. $2n = 10$.

Megoura Buckton

Aphidinae: Macrosiphini

A genus of 6 species with swollen siphunculi, associated with Leguminosae in the Old World Northern Hemisphere. Hille Ris Lambers (1949) characterized the genus and other accounts are available from Germany (Müller, 1969) and Japan (Miyazaki, 1971).

Megoura crassicauda Mordvilko (= *japonica* Matsumura)

Large, green, broadly spindle-shaped aphids with black head, prothorax, antennae, siphunculi, cauda, and legs; closely resembling European *M. viciae*, but differing in the sensoriation of the antennae (Hille Ris Lambers, 1965). In colonies on stems and at growing points of *Vicia* spp. (*faba, unijuga, cracca, angustifolia*) and *Lathyrus* (*japonicus, davidii*). In China, Korea, Taiwan, Japan, and eastern U.S.S.R. (Primorskaya). Life cycle unknown, probably monoecious holocyclic.

Megoura lespedezae (Essig & Kuwana)   Plate 119
(= *cajanae* Ghosh, Ghosh, & Raychaudhuri)

Medium-sized, green with yellowish-brown to reddish-brown head; antennae and legs dark, siphunculi black, cauda pale. On *Lespedeza* spp. (*bicolor, cyrtobotrya*) in China, Taiwan, Korea, Japan, and recorded from *Cajanus cajan* and *Desmodium trifolium* in India (Ghosh, Ghosh, and Raychaudhuri, 1971). Monoecious holocyclic on Leguminosae, males undescribed. $2n = 12$ (Shinji, 1931) or 14*.

Megoura pallipes Basu

Medium-sized, broadly spindle-shaped, green to dark green with head brown to purplish brown; dorsal abdomen with three rows of roundish or transversely oval whitish spots. Antennal segments III and IV, femora, and siphunculi all pale at base, dark distally; cauda pale brown. On both surfaces of leaves and on twigs of *Indigofera* spp. (*dosua, gerardiana, teysmanni*). Recorded from India, Kashmir, and Thailand. Life cycle unknown.

Megoura viciae Buckton   Vetch Aphid   Plate 120

Appearance in life: Large (3.0–4.3 mm), broadly spindle-shaped, dark bluish green to apple green with head, prothorax, antennae, siphunculi, cauda, and legs black. Alatae with head and whole thorax black, abdomen green.

Host plants: *Vicia* spp. (*faba, cracca, sativa, sepium*) and *Lathyrus* spp. (*montanus, pratensis*). Colonizes young, apical parts of host plants.

Virus transmission: A vector of at least 8 plant viruses including bean common mosaic and pea mosaic. Also able to transmit the persistent viruses bean leaf roll and pea enation mosaic, although conflicting reports suggest that *M. viciae* may be a relatively poor or inconsistent vector of these viruses (Cockbain and Costa, 1973; Schmütterer, 1969).

Distribution: Europe, Central Asia, Middle East, and Ethiopia.

Biology: Monoecious holocyclic on Leguminosae, with alate males. *M. viciae* has been a favourite subject for study by aphid biologists. Its polymorphism and photoperiodic responses have been studied intensively, especially by Lees (1973); Ehrhardt (1963) studied the structure and function of the digestive system; Bonner and Ford (1972) investigated the effects of crowding on development. Hille Ris Lambers (1947) describes the life cycle and discusses the damage done to *Vicia sativa* crops in Belgium; it is not an important pest of *V. faba*. $2n = 10$.

Melanaphis van der Goot   Aphidinae: Aphidini
(= *Longiunguis* van der Goot)

A genus of about 20 Old World species with short siphunculi, associated with Gramineae. A few species are known to alternate from Rosaceae (Pyroidea) like the closely related genus *Rhopalosiphum*. Three species are European but most of the remainder are native to the Far East and are associated with *Arundinaria* and other Bambusae. Accounts are available from Japan (Sorin, 1970), Taiwan (Liao, 1976), India (Raychaudhuri and Banerjee, 1974), and north-east India (Raychaudhuri, Ghosh, and Basu, 1980).

Melanaphis arundinariae Takahashi

Purplish brown, on bamboos (*Pleioblastus*) and *Yushania niitakayamensis* in Taiwan (Liao, 1976). Distinguished from *M. bambusae* by colour, by the long, fine body hairs, and by the pale wing veins of the alatae.

Melanaphis bambusae (Fullaway)   Plate 22

Appearance in life: Small black aphids, with some grey markings on dorsal abodmen, lightly dusted with white wax powder. The antennae and legs are mainly pale. Alatae have dark wing veins. Colonizes undersides of leaves. Apterae 0.8–1.4 mm, alatae 1.1–1.3 mm.

Host plants: Primary host (in Japan) is *Pourthiaea villosa* (Rosaceae). Secondary hosts are bamboos (*Arundinaria, Bambusa, Phyllostachys, Pleioblastus*).

Distribution: Of east Asian origin, now widely distributed; Japan, China,

Taiwan, Indonesia, Malaysia, Australia, India, Egypt, Italy, Spain, and the U.S.A. (Louisiana, Hawaii).

Biology: Heteroecious holocyclic in Japan, with a rosaceous primary host (Sorin, 1962), but anholocycly is also common in Japan and the sexuales have not been recorded elsewhere, so populations in many parts of the world are probably anholocyclic on bamboos. $2n = 10$ (Kuznetsova, 1974) or 8*.

Melanaphis formosana (Takahashi) (= *miscanthi* Takahashi)

Small yellowish-grey, dirty yellow or purple aphids on leaves of Gramineae (*Miscanthus* spp., *Sorghum vulgare, Zea mays*) in Japan and Taiwan, attended by ants. Monoecious holocyclic in Japan with alate males; the sexual morphs are described by Sorin (1970), who gives additional characters for separating this species from *M. sacchari*. Probably anholocyclic in Taiwan (Takahashi, 1923).

Melanaphis pyraria (Passerini) Pear–Grass Aphid Plate 24

Small (1.3–2.1 mm) dark brown aphids on undersides of young pear leaves in spring, rolling the leaves transversely or diagonally to the mid-rib. Attended by ants. Heteroecious holocyclic between pear (*Pyrus communis*) and Gramineae (*Arrhenatherum, Brachypodium, Poa, Holcus, Triticum*). The appearance of the aphid differs remarkably on its secondary host plants according to the species of grass colonized. On *Arrhenatherum* it is reddish purple, hidden under deformed leaves; on *Poa, Brachypodium* and *Triticum* it is a smaller, yellowish aphid. $2n = 8$.

Melanaphis sacchari (Zehntner) Plate 23

Appearance in life: Small, ant-attended aphids, very variable in colour according to host plant and environmental conditions; pale yellow, yellow brown, purple, or even pinkish. Alatae (and to a variable extent the apterae also) have dark sclerotic dorsal markings. Apterae and alatae 1.1–2.0 mm.

Host plants: Gramineae, especially *Sorghum* and *Saccharum*; also sometimes on *Echinochloa, Oryza, Panicum,* and *Pennisetum*. Varietal differences in susceptibility of *Sorghum* in Japan were studied by Setokuchi (1976).

Virus transmission: A vector of the persistent virus millet red leaf, but apparently unable to transmit sugarcane mosaic.

Distribution: Distributed throughout tropical and subtropical regions (Middle East, Africa, India, China, Thailand, Japan, Malaysia, Indonesia, Philippines, Australia, Hawaii, Caribbean, Central and South America). C.I.E. Distribution Map **A420** (1981).

Biology: Probably anholocyclic almost everywhere, but David (1977) records oviparae on *Sorghum* in February to March in north-west India. Males are not recorded. Zimmerman (1948) gives a general account of the biology, including natural control by parasites and predators, in Hawaii, and Setokuchi (1976) has studied its ecology on *Sorghum* in Japan. *M. sacchari* is very closely related to the host-alternating *M. pyraria* in Europe, and is apparently a complex of mainly anholocyclic races or subspecies (discussed by Eastop, 1966, and David, 1977). A dark brown form (*indosacchari* David) feeding on the exposed upper leaves of sugarcane may be distinguished from *M. sacchari s. str.* on lower leaves sheltered from the sunlight, and from the form usually found on *Sorghum*, by its shorter body hairs, longer terminal process to the antenna, and reduced lateral abdominal tubercles. The ecology and control of this form was studied by Varma *et al.* (1978). Whether these morphological differences are environmentally induced or genetically determined is still uncertain. $2n = 8$.

Melanocallis Oestlund   Drepanosiphinae: Phyllaphidini

A monotypic genus near *Tinocallis* Matsumura but with differences in the ventral structure of the head.

Melanocallis caryaefoliae (Davis)   Plate 11 (right)
(= *fumipennellus* auct., nec Fitch)   Black Pecan Aphid

Small, squat-bodied very dark green to black, with conspicuous black dorsal paired tubercles, those on the second abdominal tergite being especially large, and small tufts of white wax. All adults alate, living on both upper and lower surfaces of leaves of *Carya* spp., especially pecan. Feeding causes yellow spots on the leaves and large populations can cause defoliation. In most of the U.S.A., except the north-east, and in Canada (Ontario). Bissell (1978) gives a full account, and Tedders and Osburn (1970) investigated methods of chemical control. $2n = 14$.

Metopolophium Mordvilko   Aphidinae: Macrosiphini

A genus of about 18 species resembling *Acyrthosiphon* but perhaps really more closely related to *Sitobion*, and like *Sitobion* associated with Gramineae and sometimes also with Rosaceae. All the 11 European species are associated with Gramineae and several overwinter as eggs on *Rosa*. Two out of 3 Indian species live on Gramineae, as do 3 out of 5 North American species. The generic position of a number of species described from dicotyledons is uncertain. Stroyan (1982) has recently revised the European members of the genus. Raychaudhuri, Ghosh, and Basu (1980) give an account of Indian species placed in *Metopolophium*.

Metopolophium alpinum Hille Ris Lambers

Medium-sized to rather large (2.6–3.7 mm), evenly dull opaque green with pale appendages; spring populations on *Rosa* spp. (*alpina, pomifera, villosa*) in French and Swiss Alps, migrating to *Poa alpina* and also recorded from *Carex*. Alatae have dusky narrow transverse bands on the dorsal abdomen. Hille Ris Lambers (1966a) gives a detailed account of the life cycle and descriptions of all morphs.

Metopolophium dirhodum (Walker)   Rose–Grain Aphid   Plate 104

Appearance in life: Apterae are elongate spindle-shaped, green or yellowish green, with a distinct brighter green longitudinal mid-dorsal stripe. Antennae mainly pale with the very apices of segments III–V, segment VI near the primary sensoria, and the terminal process dusky to black. Legs, siphunculi, and cauda pale. Alatae have a green abdomen without dorsal abdominal markings. Apterae 1.6–2.9 mm, alatae 1.6–3.3 mm.

Host plants: Primary hosts are wild and cultivated *Rosa* spp., possibly occasionally *Agrimonia* and *Fragaria* spp. Secondary hosts are numerous species of grasses and cereals (*Aira, Agrostis, Avena, Bromus, Dactylis, Festuca, Glyceria, Hordeum, Lolium, Poa, Triticum, Zea*). Occasionally also on *Iris* spp.

Virus transmission: A vector of the persistent virus barley yellow dwarf.

Distribution: Europe, the Middle East, Central Asia, Africa, North and South America, and recently introduced into New Zealand (R. Sunde, personal communication).

Biology: Heteroecious holocyclic between *Rosa* and Gramineae, but in western Europe a tendency to overwinter parthenogenetically on grasses has developed, apparently quite recently (Prior, 1976). Vickermen and Wratten (1979) review the ecology of *M. dirhodum* in Europe. Hand and Williams (1981) studied holocyclic populations on wild roses in England. Hille Ris Lambers (1947a) gives a general account. See also papers in *Bulletin S.R.O.P.* 1980(3), and numerous references annually in *Review of Applied Entomology*. $2n = 18$.

Metopolophium festucae (Theobald)   Fescue Aphid   Plate 105

Appearance in life: Apterae broadly spindle-shaped, rather shiny, evenly coloured yellowish green to green or salmon pink. Antennae darkened progressively towards their apices; legs, siphunculi, and cauda pale. Alatae have a dull green to yellow-green abdomen with a dorsal segmental pattern of dark transverse bands, sometimes weakly developed in small specimens. Apterae 1.4–2.2 mm, alatae 1.8–2.2 mm.

Host plants: Many species of grasses (*Agrostis, Bromus, Deschampsia, Festuca, Lolium, Poa*); rarely and probably only transiently on cereals.

Distribution: Europe and Iceland; there are also single records of aphids resembling *M. festucae* from Argentina and California.

Biology: Monoecious on many grasses, with a strong tendency to anholocycly. Alate males and oviparae are known, but rare. Fundatrices (or individuals with the morphology of fundatrices) occur on several genera of grasses, indicating that there is no host alternation. Prior (1976) discusses the life cycle, Hille Ris Lambers (1947a) gives a general account. References to *M. festucae* on cereals prior to 1982 refer to subspecies *cerealium*. $2n = 16*$.

Metopolophium festucae cerealium Stroyan

Appearance in life: Apterae are broadly spindle-shaped, evenly coloured pale yellowish or yellowish green. Alatae have a very well-developed dorsal abdominal pattern of dark transverse bands. Apterae 1.7–3.4 mm, alatae 1.8–3.3 mm.

Host plants: Mainly cereals (*Avena, Hordeum, Secale, Triticum*), sometimes other Gramineae especially *Lolium perenne* and *Phleum pratense*.

Distribution: Europe.

Biology: Monoecious on Gramineae, and mainly anholocyclic; alate males, oviparae and fundatrices occur rarely. Stroyan (1982) discusses the biology and host-plant relationships of this subspecies, and the difficult morphological separation from *festucae s. str.* $2n = 16*$.

Metopolophium longicaudatum (David & Hameed)

Medium-sized, oval, pale aphids described from wheat in India (David and Hameed, 1975), distinguished from other *Metopolophium* on wheat by the long terminal process to the antenna (longer than antennal segment III) and a long cauda (almost as long as the siphunculi).

Metopolophium montanum Hille Ris Lambers

Medium-sized, broadly spindle-shaped, evenly coloured green aphids on undersides of leaves of *Rosa* spp. in montane regions of Switzerland, France, and Spain. Alatae have conspicuous dark transverse bands on the dorsal abdomen. Heteroecious, migrating from roses to grasses such as *Poa alpina*. For further information see Stroyan (1982).

Metopolophium sabihae Prior

Rather small, broadly spindle-shaped, yellow-green to blue-green aphids on *Festuca rubra* and *Vulpia membranacea* growing in sand dune habitats in France and the U.K. Apparently normally anholocyclic, although oviparae and males are recorded. Able to colonize certain other grasses in the laboratory, but unlikely to occur in pastures (Prior, 1976).

Metopolophium tenerum Hille Ris Lambers

Rather small, oval aphids of colour varying from dirty greenish or pinkish white to dark green, living on *Deschampsia flexuosa, Festuca ovina*, or *F. rubra* in shady places, or sometimes in the open on elevated heath or moorland. Recorded from the Netherlands, the U.K., Sweden, and Norway. Monoecious holocyclic on grasses, with alate males (Hille Ris Lambers, 1947a). Reliable separation from *M. festucae* is difficult (see Stroyan, 1982), but *M. tenerum* is unlikely to occur in lowland pastures.

Micromyzus van der Goot   Aphidinae: Macrosiphini

A genus of about 11 species of the Old World tropics, mainly associated with ferns. The monotypic subgenus *Kugegania* Eastop has reduced wing venation and lives on etiolated grasses.

Micromyzus (Kugegania) ageni Eastop

Small aphids living on etiolated parts of Kikuyu grass (*Pennisetum clandestinum*), and possibly other grasses growing in shaded conditions, in eastern and southern Africa (Kenya, Malawi, Zimbabwe, South Africa). Mainly known from trapped alatae, which have reduced wing venation (media once-branched, radius usually absent) and no central black patch on the dorsal abdomen. Life history unknown. Because of its habitat this species is unlikely to be a pasture pest.

Monellia Oestlund   Drepanosiphinae: Phyllaphidini

A genus of 4 North American species on *Carya* (Juglandaceae), resembling *Monelliopsis* but folding their more rounded wings flat over the abdomen in repose. Bissell (1978) revised the genus.

Monellia caryella (Fitch)   Little Hickory Aphid,   Plate 12 (left)
(= *costalis* Fitch)   Black-margined Yellow Pecan Aphid

Adults all alate; small, pale lemon yellow to greenish yellow, later generations from midsummer to autumn with a broad black frontal and lateral stripe.

Wings held flat over abdomen in repose. Monoecious holocyclic on leaves of *Carya* spp., especially *C. pecan* and *C. cordiformis*. Widespread in the U.S.A., in Ontario, Canada, and introduced into Israel where it has recently become a serious pest of pecans (Mansour, 1981). For an ecological study see Edelson and Estes (1983). Tedders (1967) found 2 *Trioxys* spp. (Braconidae) preferentially attacking this aphid. According to Bissell (1978), Davidson's (1914) account of *M. caryella* in California applies to a species of *Monelliopsis*, probably *bisetosa* Richards, and Richards' (1965) description and figures under the name *caryella* apply to *Monelliopsis nigropunctata*. $2n = 18$.

Monelliopsis Richards — Drepanosiphinae: Phyllaphidini

A genus of 5 North American species associated with Juglandaceae, similar to *Monellia* but holding their wings vertically at rest (as do most aphids). Richards (1965) keyed the species then recognized.

Monelliopsis pecanis Bissell — Yellow Pecan Aphid — Plate 12 (right)

All adults alate; small, yellow, with black antennal segments I and II and apices of other segments. Usually there is a black spot near the apex of each hind femur, and small black spots at the bases of many of the dorsal abdominal hairs with a larger one anterior to each siphunculus. On *Carya pecan* in the U.S.A., Mexico, and introduced into South Africa. This species was previously confused with *M. nigropunctata* (Granovsky), which occurs on other *Carya* and *Juglans* spp. in North America, and may sometimes be collected on *C. pecan*. Distinguishing characters are described by Bissell (1983). Edelson and Estes (1983) studied its population ecology (as *M. nigropunctata*). $2n = 10$.

Moritziella Börner — Phylloxeridae

A genus of 2 species on Fagaceae, resembling *Phylloxera* but without abodominal spiracles on segments 2–5, and with body bearing numerous well-developed dorsal tubercles.

Moritziella castaneivora Miyazaki

Very small (0.9–1.3 mm), yellowish-brown to purplish-brown pyriform insects feeding on the fruits of chestnut trees (*Castanea crenata*) in Japan (Miyazaki, 1968b).

Myzaphis van der Goot — Aphidinae: Macrosiphini

A genus of 5 species associated with *Rosa* and *Dasiphora* (Rosaceae). The terminal process of the antenna is short and the apterae have a rugose dorsal cuticle. The four Old World species differ from the Canadian species in having 5 hairs on the first tarsal segments (only 2 in *canadensis*) and have alatae with a dark dorsal abdominal patch (absent in *canadensis*). Accounts are available for India (David *et al.*, 1971) and Canada (Richards, 1963).

Myzaphis bucktoni Jacob

Small, pale yellow to pale green, with light to dark brown dorsal markings consisting of a brown head, two large brown patches on the pronotum, and paired, broad stripes extending from mesothorax to base of cauda, converging at about the level of the siphunculi. Alatae have a pale green abdomen with a pale brown central dorsal patch. Generally feeding singly from the mid-rib on upper sides of leaflets of *Rosa* spp. (*villosa, canina, tomentosa*). Recorded from the U.K., Portugal, and Yugoslavia. Monoecious holocyclic, with apterous males; Jacob (1946) describes the life cycle. $2n = 13$ (for one sample from Portugal)*.

Myzaphis rosarum (Kaltenbach)   Lesser Rose Aphid   Plate 71

Appearance in life: Apterae are small, rather elongate oval, rather dorsoventrally flattened, yellow-green to green, feeding mainly along the mid-rib on either side of young leaves of the host plant. Alatae have a dark central patch on the dorsal abdomen. Apterae 1.0–2.2 mm, alatae 1.2–2.0 mm.

Host plants: Rosa spp., wild or cultivated, especially rambler roses. Also frequently on *Dasiophora* (~ *Potentilla*) spp., especially *D. fruticosa*, and occasionally recorded from *Fragaria* spp.

Virus transmission: Able to transmit strawberry mottle virus.

Distribution: Europe, the Middle East, Central Asia, Morocco, South Africa, India and Pakistan, China, Japan, New Zealand, North America, Colombia, and Ecuador. Japanese populations have rather longer cephalic hairs than European ones.

Biology: Monoecious holocyclic on Rosaceae in Europe, with very small apterous males. Anholocyclic overwintering occurs in New Zealand, and probably elsewhere. Brief accounts of *M. rosarum* are given by Cottier (1953), Mimeur (1936), and Miyazaki (1971). $2n = 4$.

Myzaphis turanica Nevsky

Small aphids on wild and cultivated *Rosa* spp., closely resembling *M. rosarum* but with a rounded median frontal projection bearing longer hairs. Alatae have a paler and more irregularly shaped dorsal abdominal patch than *M. rosarum*. Monoecious holocyclic, with alate males (Tuatay and Remaudière, 1964). Recorded from Central Asia, Middle East, India, Mongolia, and Sweden.

Myzocallis Passerini Drepanosiphinae: Phyllaphidini

A genus for 29 species with a knobbed cauda and a bilobed anal plate, mainly associated with Fagaceae (26 species); 20 of the *Quercus*-feeding species are American, including all 14 members of the subgenus *Lineo myzocallis*. The other species are mainly European except for one Middle Eastern, one from India, and another from China and Japan. Apterous viviparae are known for a few species but in most all the viviparae are alate. Accounts are available for Britain (Stroyan, 1977), Fennoscandia and Denmark (Heie, 1982), and North America (Richards, 1966; Boudreaux and Tissot, 1962).

Myzocallis castanicola Baker

Medium-sized (1.6–2.2 mm), all adult viviparae alate; yellow with a broad dark spinal band on head and thorax, and a pattern of paired dark segmental markings on the abdomen. Wing veins are dark. Immatures have long, capitate hairs. On many species of *Quercus*, and *Castanea* spp. (*sativa, vulgaris*). Probably of European origin, now widespread (Europe, Middle East, southern Africa, Australia, New Zealand, Brazil, California, British Columbia). Monoecious holocyclic. Cottier (1953) gives a general account. It is possible that the name is being applied to more than one species. $2n = 14$.

Myzocallis coryli (Goetze) Hazel Aphid

Small (1.3–2.2 mm), all viviparae alate; pale yellow with apices of antennal segments black, and usually a black spot on the pterostigma of each forewing. Immature stages have long capitate hairs. On *Corylus* spp. (*avellana, colurna, heterophylla, maxima*). Probably of European origin but now widely distributed (Europe, Turkey, Ukraine, Japan, Tasmania, New Zealand, Chile, western U.S.A. (California, Utah, Oregon) and Canada (British Columbia, Ontario)). Monoecious holocyclic. Cottier (1953) gives a general account of this aphid in New Zealand. Viggiani in Cavallaro (1983, pp. 109–13) describes the natural enemy complex in Italy. Resistance to carbamate insecticides has developed in western North America (Aliniazee 1983). $2n = 14$ (Gut, 1976).

Myzus Passerini Aphidinae: Macrosiphini

A genus for about 55 species with gibbous antennal tubercles from the Old World, three-quarters of them from Asia. The few species described from America were probably either introduced, belong to other genera, or have not been recognized again since their original description. Perhaps only the circumpolar *M. polaris* occurs naturally in America. The primary hosts of the heteroecious species are *Prunus* spp., but the secondary hosts belong to many different families. *Myzus* contains several of the most polyphagous aphid species. Three species with wide host ranges, *M. ornatus, M. ascalonicus*, and *M. (Sciamyzus) cymbalariae*, have appeared during the last 50 years and spread rapidly around the world; since the populations of all three are permanently parthenogenetic, a hybrid origin is suspected. The biologies of many of the oriental species are not fully known. Accounts of *Myzus* are available for Germany (Müller, 1969), Japan (Miyazaki, 1971), China (Tao, 1963), and India (Basu and Raychaudhuri, 1976a; Raychaudhuri, Ghosh, and Basu, 1980).

Myzus ascalonicus Doncaster   Shallot Aphid   Plate 83

Appearance in life: Apterae are shiny pale greenish brown, straw-coloured or dirty yellow, with dorsum strongly convex in comparison with related species. Appendages mainly pale except that the apex of antennal segment V, whole of VI, the apices of the tibiae, and the tarsi, are all quite black. Alatae have a black dorsal abdominal patch, dark siphunculi and cauda, and a remarkably bimodal variation in the number of secondary sensoria on the antenna, either a few near the base of segment III or large numbers covering III, IV and V. Apterae 1.1–2.2 mm, alatae 1.3–2.4 mm.

Host plants: Extremely polyphagous, colonizing plants in over 20 families, particularly Alliaceae (*Allium ascalonicum, A. cepa*); Caryophyllaceae (*Cerastium, Stellaria*); Compositae (e.g. *Lactuca, Chrysanthemum, Taraxacum, Crepis*); Cruciferae (*Brassica*, etc.); Liliaceae (*Tulipa*); and Rosaceae (*Fragaria, Dasiphora*).

Virus transmission: A proven vector of about 20 plant viruses including beet mosaic, beet yellows, cabbage black ring spot, cauliflower mosaic, cucumber mosaic, *Dahlia* mosaic, dandelion yellow mosaic, onion yellow dwarf, potato leaf roll, shallot yellows, strawberry mottle, strawberry veinbanding, tomato dwarfing, turnip crinkle, and turnip yellows. Not a vector of lettuce mosaic, potato virus Y, or the persistent strawberry viruses. Important particularly for its ability to transmit viruses from wild, overwintering hosts to crops, e.g. beet yellows from *Stellaria media* (Semal, 1956) and dandelion yellow mosaic from *Taraxacum* to lettuce (Kassanis, 1947).

Distribution: Europe, India, Japan, Australia, New Zealand, Antipodes, Auckland Isles, and North and South America. The origin of *M. ascalonicus* is a mystery; it was first collected on stored onion sets in Wyoming, U.S.A., in 1940 (Fronk, 1955), and a year later on stored shallots in Lincolnshire, U.K.

Biology: Apparently exclusively anholocyclic; no sexual morphs are known. It is frequently found in winter on stored bulbs, potatoes, or root vegetables, in glasshouses and on potted plants. See Müller and Möller (1968) for further information. $2n = 12$.

Myzus cerasi (Fabricius)   Cherry Blackfly   Plate 79

Appearance in life: Rather small to medium-sized, shiny very dark brown to black, with sclerotic dorsum. Legs and antennae bicoloured yellow and black, siphunculi and cauda wholly black. Forming dense colonies at shoot apices of cherry trees in spring, curling leaves and often doing severe damage. Commonly attended by ants. Alatae have a yellow-brown abdomen with a large black central dorsal patch. Apterae 1.5–2.6 mm, alatae 1.4–2.1 mm.

Host plants: Primary hosts are *Prunus cerasus, P. avium*, and sometimes other *Prunus* (*pennsylvanica, serrulatus, sieboldii, yedoensis, virginiana*); secondary hosts are *Galium* spp., *Veronica* spp., *Euphrasia officinalis* and, especially in North America, certain Cruciferae (*Capsella, Cardamine, Lepidium*).

Virus transmission: Able to transmit wilt and decline disease of cherries and several non-persistent viruses of non-host plants including bean yellow mosaic, celery mosaic, and onion yellow dwarf.

Distribution: Europe, Turkey, India, Pakistan, Australia, New Zealand, and North America.

Biology: Heteroecious holocyclic between *Prunus* and secondary hosts in Rubiaceae, Scrophulariaceae, and Cruciferae. Apterae on secondary hosts vary in colour from dark brown to olive green or yellowish brown. Gilmore (1960) studied populations in Oregon, U.S.A., and Dahl (1968) studied intraspecific variation in life-cycle and host-plant relations in Europe, erecting several subspecies of which one is monoecious on *Galium*. The name may be being applied to more than one species. $2n = 10$.

Myzus cerasi umefoliae (Shinji)

Shiny black aphids on *Prunus mume* in Japan, morphologically resembling *M. cerasi s. str.* but not found on cherry in Japan and migrating to *Artemisia capillaris* (Takahashi, 1965b).

Myzus certus (Walker)

Small to medium-sized, pink to dark reddish-brown aphids on pinks, carnations, Sweet William, and other Caryophyllaceae (*Cerastium, Stellaria*), and also on Violaceae. On Caryophyllaceae it causes curling and spotting of the leaves. It can transmit several non-persistent viruses including beet yellows and potato virus Y. Monoecious holocyclic, with apterous males; anholocyclic populations also occur. In Europe, Iran, and North America. This aphid when preserved closely resembles *M. persicae,* particularly in its alate morph; distinguishing characters for the two species in trap catches are given by Hille Ris Lambers (1959). $2n = 12$.

Myzus (Sciamyzus) cymbalariae Stroyan

Apterae are small (1.2–2.0 mm), oval, dull yellowish green or yellowish brown, to dark brown or crimson red (more pigmented in cold conditions); alatae, unlike other *Myzus*, have more or less separate transverse dark bands on the dorsal abdomen, although these do have some tendency to coalesce. Rather polyphagous, colonizing a similar range of hosts to *M. ascalonicus* (especially species in Alliaceae, Caryophyllaceae, Compositae, Iridaceae, Liliaceae, Scrophulariaceae, and Violaceae). Apparently entirely anholocyclic, although males have been obtained in trap catches and reared in laboratory cultures. Origin unknown; present distribution indicates recent dispersal by man (England, Scotland, Rwanda, South Africa, India, Australia, and New Zealand). For further information see Brown (1984). $2n = 12$.

Myzus dianthicola Hille Ris Lambers

Small to medium-sized, deep yellow-green aphids on *Dianthus caryophyllus*, usually in glasshouses, causing white or yellow spots or blotches on the leaves. Closely resembling *M. persicae*, but distinguishable from that species by the reaction of the host plant, and from *M. certus* by its colour in life. In North America, western Europe, and New Zealand. Apparently completely anholocyclic. $2n = 14$ (heterozygous).

Myzus hemerocallis Takahashi

Small, pale yellowish green or greenish white, attacking basal parts of young leaves of *Hemerocallis* spp. (*fulva, auranticae*) and *Agapanthus umbellatus* in China, Taiwan, Japan, India, and South Africa. Life cycle not known.

Myzus japonensis Miyazaki

Small, pale yellow or yellowish green to green, with appendages mainly pale, siphunculi slightly dusky. On undersides of leaves of *Rosa rugosa* in Japan. Males are alate, but the life cycle is uncertain (Miyazaki, 1968).

Myzus mumecola (Matsumura)

Rather small, pale green aphids colonizing *Prunus mume* and *P. armeniaca var. ansu* in Japan. Life cycle and sexual morphs are not recorded. Specimens provisionally assigned to this species have also been collected in Himachal Pradesh, India, on *P. cornuta*.

Myzus ornatus Laing   Violet Aphid   Plate 82

Appearance in life: Apterae are small or very small, oval, somewhat dorsoventrally flattened, pale yellow or green, marked dorsally with a characteristic pattern of dark green or brownish dots and transverse streaks; they live singly on the leaves of the host plants. Alatae have a black central dorsal patch on the abdomen, dark antennae, and dusky siphunculi and cauda. Apterae 1.0–1.7 mm, alatae 1.2–2.1 mm.

Host plants: Very polyphagous; on many different plant families including especially Bignonaceae, Compositae, Labiatae, Polygonaceae, Primulaceae, Rosaceae, and Violaceae.

Virus transmission: Able to transmit at least 20 plant viruses, including the persistent viruses *Malva* yellows, pea enation mosaic, potato leaf roll, and radish yellow, and non-persistent viruses of crucifers, cucurbits, *Dahlia*, onion, potato, *Primula*, soybean, strawberry, and tomato.

Distribution: Distributed throughout the world, probably on ornamental plants. Although common in India since 1956, there is still only one record from South East Asia (New Guinea).

Biology: Anholocyclic everywhere, in colder climates probably overwintering in glasshouses, on pot plants, or in sheltered situations. Oviparae are not known but the male has been described in India (Chakrabarti and Raychaudhuri, 1975). *M. ornatus* is chiefly found on ornamental plants in sheltered places and does not usually occur in significant numbers in field crops. Feeding behaviour and development of this aphid in the laboratory were studied by Lowe (1967). $2n = 12$.

Myzus persicae (Sulzer)
Green Peach Aphid, Peach–Potato Aphid   Plate 81

Appearance in life: Adult apterae small to medium-sized, whitish green, pale yellow green, grey green, mid-green, pink, or red, more deeply pigmented green or magenta in cold conditions; rather uniformly coloured, not shiny. Alatae have a black central dorsal patch on the abdomen; immatures of the alate morph, especially in autumn populations, are often pink or red. In colonies curling young leaves of peach in spring, or in more dispersed populations on many other plants, often on older leaves. Apterae and alatae 1.2–2.3 mm.

Host plants: Primary host usually *Prunus persica*, sometimes *P. nigra*, *P. tanella*, and possibly *P. serotina* and peach–almond hybrids. Secondary hosts are in over 40 different plant families, and include very many economically important plants.

Virus transmission: The most important aphid virus vector, shown to be able to transmit well over 100 plant viruses (Kennedy *et al.*, 1962). Persistent viruses transmitted include beet mild yellowing, beet yellow net, beet mild yellows, pea enation mosaic, pea leaf roll, potato leaf roll, radish yellows, tobacco vein distorting, and tobacco yellow veinbanding. The relationship with potato leaf roll has received particular attention (e.g. Ponsen, 1972; Eskanderi *et al.*, 1979).

Distribution: Probably of Asian origin, like its principal primary host; now world wide.

Biology: Heteroecious holocyclic between *Prunus* and secondary host plants, but anholocyclic on secondary hosts in many parts of the world where peach is absent, and where the climate permits active stages to survive the winter season. The literature on *M. persicae* is immense, but there have been extensive reviews of ecology (van Emden *et al.*, 1969; Mackauer and Way, 1976), as well as discussions of migration and spatial dynamics (Taylor, 1977), biological approaches to control (Blackman, 1976), and development of resistance to insecticides (Georghiou, in press). This aphid has also een the subject of much laboratory research; for example, studies of the anatomy and function of the gut (Forbes, 1964), nutritional studies using host plants (van Emden, 1977) and artificial diets (Mittler, 1976), photoperiodic responses (Takada, 1982), and genetic variation of enzyme systems (May and Holbrook, 1978; Takada, 1979). $2n = 12$; a form heterozygous for a chromosomal translocation is world-wide and common (Blackman *et al.*, 1978), and anholocyclic clones (often darker green in colour) frequently have one or two chromosomes dissociated.

Myzus varians Davidson   Plate 80

Apterae are medium-sized, pale green to green, with distal halves of siphunculi conspicuously black. Alatae have a green or (in autumn) greenish-blue abdomen with a dark dorsal central patch. In colonies tightly curling the young leaves of *Prunus persica* in spring. Heteroecious holocyclic, migrating for the summer to *Clematis* spp. It is a vector of plum pox virus (Sharka disease). Presumably native to China, and also recorded from Japan, Korea, Taiwan, and Thailand. There are North American records from California (first description, 1912) and North Carolina, but none are of sexual morphs and it is not a pest on peach, suggesting that North American populations may be anholocyclic. *M. varians* was first found in Europe in 1947 (Switzerland); since then it has gradually extended its distribution and is now also recorded from Austria, Bulgaria, Corsica, England, France, Italy, Hungary, and Yugoslavia. Arzone (1979) gives a general account of *M. varians* in Italy, and for additional information see Alma and Arzone (1983), Stroyan (1979a) and Meier (1954). Massonie *et al.* (1979) studied the resistance in two varieties of peach. $2n = 12$.

Myzus yamatonis Miyazaki

Small, black aphids severely curling the leaves of ornamental cherry trees in the Far East, and also recorded from peach and Japanese apricot (Tao, 1963, as *Prunomyzus sakurae*). Although black, the dorsal surface is not sclerotic like that of *M. cerasi*. Found on *Prunus* throughout the growing season, but a facultative migration to *Salvia nipponica* occurs at least in Japan (Miyazaki, 1971). Recorded from Japan, Korea, China, and Taiwan.

Nasonovia Mordvilko   Aphidinae: Macrosiphini

A genus of about 45 species with enlarged spiracular openings, cylindrical siphunculi, and apterae with sensoria on the third antennal segment. About two-thirds of the species are nearctic; nearly half are associated with the related families Saxifragaceae and Grossulariaceae, host-alternating species having *Ribes* as primary hosts. The Old World subgenus *Nasonovia s. str.* is largely associated with Compositae and particularly *Hieracium*. There are species on Ranunculaceae in both Old and New Worlds and species-groups on Polemoniaceae and Scrophulariaceae in North America. A recent revision is provided by Heie (1979).

Nasonovia (Kakimia) brachycyclica Holman

Apterae are medium-sized, bright green or pinkish, with central part of dorsal abdomen sclerotized, brownish or shiny black; alatae mainly dark brown to black. Monoecious holocyclic on *Grossularia* in Czechoslovakia, with an abbreviated parthenogenetic phase, oviparae and alate males appearing in June (Holman, 1972). At shoot apices, causing growth retardation and slight curling of leaves similar to the effect of *N. ribis-nigri*.

Nasonovia (Kakimia) cynosbati (Oestlund)   Dogberry Aphid   Plate 91

Apterae medium-sized, broadly spindle-shaped, pale green; siphunculi usually with dark or dusky apices. Alatae have very variable dorsal abdominal pigmentation; pale green or with variably developed transverse bands or even a solid black *Myzus*-like patch. Heie (1979) discusses the wide range of morphological variation in this species. Monoecious holocyclic, with alate males, on wild and cultivated *Ribes* spp., and also recorded from certain herbaceous Saxifragaceae (species of *Heucheria, Tellima, Borykinia*). Widely distributed in North America except in southern U.S.A. Probably partially anholocyclic in California. $2n = 10$.

Nasonovia (Kakimia) houghtonensis (Troop)   Gooseberry Witch-broom Aphid

Rather small, pale yellow-green to straw-yellow aphids severely curling and twisting the terminal leaves of *Ribes* spp. in North America. Alatae have a pale abdomen with no dorsal markings. They appear to comprise a complex of races or subspecies, which is revised and keyed by Heie (1979). *N. (K.) houghtonensis houghtonensis* attacks especially the Houghton variety of gooseberry; its life cycle is described by Baker (1919). It is recorded from New York, New Jersey,

Indiana, Illinois, Ohio, Iowa, Michigan, Colorado, and Oregon, U.S.A., and from New Brunswick, Canada. *N. (K.) houghtonensis similis* is recorded from wild and cultivated *Ribes* spp. (but not so far from cultivated gooseberry) in Canada (Alberta, Saskatchewan, Manitoba, Ontario). Both these subspecies are holocyclic monoecious on *Ribes*, and have apterous males.

Nasonovia ribis-nigri (Mosley)
Currant−Lettuce Aphid, Lettuce Aphid   Plate 90

Appearance in life: Medium-sized, spindle-shaped, rather shiny pale green to apple green aphids in small colonies at shoot tips of *Ribes* spp. in spring, causing slight leaf curl and retardation of growth, rather soon giving rise to emigrant alatae which have extensive black pigmentation including a conspicuous black sclerotic dorsal abdominal pattern. On *Lactuca* and other secondary hosts the adult apterae usually live dispersed under leaves or on flower stalks, and are shiny pale yellow to green, sometimes reddish, with a dorsal abdominal pattern of dark brown markings; the siphunculi are often at least distally dark green to black. Apterae 1.3−2.7 mm, alatae 1.5−2.5 mm.

Host plants: Primary hosts are *Ribes* spp. (e.g. *grossularia, nigrum, alpinum, ussuriense*); see Keep and Briggs (1971) for a full list of species colonized in Britain. Secondary hosts are mainly liguliflorous Compositae (*Cichorium, Crepis, Hieracium, Lactuca, Lampsana*); certain members of Scrophulariaceae (*Euphrasia, Veronica*) and Solanaceae (*Nicotiana, Petunia*) are also frequently colonized. Resistance in *Lactuca* to *N. ribis-nigri* is discussed by Dieleman and Eenink (1977).

Virus transmission: A vector of the persistent virus gooseberry veinbanding. It is also able to transmit mosaic diseases of cauliflower and cucumber. Lettuce mosaic is apparently not usually transmitted by this aphid.

Distribution: Throughout Europe, east to Ukraine; introduced to North America (north-eastern U.S.A., British Columbia, Quebec, New Brunswick), and South America (Peru, Brazil, Argentina).

Biology: Heteroecious holocyclic between *Ribes* and secondary host plants; the migration from *Ribes* occurs mainly in June, the return of gynoparae and males in September−October. For additional information see Hille Ris Lambers (1949) and Heie (1979). Griffiths (1961) studied a parasite (*Monoctonus paludum*) specific to *N. ribis-nigri*. $2n = 12$.

Nearctaphis Shaposhnikov
Aphidinae: Macrosiphini

A genus of about 12 North American species with short, imbricated siphunculi and a rather short cauda, alternating from Pomoidea mostly to Leguminosae and Scrophulariaceae. Hille Ris Lambers (1970a) revised the genus and (1974) redescribed *N. argentinaeradicis*.

Nearctaphis bakeri (Cowen) — Short-beaked Clover Aphid — Plate 47

Appearance in life: Apterae on clover small, oval, varying in colour from darkish green to salmon pink, with a rather variable development of small dark spots and patches on the dorsal abdomen. Alatae have a dark dorsal abdominal patch. Apterae 1.4–2.3 mm, alatae 1.5–2.1 mm.

Host plants: Primary hosts (in North America) are *Crataegus, Cydonia, Malus,* and *Pyrus communis*. Secondary hosts mainly in Leguminosae (*Medicago, Melilotus, Trifolium, Trigonella*), also recorded from *Castilleja, Veronica, Capsella,* and *Valeriana*. It is a principal pest of red clover (*Trifolium pratense*).

Virus transmission: Able to transmit bean yellow mosaic and lucerne mosaic viruses.

Distribution: Native to North America, where it is widely distributed. Now in Europe, Egypt, Afghanistan, Iran, India, and Japan. Recorded also in New Zealand, but this record is doubted by Cottier (1953).

Biology: Heteroecious holocyclic in North America, infesting tips of twigs, leaves, and blossom buds of *Malus*, etc. in spring, migrating to *Trifolium*, etc. On red clover, all parts of the plant may be colonized; undersides of leaves, bases of stems (often with ant shelters built around them), and in the inflorescences. Populations introduced into the Old World are apparently entirely parthenogenetic on secondary hosts. See R.H. Smith (1923) for a full account of *N. bakeri* in the U.S.A., and Leclant (1967a) and Meier and Kolar (1970) for a discussion of its biology and economic importance in Europe. $2n = 12$.

Nearctaphis californica Hille Ris Lambers

Described from *Melilotus alba* and a species of either *Medicago* or *Trifolium* in California (Hille Ris Lambers, 1970a); apterae are medium sized, colour in life unknown but distinguishable by broad transverse black bars on the abdominal tergites.

Nearctaphis crataegifoliae (Fitch) — Long-beaked Clover Aphid, Hawthorn Aphid

Small to medium-sized, oval, creamy white, greenish yellow, pink, or bright red; at base of stem or on roots of *Trifolium* spp. (*hybridum, pratense, repens*), *Melilotus albus,* and *Lathyrus odoratus*. Heteroecious holocyclic, with sexual phase on *Crataegus* sp., and sometimes on *Cydonia japonica*. Spring generations on hawthorn distort the leaves and turn them purple. Patch (1915) describes the life cycle (under the name *Aphis brevis*). In central and eastern North America. In western U.S.A. (California, Colorado, Utah) aphids with a similar biology have much darker pigmentation of head, posterior abdominal tergites, and siphunculi, and are described as a separate subspecies, *N. crataegifoliae occidentalis*, by Hille Ris Lambers (1970a).

Nearctaphis sensoriata (Gillette & Bragg)

This species overwinters on *Amelanchier* in western North America and probably migrates to secondary hosts in the Leguminosae, possibly *Medicago* spp., but the life cycle is still not clear (Hille Ris Lambers, 1970a). Apterae have a solid black dorsal abdominal patch, and this species differs from all other *Nearctaphis* in having numerous scent glands on the basal half of the hind tibia of all viviparous morphs.

Neoacyrthosiphon Tao   Aphidinae: Macrosiphini

A genus for a few oriental species with cylindrical siphunculi, first tarsal segments with 3–4 hairs, spinulose second tarsal segments, a relatively short terminal process only 2–3 times longer than the base of the sixth antennal segment, and an association with *Rhododendron* species. Similar species with clavate siphunculi, or 5 hairs on the first tarsal segments, or with body hairs arising from distinct tubercles, have been described in a number of other genera (*Chaetomyzus, Indomasonaphis, Illinoia, Ericolophium*). Chakrabarti *et al.* (1983) have revised these *Rhododendron*-inhabiting aphids related to *Chaetomyzus*.

Neochromaphis Takahashi   Drepanosiphinae: Phyllaphidini

A genus of 2 oriental species associated with Betulaceae, close to *Chromaphis* Walker but with pigmented wings and longer dorsal abdominal hairs. All viviparae are alate. Higuchi (1972) revised the genus in Japan.

Neochromaphis coryli Takahashi

Small (1.6–2.0 mm), brownish, with large dark areas on wings. Feeding (on twigs?) of *Corylus* spp. (*heterophylla, sieboldiana*) and *Carpinus* spp. in Japan and Korea.

Neotoxoptera Theobald   Aphidinae: Macrosiphini

A genus for 5 *Myzus*-like aphids with clavate siphunculi and dark bordered wing veins, probably of oriental origin. Accounts are available from Australia

(Carver, 1980) and Japan (Miyazaki, 1971; Sorin, 1971). Sexuales are unknown, but the presence of fundatrices of *N. abeliae* on *Abelia* indicates an association with Caprifoliaceae, as in the somewhat similar genus *Rhopalomyzus*.

Neotoxoptera formosana (Takahashi) — Onion Aphid

Adult apterae are small to medium-sized (1.6−2.3 mm), oval, shining magenta red to almost black, with antennae black basally and distally, femora black except at bases, and siphunculi dark but paler than body. Alatae are very dark red to black with wing veins heavily black-bordered, the borders maintaining constant width along the length of the vein. Sometimes in large numbers on leaves of *Allium* spp. (*ascalonicum, cepa, porum, schoenoprasum, sativum*, etc.), or on bulbs in store. In Japan, China, Taiwan, Korea, Australia, New Zealand, Hawaii, and North America. Life cycle unknown, apparently anholocyclic almost everywhere. $2n = 12*$.

Neotoxoptera oliveri (Essig) — Plate 84

Adult apterae are small (1.1−2.0 mm), less robust than *N. formosana*, dark red appearing almost black in life; antennae dark but with only distal segments completely black, siphunculi paler than body with dusky apices. Alatae shining very dark red to black; wing veins heavily bordered with black, the borders widening out at the base and apex of each vein. *N. oliveri* has a remarkably similar host range to *Myzus ascalonicus*; colonies most commonly occur on *Allium cepa, Viola* spp., and *Stellaria media*, but it is also recorded from plants in several other families. In Portugal, Africa, Korea (as *Myzus clavatus* Paik), Australia, New Zealand, western U.S.A. (California, Washington), Bermuda, and Brazil. Müller and Schöll (1958) discuss its occurrence in South Africa. $2n = 12*$.

Neotrama Baker — Lachninae: Tramini

A palaearctic genus of 6 species differing from *Trama* only in the presence of siphunculi, in that the apterae usually have many-faceted eyes, and in that the alatae usually have 3−4 sensoria on antennal segment III (2 or less in *Trama*).

Neotrama caudata (del Guercio)

Apterae are medium-sized to rather large (2.1−3.1 mm), broadly oval, pale greenish, brownish, or yellowish white, on roots of Compositae (e.g. *Cichorium endiva, Hypochoeris radicata, Lactuca sativa, Picris echioides, Sonchus asper, S. oleraceus, Taraxacum officinale*), attended by ants. In Europe only. Apparently completely anholocyclic, producing alatae throughout the year but most in autumn. For additional information see Eastop (1953). $2n = 9$ and 11.

Nippolachnus Matsumura   Lachninae: Lachnini

An oriental genus of 4 species characterized by the absence of an ocular tubercle (triommatidium). Associated with *Eriobotrya* with, in one species, a migration in summer to other woody Rosaceae; the only case of any kind of host alternation in present-day Lachninae. A.K. Ghosh (1982a) reviews the genus.

Nippolachnus himalayensis (van der Goot)
(= *eriobotryae* Basu & Hille Ris Lambers)

Apterae are large to very large (3.5–5.5 mm), oval-bodied, brick red with distinct segmentation, the intersegmental boundaries being whitish, and with mainly black legs. Alatae have a brick-red abdomen and brownish head and thorax. It forms large colonies on undersides of leaves of *Eriobotrya petiolata* in Bengal, India.

Nippolachnus piri Matsumura

Apterae are large, long bodied, pale green with some white wax on dorsal surface, and with pale legs and antennae. On undersides of leaves, feeding along main veins of *Eriobotrya japonica* (on which the sexual generation occurs) and *Pyrus* spp., including pear trees. It drops easily when disturbed. Also recorded from *Raphiolepis umbellata* and *Sorbus alnifolia*. In India, Japan, Korea, Taiwan, and China. Takahashi's (1950) record from Malaya on *Pyrus granulosa* apparently applies to *N. bengalensis*. Life history observations in Japan are recorded by Takahashi (1923), and in India by M.R. Ghosh and Raychaudhuri (1981).

Ovatus van der Goot   Aphidinae: Macrosiphini

A genus for about 14 species resembling *Myzus*, but the alatae have no dorsal abdominal patch and often have more secondary sensoria than in *Myzus*. Eight Old World species are associated with Pyroidea and/or Labiatae, but the 5 American species occur on a wide variety of plants. Accounts are available for Europe (Müller, 1969) and Japan (Miyazaki, 1971), and the American species are to be found in accounts of *Myzus* and *Phorodon*.

Ovatus crataegarius (Walker)   Mint Aphid   Plate 76
(= *menthae* Walker)

Appearance in life: A small, oval, yellowish-green to mid-green aphid on leaves of mint, or in spring colonies on young leaves of hawthorn, apple, or quince, where their feeding has little obvious effect on the host. Alatae produced on mint in summer are often much longer than apterae of the same generation and have a green abdomen without any black dorsal markings. Apterae 1.1–1.9 mm, alatae 1.5–2.1 mm.

Host plants: Primary hosts are pome fruits, most commonly *Crataegus* (*oxyacantha, monogyna*) but also sometimes *Malus* and *Cydonia*. Secondary hosts are *Mentha* spp. (*aquatica, piperita, longifolia, spicata*), occasionally certain other Labiatae (*Melissa, Nepeta*).

Distribution: Throughout Europe, the Middle East, Central Asia, India, Pakistan, Africa (Malawi, Zimbabwe, South Africa), Australia, New Zealand, the U.S.A., Canada (Quebec, Nova Scotia), and Brazil.

Biology: Heteroecious holocyclic between Pomoidea and Labiatae in cold temperate regions, overwintering as parthenogenetic forms on mint where winter conditions permit. On primary hosts, *O. crataegarius* is liable to confusion with the morphologically similar *O. insitus* (Walker); Müller and Hubert-Dahl (1979) discuss the taxonomic relationship and hybridization of these two species. Leonard (1963) gives a general review of *O. crataegarius*. In Japan, *O. crataegarius* is recorded from apple and quince but not from *Crataegus* or *Mentha* and there are slight morphological differences (Miyazaki, 1971), suggesting that a separate subspecies or species may be involved for which the name *malicolens* Hori is available. $2n = 12^*$.

Ovatus insitus (Walker)

Small, greenish-white, rather shiny aphids on undersides of young leaves of *Crataegus* spp. or *Mespilus germanica* in spring, sometimes on *Cydonia, Pyrus, Sorbus*, and possibly *Malus*, migrating for the summer to *Lycopus europaeus* where they live on the stem or rhizomes. Heteroecious holocyclic; throughout Europe and also recorded from Israel, Iran, and Turkey. Müller and Hubert-Dahl (1979) discuss the taxonomic relationship of this species with the closely related *O. crataegarius*. $2n = 12$.

Ovatus malisuctus (Matsumura)

Small, dark green aphids curling the young leaves of *Malus* spp. (*baccata, pumila, sieboldii*) and *Chaenomeles japonica*. Recorded from Japan, China, Korea, Taiwan, and Georgian S.S.R. Monoecious holocyclic on *Malus* in Japan, where the population ecology and feeding behaviour have been studied (e.g. Takeda, 1979).

Ovatus mentharius (van der Goot)

Small whitish-green aphids on undersides of leaves of *Mentha* spp. in Europe and Central Asia. Monoecious holocyclic on *Mentha*, with alate males. Hille Ris Lambers (1947b) gives an account of this aphid under the name *O. menthastri*.

Paracletus von Heyden   Pemphiginae: Fodini

A small palaearctic genus for 4 species resembling *Forda* except for the longer outer margin of the hind coxae. Only *P. cimiciformis* is well known, but Hille Ris Lambers (1954b) discriminates the species on the material then available.

Paracletus cimiciformis von Heyden   Plate 144

Apterae are medium-sized, shining, waxy yellowish white with body dorsoventrally flattened, especially at lateral margins. On roots of many species of Gramineae throughout Europe, the Mediterranean region including North Africa, the Middle East, Central Asia, Korea, and Japan. Holocyclic in Middle East, with *Pistacia palaestina* and *P. terebinthus* as primary hosts. The galls on *Pistacia* are flat folds of the edges of the leaves. Elsewhere it is exclusively anholocyclic on roots of Gramineae (e.g. *Agrostis, Festuca, Hordeum, Oryza, Poa, Triticum*), where it is always attended by ants, especially *Tetramorium* spp. The form on rice roots in the Far East may be a separate race or subspecies, *harukawai* Tanaka (Tanaka, 1961). For fuller information see Zwölfer (1958) and references therein. $2n = 16$.

Patchia Baker   Drepanosiphinae: Phyllaphidini

A small nearctic genus associated with Fagaceae, comprising rather stout-bodied aphids with long hairs on body and appendages, a rounded cauda but cleft anal plate, and alatae with wing veins darkly bordered.

Patchia virginiana Baker

On bark of chestnut (*Castanea* spp.) in eastern U.S.A. (Florida, North Carolina, Pennsylvania, Virginia). Apterae are small to medium-sized in life, rather uniformly velvety black, secreting wax. Alatae are brown with a large black dorsal abdominal patch and heavily bordered wing veins. Biology not known.

Patchiella Tullgren   Pemphiginae: Pemphigini

A monotypic genus associated with *Tilia* and Araceae.

Patchiella reaumuri (Kaltenbach)

In spring twisting the terminal growth of lime trees, clumping the leaves into large leaf 'nests'. Apterae (fundatrices) in galls are plump-bodied, yellowish brown, sometimes greenish; they produce alatae with broad, green to yellow-green abdomens which migrate to form colonies on root tubers of *Arum* spp. In England, Italy, Bulgaria, and Turkey. Specimens assigned to this species have also been collected on the roots of *Colocasia esculentum* (taro) in Hawaii and the Solomon Islands, suggesting that an anholocyclic form of *P. reaumuri* may be widespread, but often overlooked on the roots of Araceae. For a general account of the life cycle see Roberti (1939a). Stroyan (1979b) describes the morphs from *Arum* roots in England.

Pemphigus Hartig
Pemphiginae: Pemphigini

A genus of nominally about 65 rather elongate oval aphids mostly living on the roots of plants and/or forming galls in spring and early summer on the leaves and twigs of poplars of the subgenus *Eupopulus*. Four species are known to be monoecious on poplars, the sexuales being the progeny of alate migrants from the galls. Another 13 species migrate to known secondary hosts where they mainly feed on the roots but sometimes occur in woolly wax masses above soil level. The biologies of all the remaining nominal species are incompletely known, some being described from the gall-forming stages and others from the roots of plants. Some species may be permanently anholocyclic on roots. The species are distributed rather evenly throughout the Northern Hemisphere; 17 are described from Europe, 10 from the Middle East and Central Asia, 16 from the Orient, and 21 from North America. *Pemphigus* is taxonomically difficult, but accounts are available for Britain (Furk and Prior, 1975), Japan (Aoki, 1975), China (Chang and Zhong, 1979b), India (Ghosh, 1974), and North America (Harper, 1959); Lange (1965) summarized the biologies of the American species. Several species have been subjected to intensive biometric analyses (see Sokal *et al.*, 1980). Faith (1979) investigated strategies of gall formation.

Pemphigus species on *Beta*

Yellowish aphids secreting white wax, on roots of *Beta vulgaris*, also on *Chenopodium album* and possibly *Spinacia*. Economic damage is caused to beet crops in several parts of the world, but there is still some doubt as to the number of species of *Pemphigus* involved and their life cycles and host-plant relationships. In western North America, one or more beet-feeding species have a sexual stage on *Populus* (*angustifolia, balsamifera, trichocarpa*), in spring producing pouch-like protrusions of the mid-rib on either side of the leaf lamina, the mature galls being greenish, often tinged with red. Anholocyclic overwintering on beet roots also seems to occur widely. Grigarick and Lange (1962) confirmed a migration from *Beta* to *Populus trichocarpa* in California and suggested that the widely used name *P. betae* Doane was a synonym of *P. populivenae* Fitch. Harper (1963), studying the life cycle in Alberta, concluded that *Populus angustifolia* and *Populus balsamifera* were the principal hosts of *P. betae*; according to Harper (1959), the mid-rib galls of *P. betae* are on the upperside of the poplar leaf and those of *P. populivenae* on the underside. Whitham (1978) studied population dynamics of the gall generations of *P. betae*.

Similar root aphids have relatively recently become damaging to beets in Germany and the U.S.S.R. (Sedlag, 1953; Gaponova, 1970). Old World populations of *Pemphigus* on beet, to which the name *P. fuscicornis* (Koch) is currently applied, seem to be exclusively anholocyclic. It is not clear whether this aphid is an introduced North American species or has closer affinities with an Old World species, e.g. *P. immunis* Buckton, which normally alternates between *Populus nigra* and Euphorbiaceae.

Pemphigus bursarius (L.)
Poplar–Lettuce Aphid, Lettuce Root Aphid   Plate 136

Appearance in life: Apterae on roots of Compositae are elongate oval, yellowish white with a tuft of white wax on the posterior part of the abdomen. Alatae from root-feeding colonies (sexuparae) have a brownish-orange abdomen. Colonies are not attended by ants. Apterae 1.6–2.5 mm, alatae (sexuparae) 1.7–2.2 mm.

Host plants: Primary hosts are *Populus* spp., most frequently *P. nigra* and *P. nigra* var. *italica*. Secondary hosts are mainly Compositae (*Cichorium, Lactuca, Lampsana, Sonchus, Taraxacum, Tussilago*); records of *P. bursarius* from roots of plants in other families may be mainly referable to other *Pemphigus* spp., although it is possible that non-Composite plants are sometimes colonized. In Australia and New Zealand, *bursarius*-like aphids occur in *bursarius*-type galls on *Populus*, but root-feeding *Pemphigus* are mostly recorded from Chenopodiaceae and Polygonaceae; however, *Sonchus* roots are also often colonized, at least in New Zealand.

Virus transmission: Apparently not a vector of mosaic diseases of lettuce.

Distribution: Throughout Europe, and in the Middle East, Central Asia, Siberia, North Africa, South Africa, North America, and perhaps Autralia and New Zealand.

Biology: Heteroecious holocyclic between *Populus* and roots of Compositae. The galls are formed on the leaf petioles of *Populus* in spring; they are purse-shaped, yellowish or reddish when mature, and may contain aphids until late summer. Alatae leave through a lateral opening to found the root-feeding colonies on *Lactuca*, etc. These colonies produce sexuparae to fly back to poplar in autumn, but also often continue parthenogenetically through the winter. See Dunn (1959) and Rämert (1977) for accounts of the life cycle and biology. Dunn (1960) studied the natural enemies in England, and Judge (1968) studied the factors affecting the development of sexuparae. $2n = 20$.

Pemphigus passeki Börner

Described from roots of *Carum carvi* in Central Europe (Börner, 1952); it is very similar to *P. phenax* but separated on the basis of morphological differences in the sexuparae. Heteroecious holocyclic, probably with *Populus nigra* var. *italica* as primary host. $2n = 22$ (Gut, 1976).

Pemphigus phenax Börner & Blunck

Pale lemon yellow to yellowish-white apterae with white wax, very similar to *P. bursarius*, on roots of wild and cultivated *Daucus carota* in northern Europe and western Siberia. Alate sexuparae produced in autumn have a dull yellow abdomen; they fly to *Populus nigra* var. *italica*. The galls are formed in spring by a swelling of the mid-rib of the poplar leaf; they are usually reddish, tinged with yellow laterally, elongate spindle-shaped and rather wrinkled. Populations may also persist parthenogenetically on carrots remaining in the ground through the winter. For a more detailed account see Stroyan (1964).

Pemphigus populitransversus Riley   Poplar Leaf-stem Gall Aphid

Appearance in life: Apterae on roots of Cruciferae are medium-sized, oval, dirty pale yellow with dusky head and appendages, secreting white wax. They are not attended by ants. Alate sexuparae produced in autumn have a light greenish-yellow abdomen dusted with greyish wax powder.

Host plants: Primary hosts are *Populus* spp., especially *P. deltoides* and *P. sargentii*. Secondary hosts are Cruciferae (*Brassica, Coronopus, Lepidium, Rorippa*). Records on *P. populitransversus* from secondary hosts outside the Cruciferae are probably referable to other *Pemphigus* species.

Distribution: Widespread in the U.S.A. east of the Rocky Mountains (Sokal and Riska, 1981), extending into Canada (Saskatchewan) and Mexico; and introduced into South Africa.

Biology: Heteroecious holocyclic between *Populus* and Cruciferae, but often also overwintering parthenogenetically on roots of Cruciferae. The gall on *Populus* in spring is a green, purse-like swelling of the leaf petiole with a transverse slit, similar to that of *P. bursarius*. For a life history account see Jones and Gillette (1918) or Wene and White (1953). Morphological and biochemical variation of the gall populations have been studied intensively (e.g. Heryford and Sokal, 1971; Sokal and Riska, 1981; Setzer, 1980). $2n = 20$.

Pentalonia Coquerel   Aphidinae: Macrosiphini

A genus with 2 species of small brownish aphids with characteristically shaped symmetrically clavate siphunculi and a short pentagonal cauda. The wing venation of the alate morph is also highly characteristic, with the radius of the forewing strongly curved and fused with the media for part of its length, giving a closed cell. The hindwings are reduced and have only one oblique vein.

Pentalonia nigronervosa Coquerel Banana Aphid Plate 86

Appearance in life: Small, reddish-brown to almost black, oval aphids living under the old leaf bases of the host plant near ground level. The antennae are pale except for dark base and apex. Alatae are coloured like the apterae and have wing veins all broadly bordered with brown pigment, emphasizing the unusual pattern of venation described above. Apterae and alatae both 1.1–1.8 mm.

Host plants: Banana, Manila hemp (abaca), *Heliconia* spp., taro or coco-yam (*Colocasia*), *Caladium* spp., *Dieffenbachia,* ginger (*Zingiber* spp.), cardamon (*Elettaria*), *Alpinia* spp. and other members of the families Musaceae, Araceae, and Zingiberaceae.

Virus transmission: Particularly important as the sole vector of banana bunchy top in Australia, Africa, and Asia. Also able to transmit banana mosaic, abaca bunchy top, and cardamon 'katte' mosaic.

Distribution: Widespread through all tropical and subtropical parts of the world, and introduced into hothouses in Europe and North America.

Biology: Probably completely anholocyclic almost everwhere; sexual morphs have been recorded only from India and Nepal (Bhanotar and Ghosh, 1969; K.C. Sharma, personal communication). It is usually attended by ants. Kolkaila and Soliman (1954) studied feeding habits of *P. nigronervosa* on banana in Egypt, Ilharco (1968) covers general biology, and Eastop in Kranz *et al.* (1977) gives a brief general account. $2n = 14*$.

Phorodon Passerini Aphidinae: Macrosiphini

A genus of 4–5 *Myzus*-like species with anteriorly directed processes to the antennal tubercles and a host association with Cannabidaceae. The host-alternating species have *Prunus* as primary hosts. The hop and cannabis aphids are well known in Europe and the Middle East. The lesser known species occur in China and Japan. Other species described in *Phorodon* from Europe and America have subsequently been transferred to other genera such as *Capitophorus, Ovatomyzus,* and *Ovatus.*

Phorodon cannabis Passerini Cannabis Aphid Plate 78

Small to medium-sized (1.5–2.0 mm) aphids, shiny yellowish green with darker green longitudinal stripes, or pale to mid-green without stripes, on undersides of leaves and flower stems of *Cannabis sativa* and *C. indica*. Alatae have a dark dorsal abdominal patch. A vector of Cannabis streak virus, in central and southern Europe, central and south-west Asia, India, Nepal, Pakistan, Japan, and north Africa. Monoecious holocyclic on *Cannabis*, with alate males;

Müller and Karl (1976) give an account of the life cycle and population biology in East Germany. $2n = 12$ (Pal and Khuda-Bukhsh, 1980).

Phorodon humuli (Schrank)   Damson–Hop Aphid   Plate 77

Appearance in life: Apterae are sall to medium-sized, elongate oval, pale green to yellowish green with three darker green longitudinal dorsal stripes, rather shiny. Alatae have a dorsal abdominal black patch. Colonies occur on new growth of *Prunus* in spring, the leaves being only slightly curled; or on undersides of leaves, flowers, or fruits of *Humulus*. Apterae on *Prunus* in spring 2.0–2.6 mm, on *Humulus* in summer 1.1–1.8 mm; alatae 1.4–2.1 mm.

Host plants: Primary hosts are *Prunus* (*spinosa, insititia, domestica, mahaleb*). Secondary host is *Humulus lupulus*.

Virus transmission: A vector of hop mosaic and hop split leaf blotch viruses, and also able to transmit Arabis mosaic (= hop line pattern virus) to hops and plum pox ('Sharka disease') to plum, as well as several non-persistent viruses of other, non-host plants.

Distribution: Throughout Europe, Central Asia, north Africa (Ethiopia), and introduced into North America, where it is widespread.

Biology: Heteroecious holocyclic between *Prunus* and *Humulus*. There is an extensive literature, including population studies in Poland (Micinski and Ruskiewicz, 1975) and Norway (Stenseth, 1970), and special studies of flight activity (Taimr *et al.*, 1978) and resistance to insecticides (Lorriman and Llewellyn 1983). See also numerous papers on *P. humuli* in Hrdy and Hrdlickova (1981). $2n = 12$.

Phorodon humuli japonensis Takahashi

The south-east Asian form of *P. humuli* differs from the European form in the characters given in the key. It is heteroecious holocyclic between *Prunus mume* (and *P. salicina*) and *Humulus japonicus* in Japan, Taiwan, and Korea (Takahashi, 1965c). $2n = 12^*$.

Phorodon humulifoliae Tseng & Tao

Apterae are rather small to medium-sized, pale yellowish green or green with dark head, antennae, legs, and cauda. The siphunculi are strongly curved outward and brownish towards their apices. On leaves of *Humulus japonicus* in China (Chekiang, Szechuan). Also recorded from *Ulmus* sp. (Tao, 1963).

Phylloxera Boyer de Fonscolombe   Phylloxeridae

A genus containing nominally about 60 small oviparous aphids without siphunculi, living on Juglandaceae and Fagaceae. More than half the nominal species were described from *Carya* in North America (Pergande, 1904), and Fagaceae-feeding species occur in both Europe and North America. Accounts of the oak- and chestnut-feeding species are available for Europe (Heinze, 1962; Barson and Carter, 1972) and North America (Duncan, 1922). The status of some of the *Carya*-feeding species is at last being clarified (Stoetzel, 1981; Stoetzel and Tedders, 1981).

Phylloxera castaneae (Haldeman)

Very small, yellow insects in rows along the mid-rib on both sides of chestnut leaves, causing leaf-curl. On *Castanea* spp. (*dentata, pumila, sativa*); described from Pennsylvania, U.S.A., and also recorded from Maryland and Washington D.C. Biology is not known. Forms with long and short tubercles occur but it is not known whether they are specifically distinct (Duncan, 1922).

Phylloxera spp. on *Pecan*

Five species of *Phylloxera* are recognized on *Carya illinoiensis* in the U.S.A., and brief details are summarized below. For further information see Stoetzel (1981). Carpenter *et al.* (1979) discuss the varietal resistance to *Phylloxera* spp. in pecan.

Species	Position of galls	Generations per year	Distribution	Reference
P. devastatrix Pergande	Stem, twig, petiole, mid-rib, or nut	1	Arkansas, Florida, Louisiana, Mississippi, Oklahoma, Texas	Baker (1935)
P. notabilis Pergande	Leaflet, close to mid-rib	3–5	Arkansas, Georgia, Louisiana	Stoetzel and Tedders (1981)
P. perniciosa Pergande	Leaflet	?	Louisiana, Mississippi	Carpenter *et al.* (1979)
P. russellae Stoetzel	Leaflet, between secondary veins	1	Arkansas, Georgia, Louisiana, Mississippi, Oklahoma, Texas	Stoetzel (1981)
P. texana Stoetzel	Leaflet, between secondary veins	1	Texas	Stoetzel (1981)

Picturaphis Blanchard   Aphidinae: Macrosiphini

A genus for 4 or 5 *Myzus*-like legume-feeding aphids in South and Central America. They differ by their clavate siphunculi from the Central American legume-feeding genus *Microparsus*, and resemble the fern-feeding genus *Micromyzus* of the Old World tropics. Cermeli and Smith (1979) give keys for *Picturaphis*.

Picturaphis brasiliensis (Moreira) Plate 85

Apterae are small, shiny greenish dark brown to black in life, immature stages paler, brownish yellow. Alatae are dark brown to black with wing veins heavily black-bordered. On *Dolichos lablab* and *Phaseolus* spp. (*vulgaris, lunatus*); widely distributed in the Caribbean, Central and South America, and also recorded from Florida, U.S.A. Biology not known; presumably anholocyclic. Two other *Picturaphis* species are recorded from *Phaseolus*. *P. puertoricensis* Smith is shiny reddish brown in life and is so far only recorded from *Phaseolus adenanthus* in Puerto Rico. *P. vignaphilus* Blanchard is shiny dark brown to black in life, a rather larger species than *P. brasiliensis* with longer siphunculi. It is recorded from *Phaseolus vulgaris* and *Vigna* spp. (*luteola, sinensis*) in Argentina, Brazil, and Colombia.

Pleotrichophorus Börner   Aphidinae: Macrosiphini

A genus with 60 described species of mainly pale green to yellow aphids bearing numerous short capitate hairs and with slender, often elongate appendages. With only a few possible exceptions the species live on Compositae, with at least 44 species on members of the tribe Anthemidae, particularly *Artemisia* and *Chrysothamnus*. As in other largely anthemid-feeding genera, a few species live on members of other tribes of Compositae, particularly Astereae (*Aster, Erigeron*) and Inuleae (*Gnaphalium, Helichrysum*). The genus is predominantly American (49 species), with 7 European and only 3 central Asiatic and 1 Far Eastern species known. Many of the species were originally described in *Capitophorus*. Acounts are available for Europe (Holman, 1965; Leclant, 1968) and America (Hille Ris Lambers, 1969; Corpuz Raros and Cook, 1974).

Pleotrichophorus chrysanthemi (Theobald) Plate 69

Pale, green to yellowish, medium-sized, broadly spindle-shaped, living on undersides of leaves of *Chrysanthemum* spp., including cultivated *C. indicum*. Alatae have a pale yellow to green abdomen with widely-spaced paired dusky segmental markings. Described from South Africa, and since recorded from Europe (England, France, Ireland), Egypt, Zimbabwe, India (H.P.), Nepal, Korea, Australia, North America, and Brazil. There is a single alata in the BM(NH) collection from Japan, and possibly *Capitophorus formosanus* Takahashi, recorded from Japan and Taiwan on *Chrysanthemum*, is this species. *P. chrysanthemi* is apparently entirely anholocyclic; no sexual morphs are known. Very similar species, *P. glandulosus* (Kaltenbach) and *P. pseudoglandulosus* (Palmer), are holocyclic on *Artemisia* in the palaearctic and nearctic regions, respectively. Müller and Schöll (1958) discuss the differences between these species and *P. chrysanthemi*, and Eastop (1966) gives key characters for separating alatae of *P. chrysanthemi* from those of *P. glandulosus*.

Prociphilus Koch   Pemphiginae: Pemphigini

A genus for 50 species of rather small to very large aphids without siphunculi but with well-developed wax glands. They mainly alternate from pseudogalls on Rosaceae, Caprifoliaceae, and Oleaceae (and perhaps other families) to the roots of Coniferae. A few species go to the roots of monocotyledons, and occasionally dicotyledons. Twenty of the species are American, 20 occur in the Far East, 7 in Europe, and 3 in Central Asia. The full life cycle of many species is unknown. Accounts are available for north-west Europe (Heie, 1980), Japan (Inouye, 1956), China (Tao, 1970), and America (Smith, 1974).

Prociphilus erigeronensis (Thomas)

Appearance in life: Apterae on roots are medium sized, broadly oval, white to pale yellow with darker head, appendages, cauda, and anal plate, secreting white wax. They are usually ant-attended. Alatae, which appear rarely, have brown head and thorax and greenish-yellow abdomen. Apterae 1.7–2.3 mm, alatae about 2.2 mm.

Host plants: Roots of many plants, but with a preference for Compositae (*Ambrosia, Aster, Cichorium, Erigeron, Lactuca, Solidago, Tragopogon*), Gramineae (*Agrostis, Muhlenbergia, Poa, Triticum*), and Leguminosae (*Arachis, Phaseolus, Trifolium*). Alternation to either *Crataegus* or *Amelanchier* as primary host plant is suspected (see below).

Distribution: Widely distributed in the U.S.A., and also recorded from Ontario, Canada.

Biology: Apparently this species is predominantly anholocyclic, overwintering on the roots of plants or in ants' nests. Cutright (1925) described overwintering populations and the association with ants (*Lasius* spp.). Smith (1974) re-described the species, including the sexuparous morph, and discussed a possible holocycle involving *Crataegus* as primary host. It is possible that the name is being applied to more than one species.

Prociphilus oleae (Leach ex Risso)

Aphids living in compact colonies covered in white wax wool, on shoots near the base of the trunk of *Olea europea* in spring, and also recorded from *Phillyrea media*. Pseudogalls are not formed by this species. Originally described from the south of France, and re-described from Greece by Koronéos (1939). Heteroecious holocyclic, migrating from Oleaceae to the roots of an unknown secondary host, possibly vine roots (Barbagallo and Stroyan 1982).

Protrama Baker
Lachninae: Tramini

A genus of about 10 species with greatly elongate hind tarsi, and siphunculi present as pigmented flat cones. They mostly live on the roots of Compositae from western Europe and the Mediterranean to Central Asia and India. The species seem to be permanently parthenogenetic, at least in western Europe. Accounts are available for western Europe (Eastop, 1953), Poland (Szelegiewicz, 1977), and Central Asia (Narzikulov, 1963).

Protrama flavescens (Koch)

Apterae are medium-sized to rather large (2.7–3.8 mm), dirty white or cream with variably developed dark transverse bars on the dorsum. Hind tarsus very elongate, about two-thirds of the length of the hind tibia. Alatae and intermediates between apterae and alatae are very common. On roots of *Artemisia* spp., and sometimes on *Chrysanthemum, Cichorium,* and other Compositae, in England, Germany, Poland, and the U.S.S.R. Anholocyclic. $2n = 40-42$.

Protrama penecaeca Stroyan

Very large (3.8–5.0 mm) aphids, probably dirty greyish white in life with brown head, antennae, legs, siphuncular cones, and sclerites of posterior abdominal segments. Eyes reduced to 3–9 facets. On roots of *Helianthus tuberosus* in Jammu and Kashmir, India. Biology unknown; the apterous male of this species has been described (Verma, 1969), so there is possibly a holocycle.

Protrama radicis (Kaltenbach)
Plate 6

Apterae are medium-sized to rather large (2.5–3.4 mm), broadly oval to pear-shaped, colour in life dirty white to pale yellow or pale brownish green. The eyes are large, and small brown siphuncular cones are present. The hind tarsi are very elongate, more than three-quarters of length of hind tibiae. Ant-attended, on roots of various Compositae (*Arctium, Carduus, Centaurea, Cirsium, Cynara*) in Europe. $2n = 60$ approx.

Pseudaphis Hille Ris Lambers
Aphidinae: Macrosiphini

A genus comprising 2 East African species, differing from *Sitobion* by their completely flat frons without antennal tubercles and several other characters. Eastop (1958) distinguishes between the two species.

Pseudaphis abyssinica Hille Ris Lambers

Small, broadly oval, pale green aphids on leaves of *Rosa abyssinica* in Eritrea, and also collected from 'wild rose' in Ethiopia. Attended by ants. Life cycle is unknown. Since this aphid has now been collected more than once from roses, the earlier doubts about the correct host plant (Hille Ris Lambers, 1954a) seem to have been dispelled.

Pseudocercidis Richards — Aphidinae: Macrosiphini

A monotypic genus associated with *Rosa* in North America.

Pseudocercidis rosae Richards

Medium-sized aphids covered with white wax powder, inside curled or folded leaflets of *Rosa* spp. Recorded from *Rosa woodsii, R. acicularis*, and wild *Rosa* spp. in Alaska, Manitoba, British Columbia, and Utah, usually in shaded conditions; not yet recorded from cultivated roses. Monoecious holocyclic, with alate males. Robinson (1963) provides notes on biology. $2n = 12$.

Pseudoregma Doncaster — Hormaphidinae: Cerataphidini

A genus for 12 species of wax-dusted aphids living mostly on either bamboos or Zingiberidaceae, with a few species on other Gramineae and some other monocotyledons in South East Asia. *P. panicola* (Takahashi) is now widespread in the tropics on broad-leaved grasses. Most of the species were originally described in *Oregma*, and may be found in accounts of *Oregma* until 1966. Accounts of the bamboo-feeding species are available from Taiwan (Liao, 1976) and India (A.K. Ghosh, 1975).

Pseudoregma alexanderi (Takahashi)

Apterae are medium-sized to large (2.2–4.3 mm), rather elongate pear-shaped, pale brown to brownish black with a purple tinge, lightly dusted with wax powder which is often thicker on the posterior abdomen. Large colonies attack young shoots and leaves of bamboos (*Dendrocalamus, Miscanthus*) in Taiwan, Nepal, and north-east India. Life cycle is unknown. There is a remarkable polymorphism in the first instar, some larvae being 'pseudoscorpion-like' and apparently functioning as soldiers for the colony (Aoki and Miyazaki, 1978).

Pseudoregma bambusicola (Takahashi)
(= *bucktoni* M.R. Ghosh, Pal, & Raychaudhuri*)

Apterae are medium-sized to rather large (1.8–3.6 mm), oval bodied, grey

---

* M.R. Ghosh *et al.* (1974) proposed the name *bucktoni* to replace *bambusicola* Takahashi, because the Indian *Pseudoregma* on *Bambusa* was distinct from Japanese specimens on a slide labelled *bambusicola* by Takahashi, and because Doncaster (1966) had erected *Pseudoregma*, with type species *bambusicola* Takahashi, on the basis of specimens from Buckton's Indian material. However, material in the BM(NH) collection from Japan, India, and other countries agrees with Takahashi's *bambusicola* described (and illustrated) from Taiwan, so it seems sensible to retain this name for the widely distributed species on *Bambusa*.

brown to dark purplish brown, with rather sparse wax. In ant-attended colonies on young shoots and stems of *Bambusa* spp.; not recorded from other genera of bamboos. Alatae are rare. Found in Japan, Taiwan, Hong Kong, Thailand, Malaysia, Indonesia, India, Sri Lanka, and Fiji. $2n = 12^*$.

Pterocallis Passerini   Drepanosiphinae: Phyllaphidini

A genus for 17 small pale aphids with the terminal process usually shorter than the base of the sixth antennal segment, living on *Alnus, Carpinus, Corylus,* and rarely *Betula*. The subgenera *Mesocallis, Paratinocallis,* and *Recticallis* are known only from the Far East, India, China, and particularly Japan. Apterous viviparae occur more commonly than in most other Drepanosiphini, and particularly in the subgenus *Pterocallis s. str.*, which has 3 species on *Alnus* in Europe and another 3 species on *Alnus* in America, although the Japanese species are mostly on *Corylus*. Accounts are available for Britain (Stroyan, 1977), Scandinavia (Heie, 1982), Central Europe (Börner, 1949), Japan (Higuchi, 1972—*Mesocallis*, etc.), India (A.K. Ghosh, 1974—*Mesocallis*), the U.S.A. (Granovsky, 1928), and Canada (Richards, 1965).

Pterocallis (Paratinocallis) corylicola (Higuchi)

Adult viviparae all alate, small, pale yellow with usually only antennal segments V and VI, apices of tibiae, and tarsi, pale brown. Described from *Corylus sieboldiana* in Japan (Higuchi, 1972), and also recorded from *C. heterophylla* in Korea. Biology is unknown. $2n = 10^*$.

Pterocallis heterophyllus Quednau

Adult viviparae all alate, small, whitish to whitish yellow with apices of antennal segments dark (i.e. antennae 'banded'), and veins of forewing narrowly brown-bordered, with a black spot at the base of the pterostigma. Described from *Corylus heterophylla* in Korea (Quednau, 1979). Biology is unknown.

Pterocallis montanus (Higuchi)

Adult viviparae probably all alate, very small to small, pale yellow with apices of antennal segments brown. Described from *Corylus sieboldiana* in Japan (Higuchi, 1972). Biology is unknown. $2n = 16^*$.

Pterocallis (Mesocallis) pteleae (Matsumura)

Adult viviparae all alate, small, pale yellow with antennae mainly brown except for basal halves of segments IV and V. The forewings have the anal vein only bordered with brown. On *Corylus* spp. (*heterophylla* var. *thunbergii, sieboldiana*), *Alnus* spp., and *Betula* sp. in Japan and China. Described originally from *Ptelea*, but this may not be a true host. See Higuchi (1972) for further information. Biology is not studied.

Pterochloroides Mordvilko   Lachninae: Lachnini

A monotypic genus, characterized by the double row of large spinal tubercles on the abdomen.

Pterochloroides persicae (Cholodkovsky)   Plate 2

Appearance in life: Apterae are large, shiny, oval dark brown to black insects living along undersides of stout or medium-sized branches, and sometimes in large patches on trunk, of *Prunus* spp., and occasionally Pomoidea, attended by ants. Alatae have pigmented areas on the forewings. Apterae and alatae 2.7–4.2 mm.

Host plants: *Prunus* (especially *armeniaca, persica*, but also *amygdalus, domestica, cerasus, spinosa*); sometimes on *Malus pumila* or *Cydonia vulgaris*, and also recorded from *Citrus*.

Distribution: Mediterranean, Middle East, Central Asia, India, Pakistan; it has recently extended its range into Europe (Yugoslavia, Italy).

Biology: Monoecious holocyclic on *Prunus* in cooler regions; according to Archangelsky (1917) it only completes its life cycle on peach, apricot, and sloe. Anholocyclic in the Mediterranean, coastal areas of the Middle East, and southern Europe. Most serious damage is done to peaches, where heavy populations may kill branches. For a detailed early account, including relationships with ants and natural enemies, see Archangelsky (1917). Talhouk (1977), Velimirović (1976), and Ciampolini and Martorelli (1978) describe its population biology in Lebanon, Yugoslavia and Italy, respectively. $2n = 20^*$.

Pyrolachnus Basu and Hille Ris Lambers   Lachninae: Lachnini

A genus of 3 species of very large aphids closely related to *Longistigma* Wilson, but distinguished by the rather shorter, straight pterostigma of the forewing. A.K. Ghosh (1982) reviews the genus.

Pyrolachnus pyri (Buckton)

Apterae are very large (3.5–6.0 mm), dull yellow to dark brown aphids on bark of branches of pear trees (*Pyrus communis*), also recorded from *Pyrus malus, P. pashia* and *Eriobotrya japonica*. The wings of the alata are dark at their bases, and the large, conspicuous pterostigma of the forewing is dark brown to black. Found in Iran, India, Pakistan, Nepal, Sri Lanka, and Korea. Sexual morphs are unknown.

Reticulaphis Takahashi　　　　　　　　　Hormaphidinae: Cerataphidini

A small oriental genus of aphids producing galls on *Distylium* as primary host, one or more species apparently migrating to *Ficus*, although the life cycle has not been fully determined for any species. One species is described from *Shiia cuspidata*. The apterous viviparous forms on the secondary hosts are aleyrodiform and have a strongly reticulated dorsal cuticle.

Reticulaphis distylii (van der Goot)　　　　　　　　　Plate 132

Very small, black, oval insects, strongly sclerotized and flattened against the leaf with the legs mostly hidden; the body with or without a fringe of white wax. They live in small groups on the undersides of mature leaves of *Ficus* spp., often attended by ants. Immature stages are yellow and active. Apparently holocyclic between *Distylium* and *Ficus*, although the life cycle requires further investigation. Hille Ris Lambers and Takahashi (1959) recognized several 'subspecies' on *Ficus* in Taiwan and Java, and provided a key, but noted that the differences used could prove to be within the range of variation of one species. The most commonly encountered form on *Ficus, R. distylii fici* (Takahashi), has marginal hairs about 0.075 mm long with acute apices, and is recorded from China, Taiwan, Japan, Java, and the Philippines.

Rhodobium Hille Ris Lambers　　　　　　　　　Aphidinae: Macrosiphini

A monotypic genus, differing from *Acyrthosiphon* and *Metopolophium* in its very rough frontal tubercles, and from *Aulacorthum* in the presence of sensoria along the whole length of antennal segment III in the aptera, and the absence or only weak development of dorsal abdominal pigmentation in the alata.

Rhodobium porosum (Sanderson)　　　　　　　　　Plate 106

Appearance in life: Head of aptera brownish yellow, remainder bright yellow to yellow-green, rather shiny; living usually in young, partly unfolded leaves. Alatae have a bright green abdomen without black dorsal markings. Apterae 1.2–2.5 mm, alatae 1.4–2.2 mm.

Host plants: *Rosa* spp., especially cultivated varieties, and *Fragaria vesca*.

Virus transmission: A vector of strawberry mottle and strawberry crinkle viruses, but apparently not able to transmit rose mosaic or rose streak.

Distribution: Perhaps of North American origin in view of the existence of a holocycle in that region, but now also in Europe, the Middle East, north and east Africa, Mauritius, India, Australia, Fiji, Samoa, South America, and the Caribbean.

Biology: Monoecious holocyclic in North America, sexual morphs being found on both *Rosa* and *Fragaria* (MacGillivray, 1963). Elsewhere anholocyclic, and generally found only on cultivated roses, in glasshouses in northern Europe, or outside in warmer climates. $2n = 14*$.

Rhopalomyzus Mordvilko   Aphidinae: Macrosiphini

A genus for a few species with clavate siphunculi, small but evident antennal tubercles, and alatae with secondary sensoria on both third and fourth antennal segments. The apterae are similar to those of *Myzus, Neotoxoptera*, and *Dysaphis sorbi*. The species alternate between *Lonicera* and Gramineae in the temperate Northern Hemisphere. Accounts are available from Europe (Hille Ris Lambers, 1953a; Heinze, 1960), Central Asia (Narzikulov and Umarov, 1969), and North America (Palmer, 1952, in *Rhopalosiphum*).

Rhopalomyzus poae (Gill)

Apterae are small (1.2–2.0 mm), broadly pear-shaped, pale to dark brown or almost black, shiny and with a bluish waxy bloom. Alatae have extensive black dorsal abdominal pigmentation. This species lives on the basal parts of grasses (*Dactylis glomerata, Festuca ovina, Glyceria fluitans, Poa pratensis*, etc.) close to soil level and often on etiolated stems under stones. It is not a pasture pest, but is recorded as a vector of barley yellow dwarf and maize dwarf viruses. *R. poae* is heteroecious holocyclic, migrating to grasses from *Lonicera alpigena*, where its spring colonies cause the margins of the leaves to roll up and become spotted with red. Anholocycly on grasses is also common in regions with mild winter climates. It is not attended by ants. There are records from most European countries, and it is widely distributed in the U.S.A. Wood-Baker (1970) provides notes on the biology.

Rhopalosiphoninus Baker   Aphidinae: Macrosiphini

A genus for about 19 species with well-developed antennal tubercles, large and strongly clavate siphunculi, and a short triangular cauda, living on a variety of plants but occurring particularly in cryptic habitats in the summer. Ten of the species are from India and the Far East, 6 are European, and 3 are known only from North America. Accounts are available for Europe (Hille Ris Lambers, 1953a; Heinze, 1961), Japan (Miyazaki, 1971), India (Bhattacharya and Chakrabarti, 1982), and North America (Smith and Knowlton, 1977).

Rhopalosiphoninus latysiphon (Davidson)   Bulb-and-Potato Aphid   Plate 94

Appearance in life: Apterae are shiny dark olive green with very striking swollen shiny black siphuculi, living clustered on subterranean parts of growing plants or on bulbs or tubers in store. Immature stages are paler green with black siphunculi. Alatae have shiny olive-green to black dorsal abdominal markings. Apterae and alatae 1.4–2.5 mm.

Host plants: On bulbs (*Tulipa, Gladiolus*) and potato tubers in store, and on roots of many plants, especially in clay soils (e.g. potato crops), or on etiolated stems or runners growing in darkness under stones (e.g. *Bromus sterilis, Convolvulus arvensis, Potentilla anserina, Vinca major, Urtica* spp.)

Virus transmission: Recorded as a vector of cucumber mosaic and beet yellows viruses. It also has the ability to transmit potato leaf roll occasionally.

Distribution: Europe, Egypt, Rwanda, Kenya, South Africa, India, Nepal, Sri Lanka, Japan (?—see below), Australia, New Zealand, and North and South America.

Biology: Anholocyclic, overwintering on stored bulbs and potatoes in cold temperate regions; sexual morphs are not recorded. However, it is possible that *Rhopalosiphoninus deutzifoliae* Shinji, with spring populations on *Deutzia* in Japan, is the primary host form of *R. latysiphon*. For a summary of the biology and pest status of *R. latysiphon* see Forbes (1963). Haine (1955) studied its development on potatoes in Germany, and Gibson (1971) investigated its choice of feeding sites on potatoes. $2n = 6 (+1)$ (Gut, 1976).

Rhopalosiphoninus ribesinus (van der Goot)   Currant Stem Aphid

Apterae in life are medium sized, broadly oval, dull reddish brown to brownish black, in colonies on old branches of *Ribes* (*nigrum, rubrum*) near soil level, sometimes extending onto young shoots and undersides of leaves. It is only found in rather damp and shady situations. In Europe (the U.K., the Netherlands, Norway, Sweden, Germany, Poland, Latvia, Russia). Monoecious holocyclic, with apterous males. See Hille Ris Lambers (1953a) for further biological information.

Rhopalosiphoninus staphyleae (Koch)   Plate 95

Rhopalosiphoninus staphyleae tulipaellus (Theobald)   Mangold Aphid

Appearance in life: Apterae are broadly pear-shaped, dark olive green or brownish with very dark green or black dorsal markings; the siphunculi have the swollen part paler than the base or apex. In colonies on subterranean parts of host plant, or on bulbs or vegetables in store. Alatae have an olive-green abdomen with an extensive dark green to black sclerotic dorsal patch. Apterae and alatae both 1.5–2.4 mm.

Host plants: Primary hosts (in Europe) are *Staphylea* species (*pinnata, colchica*); secondary hosts of *staphyleae s. str.* are mainly Liliaceae and Iridaceae (*Tulipa, Hemerocallis, Crocus, Anthericum*), but also occasionally on roots or etiolated parts of plants in at least 11 other plant families (e.g. *Anemone, Capsella, Cardamine, Dentaria, Lamium, Oxalis, Vinca*). The form *tulipaellus* Theobald is particularly a pest of stored mangold beets (*Beta vulgaris*) in Europe, but is also recorded from a wide range of other plant genera including *Galium, Lycopersicon, Rumex, Tulipa,* and *Viola*.

Virus transmission: The form *tulipaellus* is able to transmit potato leaf roll, a number of non-persistent viruses including beet mosaic, cabbage black ring spot, cauliflower mosaic, potato virus Y, and tomato aspermy; it can also be an important vector of beet yellows (Tahon, 1964), which is also transmitted by the main species.

Distribution: Europe, Africa (Kenya, South Africa), Japan (?), Australia, New Zealand, North America, and Peru. The form *tulipaellus* seems to be mainly restricted to Europe, but may possibly have been introduced also to North America (see below).

Biology: Heteroecious holocyclic in Europe, alternating between *Staphylea* and the roots or concealed parts of monocotyledonous and some dicotyledonous plants, but anholocyclic populations are also common. On *Staphylea* in spring *R. staphyleae* is a much paler, yellowish aphid, causing the leaves to curl and become mottled with yellow. *R. staphyleae tulipaellus* in Europe is entirely anholocyclic on roots, or often on beets in store, but bulbs in store are apparently more commonly infested by populations representative of the main species. Elsewhere in the world *R. staphyleae* is probably entirely anholocyclic on secondary host plants. Hille Ris Lambers (1953a) gives an account of *R. staphyleae* in Europe and key characters for distinguishing the form *tulipaellus*, which is treated as a distinct species by Börner and Heinze (1957) and Heinze (1961). However, *tulipaellus* may be one of several asexual 'offshoots' of holocyclic *R. staphyleae*, and it is not clear whether one or more of such forms have been distributed outside Europe. $2n = 10$.

Rhopalosiphum Koch
Aphidinae: Aphidini

A genus for about 13 species associated with *Prunus* or Pyroidea as primary hosts, and Gramineae, Cyperaceae, or more rarely other plants in summer. Probably all the species are known from North America although *R padi* with *Prunus padus* as primary host is presumably palaearctic in origin and *R. maidis* and *R. rufiabdominalis* may have originated in the Far East. The other species described from Japan and South East Asia are probably only synonyms of more widely distributed species. Previously many other species with medium-length and slightly swollen siphunculi (e.g. *Hyadaphis, Lipaphis,*

Rhopalomyzus) were placed in *Rhopalosiphum*, but now it is restricted to species with the more dorsally placed lateral abdominal tubercles and characteristic antennal tubercles that distinguish the subtribe Rhopalosiphina. The precise limits of the genera *Rhopalosiphum, Melanaphis,* and *Schizaphis* are still uncertain. Accounts are available from Britain (Stroyan, 1972), northeast India (Raychaudhuri, 1980), Japan (Takahashi, 1965c), Australia (including most of the widely distributed Old World species; Eastop, 1966), Canada (Richards, 1960), and North America (Richards, 1962).

Rhopalosiphum insertum (Walker)   Apple Grass Aphid   Plate 16
(= *fitchii* Sanderson)

Appearance in life: Apterae in spring colonies on apple (or pear) are medium sized, oval, rather shiny, bright green to yellow green with a dark green spinal stripe, curling young leaves perpendicular to the mid-rib. Alatae are shiny with black head and thoracic lobes and green abdomen. Apterae on grass roots are smaller, yellowish green with yellowish-brown head and pronotum and dusky to dark-brown siphunculi. Apterae on Pyroidea in spring 2.1–2.6 mm, on Gramineae 1.4–2.0 mm; alatae from Pyroidea in spring 1.5–2.5 mm, from Gramineae 1.6–2.0 mm.

Host plants: Primary hosts are wild and cultivated *Malus, Pyrus communis, Cotoneaster, Crataegus, Sorbus,* and occasionally *Cydonia* and *Mespilus.* Secondary hosts are various Gramineae (*Agropyron, Agrostis, Festuca, Poa*).

Virus transmission: A vector of the persistent barley yellow dwarf virus.

Distribution: Europe eastward to Turkey and the Ukraine, and North America.

Biology: Heteroecious holocyclic between Pyroidea and the subterranean parts of Gramineae in both Europe and North America. Several names have been applied to this aphid in the literature. Prior to the biotaxonomic studies of Rogerson (1947), there was widespread confusion of the secondary host plant forms of *R. insertum* and *R. padi.* Records of the synonyms of *R. insertum* from cereals probably apply mainly to *R. padi.* In North America the name *R. fitchii* has been used until recently for *R. insertum* on its primary hosts and for *R. padi* on its secondary hosts. Vickerman and Wratten (1979) reviewed the ecology of *R. insertum* in Europe. $2n = 10$.

Rhopalosiphum maidis (Fitch)   Corn Leaf Aphid   Plate 15

Appearance in life: Apterae are rather elongate, with short antennae and short, dark siphunculi; yellow-green to dark olive green or bluish green, sometimes dusted with wax. Alatae have a yellow-green to dark green abdomen without dorsal dark markings anterior to the siphunculi. Apterae and alatae 0.9–2.4 mm.

Host plants: On young leaves of *Zea mays, Sorghum vulgare, Hordeum vulgare,* and members of more than 30 other genera of Gramineae including *Avena sativa, Secale cereale, Triticum, Oryza, Saccharum officinarum,* and occasionally Cyperaceae and Typhaceae.

Virus transmission: A vector of the persistent viruses barley yellow dwarf, maize leaf fleck, and millet red leaf, and also able to transmit several non-persistent viruses, e.g. abaca mosaic and sugar cane mosaic (including maize dwarf mosaic).

Distribution: Probably Asiatic in origin, now virtually cosmopolitan, but cannot survive outdoors in regions with severe winter climates.

Biology: Apparently entirely anholocyclic throughout the world on Gramineae; males occur sporadically, but oviparae and overwintering in the egg stage have never been observed. This is probably the most important pest of cereals in tropical and warm temperate climates, and there is an extensive literature. Hassan's (1957) studies in Egypt included accounts of the natural enemies, Muddathir (1976) studied populations on wheat in Sudan, and Wildemuth and Walter (1932) gave an account of *R. maidis* in North America. Resistance of cereals to *R. maidis* has been studied particularly in Kansas, U.S.A., by Painter and Pathak (1962); the significance of resistance-breaking biotypes of *R. maidis* is discussed by Blackman (1981). For a brief general account see Eastop in Kranz *et al.* (1977). $2n = 8, 9,$ and 10.

Rhopalosiphum nymphaeae (L.) — Water Lily Aphid

Apterae are medium-sized (1.6–2.6 mm), oval, reddish brown to dark olive, dusted with light grey wax, especially on first 4 abdominal tergites; feeding in spring on young twigs, leaf petioles, and fruit stalks of *Prunus* spp. (*domestica, persica, armeniaca, insititia, maheleb, serotina, spinosa*). Alatae are shiny black and dark brown with a whitish bloom on the ventral surface of the body. Heteroecious holocyclic, migrating for the summer to a large variety of water plants (*Alisma, Butomus, Nuphar, Nymphaea, Typha, Sagittaria, Juncus, Potamogeton,* etc.). Almost cosmopolitan. For brief general accounts see Davidson (1917—California) and Cottier (1953—New Zealand). $2n = 8$.

Rhopalosiphum padi (L.) — Bird Cherry–Oat Aphid   Plate 17

Appearance in life: Apterae on grasses and cereals are broadly oval, green mottled yellowish green or olive green, or dark olive to greenish black; often with rust-coloured patches around the bases of the siphunculi. Alatae have a pale to dark green abdomen. Apterae and alatae 1.2–2.4 mm.

Host plants: Primary host is *Prunus padus*, rarely *P. spinosus, P. tenellus,* or *P. virginiana*. Secondary hosts are numerous species of Gramineae, including all the major cereals and pasture grasses. Also sometimes recorded from Cyperaceae, Iridaceae (especially *Iris*), Juncaceae, and Typhaceae, and in England it has been found overwintering on dicotyledonous weeds (*Capsella, Stellaria*). Rautapää (1970) studied host-plant preferences in the laboratory.

Virus transmission: Able to transmit the persistent viruses barley yellow dwarf, filaree red leaf, maize leaf fleck, and rice 'giallume' (Belli *et al.*, 1975), as well as oat yellow leaf, abaca mosaic, onion yellow dwarf, maize dwarf mosaic, and several other non-persistent viruses.

Distribution: Presumably palaearctic in origin, now virtually world wide.

Biology: Heteroecious holocyclic between *Prunus padus* and Gramineae, or anholocyclic on Gramineae where winter conditions permit, and in many parts of the world where *P. padus* or alternative primary hosts are not available. There have been numerous field and laboratory studies on this aphid, but various names were used in the earlier literature; most of the records of *R. fitchii* and *R. prunifoliae* on cereals apply to *R. padi*. Ewing's (1916) classic studies on clonal inheritance were done with this aphid (as *Aphis avenae*). Dixon (1971) described the life cycle and discussed host alternation. Vickerman and Wratten (1979) reviewed the ecology of *R. padi* in Europe. Leather and Dixon (1981) studied the effect of cereal growth stage and feeding site on development and reviewed other work on host-plant relations of *R. padi*. Bode (1980) studied populations on *R. padus* and their natural enemies in Germany. $2n = 8$.

Rhopalosiphum padiformis Richards

Apterae are small to medium-sized (1.4–2.2 mm), green, with large red patches around bases of siphunculi; distinguished from *R. padi* and *R. insertum* by the long, pointed hairs on the posterior abdominal tergites (Richards, 1962). Described from the flower heads of *Poa* sp. in British Columbia, but recently also collected on *Triticum* in Montana, U.S.A. Biology is unknown, presumably heteroecious holocyclic with a primary host in the woody Rosaceae; alate males were obtained in a culture of the Montana population by T.W. Carrol and D. Yount (personal communication). $2n = 10*$.

Rhopalosiphum rufiabdominalis (Sasaki)   Rice Root Aphid   Plate 18

Appearance in life: Apterae on roots of secondary host plants are dark green or olive with usually a reddish area at the posterior end of the abdomen between and around the siphunculi. The abdomen of the alata is similarly coloured. Apterae and alatae 1.2–2.2 mm.

Host plants: Primary hosts are *Prunus* spp. (*yedoensis, mume, donarium* varieties, *persica*, etc.). Secondary hosts are numerous species of Gramineae,

Cyperaceae, and some dicots, particularly Solanaceae (potato, tomato). It is especially damaging to *Oryza sativa*; incidence of *R. rufiabdominalis* on different rice cultivars in India was studied by Dani and Majumdar (1978).

Virus transmission: A vector of barley yellow dwarf virus (Paliwal, 1980).

Distribution: Virtually world wide, but of economic importance chiefly in warmer climates.

Biology: Heteroecious holocyclic in Japan between *Prunus* and roots of many plants, especially Gramineae. Tanaka (1961) described the life cycle in Japan and Hsieh (1970) studied the population biology on rice in Taiwan. In most parts of the world, *R. rufiabdominalis* is anholocyclic on roots of secondary host plants. Doncaster (1956) and Yano *et al.* reviewed the information on this aphid. $2n = 8$ (Gut, 1976).

Sappaphis Matsumura   Aphidinae: Macrosiphini

A genus of 3 or 4 species of hairy aphids with short cauda, having *Pyrus* as primary host in the Far East. Accounts are available from Japan (Miyazaki, 1971) and China (Zhang and Zhong, 1980). The name *Sappaphis* was often used from 1952 to 1965 in a broader sense to include species now placed in *Dysaphis*.

Sappaphis piri Matsumura

Apterae are broadly oval, small to medium-sized, reddish brown or yellowish brown, covered with white wax; living on undersides of leaves of *Pyrus communis* in China, Korea, and Japan. Heteroecious holocyclic, migrating to roots of *Artemisia* spp. Also recorded from other *Pyrus* spp., and from *Prunus serotina*. Takada and Shiga (1974) described a specific Aphidiid parasite. $2n = 12$.

Schizaphis Börner   Aphidinae: Aphidini

A genus for 40 species resembling *Rhopalosiphum* but with more cylindrical or tapering siphunculi and the media of the forewing only once-branched. Probably about half the species are holocyclic on Gramineae and most of the rest are on Cyperaceae, but a few overwinter as eggs on *Pyrus*. More than half the species are European, and the others occur in Central Asia and the Middle East (3 species), Africa (1), the Far East and South East Asia (7), and North America (7 species). Eastop (1961b) provided a key to the world fauna then known, and more detailed accounts are available for Europe (Hille Ris Lambers, 1939, 1947b), China (Tao, 1962), and Australia (Eastop, 1966).

Schizaphis graminum (Rondani)   'Greenbug'   Plate 19

Appearance in life: Apterae are small, rather elongate oval, with head and prothorax yellowish or greenish straw-coloured, rest of thorax and abdomen yellowish green to bluish green with a darker spinal stripe; on leaves of grasses and cereals, their feeding often causing yellowing and other phytotoxic effects. Alatae have a brownish-yellow head and prothorax, black thoracic lobes, and yellowish-green to dark green abdomen. In both morphs the siphunculi are pale but usually have dark apices. Apterae and alatae 1.3–2.1 mm.

Host plants: Many species of grasses and cereals including *Agropyron, Avena, Bromus, Dactylis, Eleusine, Festuca, Hordeum, Lolium, Oryza, Panicum, Poa, Sorghum, Triticum*, and *Zea*. Several subspecies or species closely related to *S. graminum* have been described from particular grasses, especially *Phleum pratense* (see Orlob, 1961).

Virus transmission: A vector of barley yellow dwarf, millet red leaf, sugar cane mosaic (including maize dwarf mosaic), and western wheat mosaic viruses; and of a virus disease of rice (rice yellow mottle?) in the U.S.S.R. (Luk'yanchikov and Sorokina, 1979). Rochow (1960) studied variation between clones in the ability to transmit barley yellow dwarf virus.

Distribution: Southern Europe, the Middle East, Central Asia, Africa, India Nepal, Pakistan, Thailand, Korea, Taiwan, Japan, and North, Central, and South America. Records from Australia and Philippines appear to be *S. hypersiphonata*. Records of *S. graminum* from grasses in western Europe are now thought to apply to other species (Tambs-Lyche, 1959; Stroyan, 1960; Pettersson, 1971).

Biology: Monoecious holocyclic on Gramineae, with alate males, in cold temperate climates, e.g. in the northern U.S.A. where overwintering in the egg stage occurs predominantly on *Poa pratensis*. Anholocyclic wherever winter conditions permit. Severe outbreaks of this aphid on wheat and barley, especially after its first introduction into North America in about 1882, and after the appearance of a sorghum-preferring biotype there in 1968 (which subsequently acquired insecticide resistance), have generated much published work. Starks and Burton (1977) described methods of identifying and studying greenbug biotypes in the mid-western U.S.A. Biological control has been attempted in Oklahoma with introduced parasites (Eikenbary and Rogers, 1974) and predators (Cartwright *et al.*, 1977). Muddathir (1976) studied populations on wheat in Sudan. Walker *et al.* (1972) provided a bibliography of English language papers on *S. graminum*, and for a brief general account see Kranz *et al.* (1977, p. 339). $2n = 8$.

Schizaphis hypersiphonata Basu

Small, oval, yellowish-green aphids feeding on leaves of Gramineae, similar to

S. graminum but apparently not having the same phytotoxic effects. The terminal process of the antenna is longer (always more than 1.5 times the length of antennal segment III in alatae), and the siphunculi and hairs on the eighth abdominal tergite are also longer than in *S. graminum*. Recorded from West Bengal, Sri Lanka, Thailand, the Philippines, and Australia (referred to by Eastop, 1966, as a form of *graminum*). It appears to occur especially on *Digitaria*, but is recorded from wheat in the Philippines. It is a vector of abaca mosaic virus. This may be the species to which Mordvilko's (1921) description of *S. graminum* applies.

Schizaphis rotundiventris (Signoret) (= *cyperi* van der Goot)

Small (1.2–1.7 mm), dark green to almost black aphids sometimes forming small colonies in the young, unfolding leaves of Palmae, especially *Elaeis guineensis*, although the more normal hosts are *Cyperus* species. Probably this is the secondary host form of a species using *Pyrus communis* as primary host in Asia (see below), with one or more anholocyclic races distributed widely on Cyperaceae in Old World tropical and warm temperate regions. Recorded (mainly on *Cyperus*) from southern Europe, the Middle East, Africa, Mauritius, India, Pakistan, Nepal, Sri Lanka, Taiwan, Japan, the Philippines, Malaysia, Indonesia, Australia, New Zealand, and Hawaii. Lecland (1967b) provided an account of the '*S. cyperi* group'.

Schizaphis (Paraschizaphis) scirpi (Passerini)

Dark green or brown to almost black shiny aphids normally colonizing the leaf bases of Cyperaceae (*Carex, Scirpus*) in Europe and Central Asia; recorded from *Iris* in Sweden (Ossiannilsson, 1959).

Schizaphis species on *Pyrus*

Three species of *Schizaphis* have been described from *Pyrus; S. piricola* (Matsumura) from Japan, *S. pyri* Shaposhnikov (= *S. fritzmuelleri* Leclant) from European U.S.S.R., and *S. punjabipyri* (Das) from Pakistan. Their life cycles presumably all involve alternation to Gramineae or Cyperaceae, but have not been properly investigated. *S. fritzmuelleri* Leclant, described from *Carex divulsa* in France, seems to be the secondary host form of *S. pyri*, and *S. rotundiventris* (see above) could be the secondary host form of *S. punjabipyri*, but experimental work is necessary to confirm this. *S. piricola*, recorded from Japan and Korea on *Pyrus sinensis* and *Pyrus* sp., is yellowish brown or green, with siphunculi concolorous with body except for their darker apices. It forms pseudogalls by folding the edges of the leaves of *Pyrus* (Essig and Kuwana, 1918).

S. punjabipyri is a darker, greyish or blackish aphid dusted with white wax, with dark siphunculi and cauda, living inside a slightly inflated pod-shaped pseudogall, formed by preventing the young pear leaves from unfolding, so that the edges of the leaf remain stuck together even when it is mature (Das, 1918). It is recorded from *P. communis* in India and Pakistan. *S. pyri* is green with dark siphunculi and cauda, producing a similar pod-like gall to that of *S. punjabipyri*. Alatae of *S. pyri* have 1–4 secondary sensoria on antennal segment V as well as sensoria on III and IV, whereas alatae of *S. punjabipyri* have sensoria on III and IV only. In southern and south-eastern Europe, and probably further east into Central Asia.

Semiaphis van der Goot Aphidinae: Macrosiphini

A palaearctic genus of about 14 species similar to *Hyadaphis* but with very short siphunculi. The species are heteroecious between *Lonicera* and Umbelliferae, or live without host alternation on one or other of these host-plant groups. A few species occur on other plants, including two on *Impatiens*. Accounts are available from Central Europe (Heinze, 1960), European U.S.S.R. (Shaposhnikov, 1964), Siberia (Ivanoskaya, 1977), and Japan (Miyazaki, 1971).

Semiaphis dauci (Fabricius) Plate 66

Apterae are small (1.3–1.6 mm), pale blue-green with dark brown head and legs and very short, dark brown siphunculi; the body has a waxy bloom. Alatae are similarly coloured with black head and thorax. Colonies occur on upper sides of rolled leaves and leaflets, or later in umbels of wild and cultivated carrots; its feeding can cause severe stunting and rolling of young shoots. It is able to transmit celery (western) mosaic virus, and several other non-persistent viruses. Monoecious holocyclic on *Daucus carota*, with alate males, in Europe and the Middle East. Anholocyclic populations apparently occur in southern England and probably elsewhere. Prior (1971) gives an account of this species.

Semiaphis heraclei (Takahashi)

Apterae are small, oval, mainly pale green to yellowish green with a waxy bloom, feeding on stem or on curled leaves of Umbelliferae (*Apium graveolens*, *Daucus carota*, *Coriandrum sativum*, *Heracleum*, *Cryptotaenia*, *Angelica*, etc.). Alatae are similarly coloured but with black head and thorax. *S. heraclei* is probably heteroecious holocyclic; *Brachycolus lonicerae* Shinji is believed to be the primary host form on *Lonicera gracilipes*. It is recorded from Japan, Ryuku, Korea, China, Taiwan, Sumatra, India, and Hawaii. $2n = 8^*$.

Sinomegoura Takahashi Aphidinae: Macrosiphini

A genus for about 5 species with the head spinulose ventrally and with weakly clavate, calf-shaped siphunculi and an elongate cauda, living on shrubs in eastern and south-east Asia. Accounts are available for Japan (Miyazaki, 1971), China (Tao, 1963), and India (Raychaudhuri, 1980).

Sinomegoura citricola (van der Goot) Plate 121

Appearance in life: Apterae are medium-sized, broadly spindle-shaped, shiny dark brown, living on undersides of leaves or young growth of host plants. The antennae and legs are dark, the siphunculi are basically pale, dark on apical one-third, and the cauda is conspicuously large and dark. Apterae 1.4–2.7 mm, alatae 1.8–2.7 mm.
Host plants: Numerous species of tropical shrubs in about 20 families including *Bridelia, Camellia, Cinnamomum, Citrus, Eurya, Ficus, Mangifera, Murraya, Musa, Persea*, and *Thea*.
Distribution: India, Sri Lanka, China, Thailand, Taiwan, Malaysia, Indonesia, the Philippines, and Australia.
Biology: Life cycle is not known; no sexual morphs are recorded and the species is apparently mainly or entirely anholocyclic. It is not very common on *Citrus* and not known to be a vector of tristeza virus.

Sinomegoura rhododendri (Takahashi)

Apterae are small to medium-sized, green, broadly spindle-shaped with pale legs and antennae; siphunculi are pale basally, darker at apices, and the cauda is dark. On undersides of leaves of *Rhododendron* spp. (*arboreum, oldhami*), and also reported from *Gardenia, Wendlandia*, and *Symplocos* (Raychaudhuri, 1980). In India, Taiwan, and Japan.

Sipha Passerini Chaitophorinae: Siphini

A genus for 12 spiny-haired, Gramineae-feeding aphids with short siphunculi and 5-segmented antennae with a terminal process less than 2.5 times the base of the fifth segment. The two North American species have a smooth dorsum and a knobbed cauda and two of the Old World species have a spinulose dorsum and knobbed cauda (*Sipha s. str.*), while the remainder have a smooth dorsum and a broadly rounded cauda (subgenus *Rungsia*). Accounts are available for Britain (Stroyan, 1977), Scandinavia (Heie, 1982), European U.S.S.R. (Shaposhnikov, 1964), and the Rocky Mountain Region, U.S.A. (Palmer, 1952).

Sipha (Rungsia) elegans del Guercio
 (= *agropyrella* Hille Ris Lambers
 = *kurdjumovi* Mordvilko)

Appearance in life: Apterae are rather small, elongate oval, rather dorsoventrally flattened, brownish yellow, or yellowish brown with a distinctly paler spinal stripe. Alatae have a yellowish-brown abdomen with black sclerites which are separate on anterior tergites and fused into transverse bars on posterior tergites. This aphid usually forms colonies on the upper surfaces of the leaf blades of grasses and cereals, often causing them to roll upwards and develop yellow patches. Apterae and alatae 1.4–2.1 mm.

Host plants: mainly grasses (*Agropyron* spp., *Arrhenatherum elatius, Festuca pratensis, Hordeum murinum*) but also sometimes on wheat and barley. Robinson and Hsu (1963) list 12 species on which colonies were found in Manitoba.

Distribution: Europe, Central Asia, and introduced into North America.

Biology: Monoecious holocyclic on Gramineae, with apterous males, in both Europe and North America. Dixon and Shearer (1974) investigated the factors influencing its choice of feeding site. Bode (1977) gives an account of this species on wheat in Germany. $2n = 6$.

Sipha flava (Forbes)   Yellow Sugar Cane Aphid   Plate 8

Appearance in life: Apterae are small, oval, yellow (green at low temperatures), with numerous long bristle-like hairs on the body; often there are paired, dusky transverse intersegmental markings on the dorsum. Alatae have a yellow abdomen with a variably developed dorsal pattern of dark markings. They occur in large colonies on leaf blades of grasses and cereals. Apterae and alatae 1.3–2.0 mm.

Host plants: Numerous species of Gramineae including *Digitaria, Hordeum, Panicum, Paspalum, Pennisetum, Saccharum, Sorghum,* and *Triticum.* Also recorded from Cyperaceae (*Carex, Cyperus*). Starks and Mirkes (1979) studied resistance of cereal crops to *S. flava.*

Virus transmission: A vector of sugar cane mosaic virus.

Distribution: North America, north as far as New York State in the east and Washington in the west; Caribbean, and Central and South America.

Biology: Monoecious holocyclic on Gramineae, with apterous males, in areas with cold winters. Anholocyclic wherever winter temperatures permit. Hayward (1944a) gives an account of this aphid on sugar cane in Argentina, and (1944b) a bibliography of the literature to that date. Smith *et al.* (1963) describe it in Puerto Rico.

Sipha glyceriae (Kaltenbach)

Apterae are rather small (1.5–2.4 mm), oval, somewhat flattened dorsoventrally; in life dull, pale green to mid-green, sometimes reddish, with paler legs, and often with a paler spinal stripe. Dorsal surface of body with numerous spine-like hairs. Monoecious holocyclic, with dark brown apterous males, on many species of Gramineae, especially those growing in marshy habitats (*Agropyron, Aira, Alopecurus, Glyceria, Holcus, Hydrochloa, Oryza sativa, Phalaris arundinaceae, Poa*); occasionally also found on Cyperaceae and Juncaceae. It occurs throughout Europe and eastward to Ukraine, Turkey, and Iran. *S. glyceriae* can be injurious to rice crops, although not a vector of rice yellows virus (Osler and Longoni, 1975). $2n = 12$. (Gut, 1976).

Sipha (Rungsia) maydis Passerini   Plate 9

Appearance in life: Apterae are rather small, pear-shaped, somewhat dorsoventrally flattened, shining dark brown to almost black on dorsal surface, which is fully sclerotized. They usually live on upper sides of leaf blades near the bases, and sometimes on stems or flower heads, of grasses and cereals, often attended by ants. Heavily infested leaves may become yellowed, rolled into tubes, and desiccated. Alatae have a solid black patch extending over abdominal tergites 4–7, with separate dark transverse bands on tergites 1–3. Apterae 1.0–1.9 mm, alatae 1.3–2.0 mm.

Host plants: Recorded only from Gramineae, but feeding on numerous species within this family including, in drier climates, all the economically important cereal crops. Not usually found on cereals in northern and western Europe.

Virus transmission: Able to transmit cucumber mosaic virus.

Distribution: Europe, the Mediterranean, the Middle East, Central Asia, India, Pakistan, and South Africa.

Biology: Holocyclic monoecious on Gramineae, with apterous males, in Turkey (Tuatay and Remaudière, 1964). The sexuales have not been recorded elsewhere, and overwintering is probably usually anholocyclic. $2n = 12$.

Sitobion Mordvilko   Aphidinae: Macrosiphini

A genus for about 75 species of *Macrosiphum*-like aphids but with less well-developed antennal tubercles and often with a more sclerotic dorsum which bears shorter hairs. Seven of the species live on pteridophytes and the remainder are about equally distributed between monocotyledons, particularly Gramineae, and dicotyledons. Several species are known to alternate from dicotyledons (*Rosa, Rubus, Akebia*) to Gramineae, and some of the other dicotyledon-feeding species may prove to be the primary host forms of species on Gramineae. The 6 European species are associated with Gramineae (4) and pteridophytes (2). The 24 African species include 9 grass-feeders and 13 from dicotyledons. The 30 Asiatic species include only 6 from Gramineae, but 10 from orchids and other monocotyledons and 13 from dicotyledons. The 14 American species include 5 from ferns and 5 from dicotyledons. Accounts are available for Europe (Holman, 1961, added to by Stroyan, 1969b), East Africa (Eastop, 1958), West Africa (Eastop, 1961a), Japan (Miyazaki, 1971, in *Macrosiphum*), China (Tao, 1963), India (David, 1976), north-east India

(Raychaudhuri, 1980, in *Macrosiphum*), and California (Hille Ris Lambers, 1966c).

Sitobion africanum (Hille Ris Lambers) Plate 112

Apterae are medium-sized (1.6–3.1 mm), broadly spindle-shaped, variably coloured yellowish green to green, reddish, yellowish, or greenish brown; reddish or brownish forms are usually densely wax-covered. There is also a very variably developed dorsal abdominal pattern of dark pigmentation; there may be an extensive, broken pattern of sclerotization of the whole dorsal surface or, at the other extreme, the cuticle of the paler green forms may be almost completely membranous. Alatae have a pattern of transverse, narrow, often interrupted dark bars on the dorsal abdomen.

S. *africanum* occurs on Gramineae (*Brachyaria, Bromus, Cynodon, Eragrostis, Panicum, Setaria, Zea*) in Africa (Angola, Botswana, Cameroun, Eritrea, Ghana, Kenya, South Africa), and it is also recorded from Syria. Aphids collected from wheat in Chile have also been assigned to this species. In Africa, S. *africanum* or very similar species have also been collected quite frequently from various dicotyledons (*Euphorbia, Ficus, Pavetta*, etc.). Müller and Schöll (1958) give an account of the variation and host plants of S. *africanum* in South Africa. It may eventually transpire that this name is being applied to a complex of species. The biology of the group is unstudied; Müller and Schöll (1958) describe oviparous females from *Ficus*.

Sitobion akebiae (Shinji)

Apterae are medium-sized to rather large, broadly spindle-shaped, variably coloured green, yellowish green or orange red, with dorsal abdomen pale or variably pigmented. In Japan and Korea, Takahashi (1964) states that it is holocyclic heteroecious between *Akebia* and secondary hosts mainly in the Gramineae, but Miyazaki (1971) found fundatrices on *Stellaria* and *Platanus*, so overwintering eggs may be laid on various plants. Summer host plants include *Agropyron, Festuca, Oryza, Poa,* and *Triticum*; also *Gladiolus, Plantago, Ranunculus,* and *Trifolium*. The more widely distributed S. *miscanthi* is very difficult to separate from S. *akebiae* and may perhaps be the same species. For further information see Yano et al. (1983).

Sitobion anselliae (Hall) (= *schoelli* Müller)

Apterae are small to medium-sized (1.4–2.2 mm), broadly spindle-shaped, grass green to yellow green with black siphunculi. Alatae have a green or yellow-green abdomen with dark marginal sclerites but no dorsal abdominal sclerotization anterior to the siphunculi. On wild and cultivated orchids (*Ansellia, Cattleya, Epidendrum*, and also collected from *Kniphofia* (Liliaceae); in Angola, Kenya, South Africa, and Zimbabwe.

Sitobion avenae (Fabricius) — Grain Aphid — Plate 113

Appearance in life: Apterae on grasses and cereals are medium-sized, broadly spindle-shaped, yellowish green or dirty reddish brown, sometimes rather shiny, with black siphunculi not much longer than the pale cauda. The dorsal cuticle is distinctly and uniformly sclerotic, but variably pigmented, from colourless with faint intersegmental markings to dark brown. Alatae are similarly coloured but with more distinct dark dorsal intersegmental markings. Apterae 1.3–3.3 mm, alatae 1.6–2.9 mm.

Host plants: Numerous species of Gramineae, including all the cereals and pasture grasses of temperate climates, and many other monocotyledons. Lowe (1981) studied resistance and susceptibility of spring and winter wheats to *S. avenae*. Rautapää (1970) studied preferences among a wider range of host plants.

Virus transmission: A vector of barley yellow dwarf virus, and also able to transmit bean yellow mosaic, pea mosaic, and radish yellows.

Distribution: Europe, the Mediterranean, the Middle East, Central Asia, India, Nepal, Pakistan, Africa (Ethiopia, Libya, Morocco, Zimbabwe, South Africa); North, Central, and South America. (Oriental records refer to *S. miscanthi*.)

Biology: Monoecious holocyclic, with alate males, on many species of Gramineae; the fundatrix and the holocycle are described by Müller (1977b). Anholocyclic overwintering is common in regions with mild winters. On cereals the aphids prefer to feed on the upper leaves, and on the ears once these have emerged. Vickerman and Wratten (1979) review the ecology of *S. avenae* on cereals in Europe. Population development has been modelled in western Europe (Carter and Rabbinge, 1980). Factors determining alata production were studied by Ankersmit and Dijkman (1983). Enzymes of *S. avenae* and its parasites in U.K. were characterized by Castanera *et al.* (1983). Biological control has been attempted in Chile (van den Bosch, 1977). $2n = 18$.

Sitobion chanikiwiti (Eastop)

Apterae are rather small (1.5–2.0 mm), green, with usually pale siphunculi dusky only at their apices. Recorded from *Cynodon dactylon, Eleusine indica, Pennisetum purpureum*, and other tropical grasses in Africa (Kenya, Malawi, South Africa, Uganda, Zimbabwe). Biology is not known.

Sitobion fragariae (Walker) — Blackberry–Cereal Aphid — Plate 114

Appearance in life: Apterae on *Rubus* are rather large, broadly spindle-shaped, yellowish green with small brown intersegmental dorsal markings; the siphunculi are wholly dark or pale basally and brownish apically. Apterae on grasses are smaller, dirty greenish yellow with brown intersegmental markings and black siphunculi; but also with a tendency for the whole dorsal abdomen to be tanned to a smoky or even blackish brown. Alatae have a yellowish-green abdomen with very distinct dark intersegmental ornamentation. Apterae (on *Rubus*) 2.0–3.0 mm, (on grasses) 1.6–2.9 mm; alatae 2.0–3.0 mm.

Host plants: Primary hosts are *Rubus* spp., especially *R. fruticosus*, and probably overwintering in the egg stage also occurs sometimes on *Fragaria, Rosa,* and *Geum,* as oviparae are often found on these plants. Secondary hosts are numerous species of Gramineae (particularly *Avena, Bromus, Dactylis, Holcus, Hordeum (murinum), Phleum*), and occasionally Juncaceae. Only small populations occur on wheat and barley.

Virus transmission: A vector of barley yellow dwarf virus.

Distribution: Europe, the Mediterranean, the Middle East, South Africa, Australia, New Zealand, Antipodes, and western North America.

Biology: Heteroecious holocyclic between *Rubus* and Gramineae; the life cycle and ecology of *S. fragariae* in Europe are reviewed by Vickerman and Wratten (1979). $2n = 18$.

Sitobion graminis Takahashi

Apterae in life are medium sized (1.8–2.9 mm), dark reddish brown covered with grey wax dust, often of a bluish appearance; in alcohol they are red, with a variably developed pattern of dark sclerites on the dorsal abdomen. Antennae, legs, siphunculi, and cauda are all very dark. On many species of Gramineae (*Bouteloua, Chloris, Cynodon, Digitaria, Eragrostis, Glyceria, Oryza, Setaria, Sporobolus, Triticum*), usually feeding on the flower stems. In Africa (Angola, Cameroun, Kenya, Madagascar, South Africa), India, Malaya, New Guinea, and the Philippines. Biology is not known.

Sitobion howlandae (Eastop)

Rather small (1.7–2.2 mm), bright yellow aphids with dark apices to the antennal segments, and siphunculi which are basally pale and apically dark. Recorded from *Zea mays* and *Sorghum arundinaceum* in Africa (Cameroun, Kenya, Malawi, Nigeria, South Africa, Zimbabwe). Biology is not known.

Sitobion ibarae (Matsumura)

Medium-sized (2.1–3.1 mm), yellow, yellowish-green, or yellowish-brown aphids with rather long, tapering dark siphunculi and a long pale cauda; living on young leaves and flower buds of wild and cultivated *Rosa* spp. In summer dwarf forms may occur in which the dorsal abdominal cuticle is tanned dark brown. Monoecious holocyclic, or often anholocyclic, on roses; in Japan, Ryuku, Korea, China, Taiwan, Sumatra, Malaya, and the Philippines. This species was sometimes confused with *Macrosiphum rosae* in the earlier oriental literature.

Sitobion indicum Basu

Apterae are medium-sized, spindle-shaped, with a brown to blackish irregularly shaped and perforated sclerotic area on abdominal tergites 1–5, black siphunculi, and a pale brown cauda. Body colour in life is not recorded, but according to the original describer, probably yellowish green. On *Cymbidium* spp. (Orchidaceae) in hilly regions of north-east India. Apparently anholocyclic, and collected usually in glasshouses. This aphid is clearly closely related to *S. luteum*; in addition to the key characters, the dorsal body hairs are longer than in *S. luteum* (see David, 1976), and there is also a difference in chromosome number. $2n = 18$ (Kurl, 1980).

Sitobion lambersi David

Apterae are medium-sized, spindle-shaped, pale green, without any dorsal sclerotization. The siphunculi are long and brown, darker apically; the cauda almost colourless. Alatae have a pale green abdomen with only very small dusky intersegmental sclerites. On stems and in flower heads of tropical grasses (*Bothriochloa pertusa, Chloris barbata, Cynodon dactylon, Digitaria marginalis, Panicum maximum, Paspalum conjugum*, etc.), and occasionally on Cyperaceae. Recorded from southern India, Sri Lanka, Cameroun, Kenya, Nigeria, Zimbabwe, and Fiji. *S. lambersi* is recognized easily by the presence of only 2 long hairs on the cauda, but the name is possibly being used for a complex of species with this character in the Old World tropics. Biology is unstudied and sexual morphs are unrecorded.

Sitobion leelamaniae (David)   Banded Cereal Aphid

Apterae are small to medium-sized, spindle-shaped, pale yellowish green. The antennae have brown joints giving them a banded appearance, with the whole of segment VI dark; the siphunculi are pale brown, the legs and cauda are pale. Alatae have a pale green abdomen, and dark brown antennae and siphunculi. Unlike most grass-feeding *Sitobion*, it lives on the undersides of the leaves and can severely damage seedling plants. Host plants include *Eleusine coracana, Pennisetum typhoides*, and *Sorghum vulgare*; it is recorded from southern India, Sri Lanka, and Cameroun.

Sitobion luteum (Buckton)   Yellow Orchid Aphid

Apterae are small to medium-sized (1.3–2.1 mm), spindle-shaped, bright greenish yellow to pale yellowish green, with a black oval sclerotic area extending over abdominal tergites 1–5, and a narrow black transverse band on the metathorax. The antennae are black except at their bases, the siphunculi are black and the cauda pale yellow. Alatae do not have the black oval patch on the dorsum, only dark intersegmental markings. On orchids (*Dendrobium, Epidendrum*, etc.), usually under glass. Recorded from Europe, Mauritius, India, Singapore, Australia, Fiji, Tahiti, the U.S.A. (New Mexico, New York), and Central and South America. Apparently entirely anholocyclic; no sexual morphs have been recorded. $2n = 12$.

Sitobion miscanthi (Takahashi)

Appearance in life: Apterae are medium-sized, broadly spindle-shaped, variably coloured green, reddish brown to dark brown with abdominal tergites variably tanned from pale with only very small intersegmental sclerites to very dark; in the latter case the pigmented area may be entire or segmentally divided. Siphunculi are shiny black, cauda relatively pale. Alatae are similarly coloured but with larger and darker dorsal abdominal markings. Apterae 1.7–3.0 mm, alatae 1.9–3.1 mm.

Host plants: Many species of Gramineae including *Avena, Bromus, Chloris, Cynodon, Hordeum, Oryza, Paspalum, Pennisetum, Secale, Setaria,* and *Triticum.* Sometimes *S. miscanthi* is also found in large numbers on dicotyledons, particularly semi-aquatic species, e.g. *Polygonum hydropiper* (Eastop, 1966).

Virus transmission: A vector of barley yellow dwarf and millet red leaf persistent viruses.

Distribution: India, Pakistan, Bangladesh, Nepal, Sri Lanka, Vietnam, China, Taiwan, Borneo, Malaya, Australia, New Zealand, Fiji, Tahiti, Tonga, Cook Islands, and Hawaii. Dispersal of *S. miscanthi* by flight from Australia to New Zealand in 1967–8 has been suggested (Close and Tomlinson, 1975).

Biology: Probably *S. miscanthi* is anholocyclic almost everywhere. An ovipara collected on *Polygonum chinense* in India may be this species, and alate males collected from *Poa annua* and an unidentified fern in India have also been assigned to *S. miscanthi* (see discussion by David, 1976). *S. akebiae* in Japan and Korea is closely related to *S. miscanthi,* and may be a synonym, in which case one would expect any holocycle to involve overwintering as eggs on various plants, as in Japan. The taxonomy of the *miscanthi/akebiae* group requires clarification by biological studies.

Sitobion nigrinectaria (Theobald)

Medium-sized (1.8–2.9 mm), green aphids with black siphunculi and long pale cauda living on *Cajanus indicus*, and occasionally found on other Leguminosae (*Glycine javanica, Crotalaria, Dolichos, Rhynchosia*); in Africa (Angola, Cameroun, Ethiopia, Kenya, Malawi, Madagascar, Rwanda, Uganda, Zimbabwe) and also recorded from Mauritius. Apparently not a vector of groundnut rosette virus (Eastop, 1958). Life cycle is unknown. Odebiyi (1979) studied field populations on *Cajanus* in Kenya. $2n = 18$.

Sitobion rosaeiformis (Das)

Medium-sized to rather large (2.1–3.3 mm), narrowly spindle-shaped, bright yellowish-green, green, or pinkish-brown aphids on new growth of wild and cultivated roses in India and Pakistan. The dorsal abdomen is normally pale except for small brown intersegmental sclerites, and the siphunculi are basally pale and apically dark in apterae, wholly dark in alatae. Holocyclic, and probably monoecious on roses, with alate males, but overwintering of parthenogenetic forms also occurs. For additional information see David (1976). $2n = 18$ (Kurl and Misra, 1979).

Sitobion rubiphila (Takahashi) (= *kamtshaticum* Mordvilko?)

Green aphids with black siphunculi and a pale cauda, on young shoots of *Rosa multiflora* and *Rubus* in Japan. Alatae are unusual for *Sitobion* in having secondary sensoria on antennal segment IV as well as III. Heteroecious holocyclic, migrating from *Rosa* and *Rubus* to Gramineae (*Microstegium, Digitaria*). For further information see Miyazaki (1971). *Anameson kamtshaticum* Mordvilko, described from Kamtshatica, U.S.S.R., may be this species.

Slavum Mordvilko   Pemphiginae: Fordini

A small palaearctic genus of 4 species closely related to *Aploneura*. They produce sac-like or coral-like galls on *Pistacia* spp. in the Middle East and Central Asia. Three species produce emigrant alatae from the galls and are apparently heteroecious, but the identities of the secondary host forms are not established. Davatchi (1958) revised the genus. Two species form galls on *Pistacia vera:*

Slavum lentiscoides Mordvilko

This is possibly the primary host form of *Aploneura ampelina* (q.v.). The galls are bag-like outgrowths of the mid-rib of the leaf, similar in position to those of *Geoica*, but with pointed apices. On *Pistacia vera* and *P. mutica* in Central Asia, Iran, and Crimea.

Slavum mordvilkoi Kreutzberg

This species produces branched, coral-like galls on wild *Pistacia vera* in Turkmeniya, Afghanistan, and eastern Iran. Only the primary host part of the life cycle is known.

Smynthurodes Westwood   Pemphiginae: Fordini

A monotypic genus, characterized by the elongate second antennal segment and the thick sclerotized rims of the primary sensoria on the last two antennal segments.

Smynthurodes betae Westwood   Bean Root Aphid   Plate 141
(= *Trifidaphis phaseoli* Passerini)

Appearance in life: Apterae on roots are small to medium-sized, almost globular, dirty yellowish white, wax-dusted, with light brown head, prothorax, antennae, and legs; the whole insect is clothed with numerous fine hairs. Alatae have dark transverse bars on the abdominal tergites. Apterae 1.6–2.7 mm, alatae 1.9–2.9 mm.

Host plants: Primary hosts are *Pistacia atlantica* and *P. mutica*. Secondary hosts are numerous dicotyledons, particularly Compositae (*Artemisia, Arctium*), Leguminosae (*Phaseolus, Vicia, Trifolium*), and Solanaceae (*Solanum tuberosum, S. nigrum, Lycopersicon esculentum*); also sometimes on *Beta, Brassica, Capsella, Gossypium, Heliotropum, Rumex*, etc. Rarely on monocotyledons (Gramineae, Cyperaceae).

Distribution: Virtually world wide on secondary host plants.

Biology: Heteroecious holocyclic with a two-year cycle in the Mediterranean region; the galls on *Pistacia* are yellow-green or red, spindle-shaped, formed by rolling of the edge of the leaflet near its base. Alatae from the galls migrate to the roots of numerous dicotyledonous plants (see above). Anholocycly is probably common even where primary hosts are available, and elsewhere reproduction is exclusively parthenogenetic, with overwintering in ants' nests common. See Mordvilko (1935) for fuller information, or Heie (1980) for a recent brief account. $2n = 8$.

Subacyrthosiphon Hille Ris Lambers   Aphidinae: Macrosiphini

A genus for 1 species living on *Trifolium repens* in north-west Europe, with characters between *Acyrthosiphon* and *Aulacorthum*. The peculiar habitat, the presence of distinct antesiphuncular sclerites in the aptera, and the unusual sensoriation of the antennae in the alatae which have relatively few secondary sensoria usually restricted to the basal half of segment III, are distinctive features.

Subacyrthosiphon cryptobium Hille Ris Lambers

Apterae are medium-sized, broadly oval, pale olive green with the head sometimes faintly reddish. The antennae are apically blackish, but otherwise the appendages are concolorous with the body. Alatae have a pale olive green abdomen with dark marginal and antesiphuncular spots. Monoecious holocyclic, with apterous males, living hidden on the older parts of prostrate stems of *Trifolium repens*, and dropping when disturbed so that they are rarely observed or collected. The plant shows no reaction. Recorded from England, the Netherlands, Norway, Sweden, and the U.S.A. For further information see Hille Ris Lambers (1947a).

Takecallis Matsumura   Drepanosiphinae: Phyllaphidini

A genus for 4 species resembling *Myzocallis* but with the clypeus bearing an anteriorly directed tubercle. Usually all viviparae are alate, and all the species live on bamboos. The genus is oriental in origin and accounts are available by Higuchi (1968, 1972) and Stroyan (1977).

Takecallis arundicolens (Clarke)

Adult viviparae are medium-sized (1.8–2.8 mm), all alate; whitish, pale yellow or greyish yellow, without dark dorsal abdominal markings, but with a black cauda. They live under the mature leaves of bamboos (*Arundinaria, Bambusa, Phyllostachys, Pleioblastus, Sasa*) in China, Japan, and Korea, and introduced into Europe and California. Life cycle in the Far East is not known; anholocyclic where introduced, although Leclant (1966) has reported oviparae in southern France. $2n = 18^*$.

Takecallis arundinariae (Essig)

Alatae (1.7–2.4 mm) are whitish, pale yellow or greyish yellow, with a paired series of black elongate spots on abdominal tergites 1–7, and a pale cauda. An apterous morph is described, but is found only at high altitudes in Taiwan. This aphid lives under the mature leaves of bamboos (*Arundinaria, Bambusa, Dendrocalamus, Phyllostachys, Pleioblastus*). It occurs in India, China, Taiwan, Korea, Japan, and introduced into England and North America. Life cycle in the Far East is unknown; sexual morphs are not recorded, and the aphid is anholocyclic where introduced. $2n = 18$.

Takecallis taiwanus (Takahashi)

Adults all alate, 1.4–2.3 mm, pale green with a pale cauda, living in young, still unrolled leaves and on new shoots of bamboos (mainly *Arundinaria* spp. and *Phyllostachys* spp., occasionally *Bambusa, Sasa*). In Taiwan and Japan, and introduced into Europe, South Africa, New Zealand and the U.S.A. Cottier

(1953) gave an account of this species under the name *T. arundinariae*. Life cycle and sexual morphs unknown, probably anholocyclic everywhere. $2n = 16*$.

Tetraneura Hartig   Pemphiginae: Eriosomatini

A genus for about 20 species in which the apterae have only 1-segmented tarsi, and the alatae have a simple media in the forewing and usually the fourth antennal segment much shorter than the fifth. At least 7 species alternate between galls on *Ulmus* and the roots of Gramineae, and most of the lesser known species are described either from *Ulmus* or from grass roots. *Tetraneura* is predominantly oriental and eastern Palaearctic with only *T. ulmi* in north-west Europe, *T. caerulescens* in south-east Europe and *T. africana* in the Mediterranean and Middle East. Hille Ris Lambers (1970b) revised the world fauna, and Chakrabarty and Maity (1978) and Raychaudhuri Pal and Ghosh, (1978) give accounts of the Indian species.

Tetraneura africana van der Goot (= *cynodontis* Theobald)

Rather large (2.0–3.3 mm), very globose, buff-coloured to pale brownish with a dark brown head and prothorax, in root-feeding colonies on *Cynodon dactylon*, and also recorded from *Sorghum halepense, Calamagrostis* spp., and rarely on cereals. Willcocks (1925) provides a colour plate. Recorded from Italy, Morocco, Egypt, Iraq, Transcaucasia, and Turkestan. A smaller aphid collected on the roots of *Calamagrostis epigeios* in Poland may also be this species (Hille Ris Lambers, 1970b). The galls of *T. africana* on *Ulmus* have not been identified, although Mordvilko (1935) described sexuparae collected from *Ulmus campestris* in Italy. Populations on *Cynodon* roots are probably mainly or entirely anholocyclic.

Tetraneura akinire Sasaki

Apterae are small to medium-sized, very globose; pale yellowish brown, lightly covered with wax. In root-feeding colonies on Gramineae (*Cynodon, Digitaria, Echinochloa, Oryza, Setaria*), and in hairy, stalked galls formed as outgrowths of the leaves of *Ulmus* spp. (*campestris, foliacea, parvifolia, davidiana* var. *japonica*) in spring. Recorded from Czechoslovakia, Hungary, Italy, Yugoslavia, Georgian S.S.R, Japan, and Korea. In Europe, *T. akinire* has rarely been encountered on Gramineae. This aphid has been confused with the very similar *T. nigriabdominalis* and Tanaka (1961) may have had both species. Hille Ris Lambers (1970b) discusses their separation. A detailed account of this aphid in Italy is given by Roberti (1972).

Tetraneura basui Hille Ris Lambers

Rather small (1.5–1.8 mm), globose; colour in life not recorded. On roots of

Gramineae (*Echinochloa, Eleusine, Oryza, Pagonantherum, Polypogon, Setaria*, and others) in north-east India (Raychaudhuri, 1980). It is not known whether there is a gall generation on *Ulmus*.

Tetraneura javensis van der Goot

Small to medium-sized, globose, whitish to yellowish-white aphids on roots of sugar cane (*Saccharum officinale*), known so far from Java, New Guinea, Pakistan, and India (Maharashta, Mysore). Probably heteroecious holocyclic in Pakistan between *Ulmus wallichii* and *Saccharum* (Hille Ris Lambers, 1970b), although the galls have not been described and the life cycle needs to be confirmed by host-plant transfers. *T. javensis* has also been found on roots of *Neyraudia arundinaceae* in Pakistan, and possibly occurs on other Gramineae.

Tetraneura nigriabdominalis (Sasaki)   Plate 139
(= *hirsuta* Baker)

Appearance in life: Rather small, greenish- or brownish-white, plump, oval-bodied apterae clustered on roots of Gramineae; body is not so globose as in most other *Tetraneura* species. Its presence on the roots of some hosts is indicated by a reddish-purple discolouration of the leaves. Alatae have a shiny black head and thoracic lobes and a brown abdomen. Apterae 1.5–2.5 mm, alatae 1.5–2.3 mm.

Host plants: Primary hosts are *Ulmus* spp. (usually *U. davidiana* var. *japonica*, and also recorded from *U. parvifolia*). Secondary hosts are numerous species of Gramineae (*Agropyron, Cenchrus, Chloris, Cynodon, Echinochloa, Eleusine, Eragrostis, Oryza, Panicum, Paspalum, Saccharum, Setaria* and others). Tanaka (1961) lists the degrees of preference for particular species. It is known particularly as a pest of rice; field resistance of rice varieties in India was studied by Dani and Majumdar (1978).

Distribution: Africa, India, Nepal, Bangladesh, Pakistan, Sri Lanka, Japan, Korea, Indonesia, Malaysia, the Philippines, Australia, New Zealand, Fiji, Tonga, Central America, Caribbean, the U.S.A. There are also single alatae in the BM(NH) collection from France, Greece, and Iran.

Biology: Heteroecious holocyclic in Japan between *Ulmus* and roots of Gramineae. Tanaka (1961) gives an account of the biology in Japan. The galls on *U. davidiana* var. *japonica* are elongate protrusions of the upper surface of the leaf, shaped like narrow-necked inverted flasks and covered with short hairs. Throughout most of its range *T. nigriabdominalis* is anholocyclic on the roots of Gramineae. Hille Ris Lambers (1970b) distinguished several forms on the basis of abdominal chaetotaxy, and erected a subspecies, *bispina*, for his material from Africa and North America. However, most of the African material in the BM(NH) collection does not conform to *bispina* and probably there are many distinct clonal populations distributed around the world. Raychaudhuri (1980) distinguished two populations in north-east India. Gadiyappanavar and Channabasavanna (1973) studied the population biology on *Eleusine coracana*. $2n = 17^*$ (one sample).

Tetraneura radicicola Strand

Tetraneura yezoensis Matsumura

These two species are taken together since it is not possible to separate them reliably on their secondary host plants. They are medium-sized, pale yellowish, brownish or pinkish white, very globose, with brown head and prothorax, on roots of various Gramineae. Members of the *radicicola/yezoensis* group are recorded from roots of *Echinochloa, Eleusine, Eragrostis, Imperata, Miscanthus, Oryza, Saccharum, Setaria, Triticum,* and other Gramineae, and from India, Nepal, Sri Lanka, Japan, Taiwan, Malaysia, the Philippines, and Australia. Presumably anholocyclic on grass roots everywhere, but in Japan at least there is an alternation to *Ulmus davidiana* var. *japonica*; the galls produced on *Ulmus* in spring are similar to those of *T. nigriabdominalis*, but without hairs. For further information see Hille Ris Lambers (1970b) and Raychaudhuri (1980). $2n = 14^*$ (*radicicola*), $2n = 12^*$ (*yezoensis*); Akimoto, personal communication.

Tetraneura ulmi (L.) Elm−Grass Root Aphid Plate 140

Appearance in life: Apterae on roots are medium-sized, globose, pale orange yellow, yellowish white or reddish, with head, prothorax, and appendages brown; the body lightly dusted with wax. Ant-attended, and often in ants' nests. Alatae have a shiny black head, prothorax and appendages, and a dusky abdomen. Apterae 1.7−2.8 mm, alatae 1.7−2.5 mm.

Host plants: Primary hosts are *Ulmus* spp. (*campestris, glabra*). Secondary hosts are numerous species of Gramineae (especially *Agropyron, Bromus, Dactylis, Deschampsia, Festuca, Holcus, Hordeum, Lolium, Poa, Zea*).

Virus transmission: A vector of maize dwarf virus.

Distribution: Europe, Central Asia, the Middle East (Iran, Iraq, Syria, Turkey), and introduced into North America.

Biology: Heteroecious holocyclic between *Ulmus* and roots of Gramineae; the gall on *Ulmus* in spring is bean-shaped, higher than wide, smooth and shiny red, green or yellow, with a stalk-like connection to the upper surface of the leaf. However, anholocyclic overwintering on grass roots or in ants' nests predominates, root-feeding colonies producing rather few sexuparae in autumn. Zwölfer (1957) has made a detailed study, and for a brief account see Heie (1980). $2n = 14$ and 16^*.

Thelaxes Westwood   Thelaxinae

A genus for 3 or 4 species with spinulose 5-segmented antennae with a short terminal process, and with a knobbed cauda. The winged forms hold the wings flat over the back at rest, and the media of the forewing is only once-forked. Adult viviparae are conspicous on the shoots and young leaves of oaks until early summer when aestivating larvae are produced and conceal themselves in crevices, under the marble galls of cynipids, etc. Ilharco (1967) keys the European species and compares their first instar larvae with those of the single American species.

Thelaxes suberi (del Guercio)

Small pale green aphids on leaves and young shoots of Fagaceae; usually found on *Quercus* spp., but del Guercio (1914) described two aphids, *Vacuna castaneae* and *V. carlucciana*, from *Castanea sativa* in Italy, which are believed to be synonymous with *T. suberi*. Ilharco (1966) gives an account of this species on *Quercus* in Portugal. *T. suberi* occurs in the Mediterranean region (Portugal, Italy, Sardinia, Madeira, Algeria, Morocco and the Middle East, Israel, Iraq, Turkey), and Ilharco records that he has seen specimens from the U.S.A. $2n = 8^*$.

Therioaphis Walker   Drepanosiphini: Phyllaphidini

A genus for 24 maculate aphids with knobbed cauda living on Leguminosae of the tribes Trifolieae, Loteae, Galeagneae, and Coronilleae. Half the species are known only from south-east Europe and the Middle East, 4 extend into north-west Europe and 4 or 5 extend across Russia, 1 as far as China. The world fauna was revised by Hille Ris Lambers and van den Bosch (1964), and Szelegiewicz (1969b) gives an account of the Hungarian species.

Therioaphis luteola (Börner)

Apterae are medium-sized, pale yellow with paired, faintly dusky dorsal patches on each segment; sometimes white and without dusky patches. Alatae have a similarly coloured abdomen with only the dorsal hair-bearing tubercles pigmented, forming 4 rows of small dark spots. Monoecious holocyclic, with alate males, and apparently specific to *Trifolium pratense*. Recorded from Austria, Denmark, England, Germany, Italy, the Netherlands, Sweden, and Yugoslavia. Alatae are rather regularly obtained by sweeping clover, or in yellow traps, but apterae are encountered more rarely.

Therioaphis subalba Börner

Small, pale yellow, whitish anteriorly; living on *Trifolium* spp. (*alpestre, medium*, apparently not *pratense*), and recorded from Austria, Germany, Sweden, and Yugoslavia. Aphids found on *Melilotus sauveleons* in Korea have also been provisionally assigned to this species. Biology is unknown; presumably monoecious holocyclic.

Therioaphis trifolii (Monell)   Yellow Clover Aphid
Therioaphis trifolii forma maculata (Buckton)
Spotted Alfalfa Aphid   Plate 13

Appearance in life: Apterae are distinctive, pale yellow, greenish white to almost white, rather shiny, with rows of dorsal light or dark brown pigmented tubercles bearing capitate hairs. Alatae also have rows of pigmented tubercles on the dorsal abdomen, varying in extent and depth of pigmentation. Apterae and alatae both 1.4–2.2 mm.

Host plants: Many species of Leguminosae in the genera *Astragalus, Lotus, Medicago, Melilotus, Onobrychis, Ononis,* and *Trifolium*. Recognizable forms with more specific host-plant relationships occur within the species, the best known being the Spotted Alfalfa Aphid. Various resistance-breaking genotypes of the Spotted Alfalfa Aphid have been described (Nielson and Don, 1974).

Virus transmission: Able to transmit lucerne mosaic and clover (red) vein mosaic viruses.

Distribution: Europe, the Mediterranean, North Africa, the Middle East, India, and Pakistan. The following introductions into other parts of the world seem to have occurred:

(1) To eastern U.S.A., about 1882; a form (the Yellow Clover Aphid) feeding almost exclusively on *Trifolium pratense*. Now widely distributed in North America, but particularly in central and eastern U.S.A.

(2) To south-western U.S.A., about 1953; a form (the Spotted Alfalfa Aphid) found feeding mainly on, and very injurious to, *Medicago sativa*. Now distributed throughout the U.S.A., wherever its host plant is grown, and in Mexico. In about 1977, this form was introduced into Australia (New South Wales, Queensland, Victoria), and by 1982 it had also reached New Zealand.

(3) To South Africa, in about 1980; a form capable of feeding on both *Medicago* and *Trifolium* (H.J. Dürr, personal communication).

(4) To Japan, possibly a recent (about 1980) introduction, as *T. trifolii* has only been found in Japan once previously, in 1944; collected from *Medicago* spp. (*lupulina, denticulata, sativa*), and not yet found on *Trifolium* (Yano et al., 1982).

Biology: Monoecious holocyclic, with alate males, on Leguminosae in cold temperate climates of northern and central Europe and more northerly U.S.A.; anholocyclic in warmer regions. The Yellow Clover Aphid in North America is holocyclic. The Spotted Alfalfa Aphid was anholocyclic when first introduced into southern U.S.A., but acquired the ability to produce viable sexuales and over-winter as eggs in about 1960, in the course of its spread to more northerly latitudes (Manglitz *et al.*, 1966). Cross matings between the two North American forms have been attempted (Manglitz and Russell, 1974). Hille Ris Lambers and van den Bosch (1964) discussed the morphological variation in the species and its taxonomic significance. Lehane (1982) reviewed biological control measures against the Spotted Alfalfa Aphid in the U.S.A. and in Australia. $2n = 16$.

Tiliaphis Takahashi — Drepanosiphinae: Phyllaphidini

A genus of about 4 oriental species very similar to *Eucallipterus*, but with longer, pale siphunculi and more extensive pigmentation of the forewings. Higuchi (1972), Quednau (1979), and Zhang and Zhong (1982) have described and compared species.

Tiliaphis shinae (Shinji)

Adult viviparae are all alate, small, yellow with brown markings on the wings and on the sides of the head and prothorax. Monoecious holocyclic on *Tilia* spp. (*japonica, maximowicziana*, etc.) in Japan and Korea (Higuchi, 1972). Shinji (1931) gives $2n = 14$ for *T. shinae*, but since he did not distinguish between *T. shinae* and *T. shinjii* this number could apply to either species.

Tiliaphis shinjii Higuchi

Adult viviparae all alate, rather small, colour in life not recorded. Forewings and sides of head and thorax marked with brown, and a pair of brown patches on each abdominal tergite. On *Tilia japonica* and *Tilia* sp., in Japan.

Tiliphagus Smith — Pemphiginae: Pemphigini

The single species differs from *Prociphilus* in having more numerous sensoria on the antennae.

Tiliphagus lycoposugus Smith

Medium-sized to rather large aphids producing large leaf-nest galls in spring on *Tilia americana* in the U.S.A. (North Carolina, Maryland). The terminal leaves are clumped and cupped so as to resemble a poorly-formed head of lettuce,

10-30 cm in diameter. The fundatrices inside the galls are large, brown, globose, producing reddish-brown alatae which fly to form summer colonies on the roots of *Lycopus virginicus*. For further information see Smith (1965).

Toxoptera Koch   Aphidinae: Aphidini

A genus for about 4 species resembling *Aphis* but with a stridulatory apparatus consisting of latero-ventral ridges on the abdomen and peg-like hairs on the hind tibiae (Fig. 8). The group is Far Eastern in origin although 2 species are now widely distributed on *Citrus* and other shrubs. Tao (1961) provided a key to the alatae, and the more widely distributed species were keyed and illustrated by Tao and Tan (1961) and Eastop (1966).

Toxoptera aurantii (Boyer de Fonscolombe)   Black Citrus Aphid   Plate 43

Appearance in life: Apterae are rather small, oval, shiny, reddish brown, brown-black or black, with black-and-white banded antennae and black siphunculi and cauda. Immature stages are brownish. Alatae have a dark brown to black abdomen; the forewing has a black pterostigma and normally a once-branched media, which is unusual for Aphidinae. It lives often in dense colonies on young shoots and undersides of young leaves of host plants, causing slight rolling, twisting, or bending of the mid-rib. Ant-attended. Large colonies produce an audible scraping sound when disturbed. Apterae and alatae both 1.1 mm-2.0 mm.

Host plants: Recorded from more than 120 plant species, especially in the families Anacardiaceae, Anonaceae, Araliaceae, Euphorbiaceae, Lauraceae, Moraceae, Rubiaceae, Rutaceae, Sterculiaceae, and Theaceae. It is particularly damaging to young *Citrus*, and other hosts of economic importance include *Theobroma cacao, Coffea, Thea, Ficus, Mangifera indica, Anona, Camellia, Gardenia, Cydonia japonica, Cinchona,* and *Magnolia*.

Virus transmission: Able to transmit *Citrus* tristeza virus. It is also a vector of little leaf and lemon-ribbing virus of lemon, and of a virus disease causing leaf mottle on *Citrus vulgaris* (citrus infectious mottling virus). Two coffee viruses, blister spot of arabic coffee and ringspot of excelsa coffee, are also transmitted by *T. aurantii*.

Distribution: Distributed throughout the tropics and subtropics including the Pacific Islands, and in glasshouses in temperate climates.

Biology: Apparently entirely anholocyclic; no sexual morphs have ever been observed in the field. Broughton and Harris (1971) analysed the sound produced by *T. aurantii*, which is the only aphid with audible stridulation. Studies of population dynamics include those of Rivnay (1938) on citrus in Israel, and Firempong (1976) on cocoa in Ghana. Wilson (1960) reviewed natural enemies and biological measures in Australia. For a brief general account see Carver (1978) or Kranz *et al.* (1977, p. 342). $2n = 8$.

Toxoptera citricidus (Kirkaldy)   Tropical Citrus Aphid   Plate 44

Appearance in life: Apterae are medium-sized, shiny, very dark brown to black; immature stages are brown. The antennae, in addition to having much longer hairs than those of *T. aurantii*, are not so conspicuously black-and-white banded. Alatae have a shiny black abdomen, a forewing with a pale pterostigma and a normally twice-branched media, and a dark antennal segment III. *T. citricidus* lives in colonies on young growth of host plants, rolling leaves and stunting shoots, usually visited by ants. When disturbed, the aphids may make stridulatory movements with their hind legs, but produce no sound audible to the human ear. (Specimens in alcohol colour the preserving fluid deep red.) Apterae and alatae both 1.5–2.4 mm.

Host plants: *T. citricidus* has a much narrower range of host plants than *T. aurantii*, living almost exclusively on Rutaceae, especially *Citrus*. Very occasionally quite large colonies may occur on the young growth of plants in other families.

Virus transmission: It is the principal vector of citrus tristeza virus (see Costa and Grant, 1951), and citrus vein enation is also transmitted. Other citrus viruses or suspected virus diseases in which *T. citricidus* has been implicated are 'stem pitting' virus, Eureka-seedling virus, and 'bud union decline' of lemon and orange (Heinze in Kranz *et al.*, 1977). Also able to transmit mosaic viruses of abaca, pea, and yam, and chili veinal mottle virus.

Distribution: Widespread in Africa south of the Sahara, South East Asia, Australia, New Zealand, the Pacific Islands, and subtropical and warm temperate parts of South America. Absent as yet from important citrus-growing areas in the Caribbean north of Trinidad, Central and North America, the Mediterranean region, and the Middle East.

Biology: *T. citricidus* is entirely anholocyclic through most of its range, although an abortive holocycle on *Citrus* is reported from Japan (Komazaki *et al.*, 1979). It thrives in moist warm climates and can apparently tolerate colder conditions than *T. aurantii*, for example occurring at higher altitudes, but it is not found in regions with long hot dry seasons. Schwarz (1965) studied seasonal variation in flight activity. For a brief general account see Carver (1978) or Kranz *et al.* (1977, p. 343). $2n = 8$ (Kurl, 1980).

Toxoptera odinae (van der Goot)   Plate 45

Appearance in life: Apterae are small to medium-sized, grey brown to reddish brown, on undersides of leaves of host plants along the main veins, or in dense colonies on young shoots, ant-attended. Alatae have a reddish-brown to dark brown abdomen. Apterae and alatae both 1.3–2.4 mm.

Host plants: Rather polyphagous on tropical shrubs, especially the families Anacardiaceae (*Anacardium, Mangifera, Rhus*), Araliaceae (*Aralia, Polyscias, Kalopanax*), Caprifoliaceae (*Viburnum*), Ericaceae (*Rhododendron*), Pittosporaceae (*Pittosporum*), Rubiaceae (*Coffea, Mussaenda*), and Rutaceae (*Citrus*).

Virus transmission: T. odinae has not yet been implicated in the transmission of any plant viruses.

Distribution: India, Nepal, Sri Lanka, Thailand, Laos, Korea, China, Taiwan, Japan, Malaysia, Indonesia, the Philippines, and recently reported from South Africa.

Biology: Apparently entirely or predominantly anholocyclic; sexual morphs have not been recorded. Mondal *et al.* (1976) provided a description and biological notes. $2n = 8$ (Kurl, 1980).

Trama von Heyden   Lachninae: Tramini

A genus of medium-sized to rather large, hairy whitish aphids with very long hind tarsi and without siphunculi, mostly living on the roots of Compositae in association with ants. Fifteen nominal species have been described from Europe and western Asia. *T. oculata*, described from North America, is a synonym of the Old World *T. rara*, probably introduced to America on the roots of *Taraxacum*. No satisfactory account of the world fauna is available, partly at least because as far as is known the group is entirely parthenogenetic. Heinze (1962) gave a key to the central European species.

Trama troglodytes von Heyden   Plate 5

Large, plump, white aphids on the roots of Compositae (*Achillea, Artemisia, Centaurea, Cirsium, Cynara, Helianthus, Lapsana, Sonchus, Taraxacum*), invariably attended by ants. In Europe and western Siberia. Eastop (1953) gives a full account of this species which is probably a complex of anholocyclic races; host-plant transfers did not provide any clear evidence of host-plant specificity within the group. Blackman (1980b) describes variation in the chromosomes of U.K. populations; '$2n$' = 14–22.

Tuberocephalus Shinji
Aphidinae: Macrosiphini

A genus for about 7 *Myzus*-like aphids often with hair-bearing siphunculi, alternating between *Prunus* and *Artemisia* in the Far East. The generations on *Artemisia* may look very different from the spring forms on *Prunus*, and this has contributed to the taxonomic confusion in the group. Accounts are available from Japan (Miyazaki, 1971) and China (Chang, now Zhang and Zhong, 1976).

Tuberocephalus momonis (Matsumura)
Plate 75

Rather small, yellowish-brown to dark brown or dark green, broadly oval apterae with short dark siphunculi and a short cauda, in rolled leaves of *Prunus persica*. Alatae have a pale yellowish-brown abdomen and paler, thinner siphunculi. In Japan, China, Taiwan, and Korea. Life cycle is not known. According to Miyazaki (1971), the aphid studied by Moritsu (1947) as *Myzus momonis* was actually *Tuberocephalus sakurae*, and the true *T. momonis* does not colonize *Prunus* species other than *P. persica*.

Tuberolachnus Mordvilko
Lachninae: Lachnini

A genus for a very large, well-known, and widely distributed aphid on *Salix*, *T. salignus* (Gmelin), characterized by the single large tubercle on abdominal tergite 4. A second species on *Eriobotrya* in India with a similar tubercle, but on a sclerotic tergum, has been placed in a separate subgenus. Raychaudhuri (1980) revised the genus.

Tuberolachnus (Tuberolachniella) sclerata Hille Ris Lambers & Basu

Very large (4.0–5.4 mm) blackish aphids in large colonies on undersides of leaves, along main veins, on petioles and young shoots of *Eriobotrya* spp. As yet this species is only recorded from *E. dubia* and *E. peltata* in West Bengal and Meghalaya, India.

Uroleucon Mordvilko (= *Dactynotus* Rafinesque)
Aphidinae: Macrosiphini

A genus for about 180 species with a band of rather angular polygonal reticulation at the apices of the elongate siphunculi, usually with 5 hairs on the first tarsal segments and often with dorsal abdominal hairs arising from pigmented spots (scleroites). Many of the species have dark bronzy red, almost black, viviparae and live on the stems of Compositae, but some live under the rosette leaves. At least 160 species live on Compositae, 11 species live on Campanulaceae in the Old World, and 9 species are described from other plant families. The males of many species are green as are the females of some, particularly members of the subgenus *Lambersius* (30 species) which live mostly on the tribe Astereae in both North and South America. The other American species live mostly on Heliantheae (16 spp.) or Astereae (13 spp.), while the Old World species (100+) more frequently colonize Cynareae (20 spp.) and Cichoreae (23 spp.). Only 7 species are described from Anthemideae, the hosts of most members of the related genus *Macrosiphoniella*. From about 1935 to 1975 the genus was commonly called *Dactynotus*. Accounts are available for western Europe (Hille Ris Lambers, 1939), European U.S.S.R. (Shaposhnikov, 1964), Siberia (Ivanovskaya, 1977), palaearctic *Inula*-feeders (Holman, 1981), Japan (Miyazaki, 1971), Korea (Paik, 1965), China (Tao, 1963), north-east India (Raychaudhuri, 1980), Argentina (Blanchard, 1939, in *Macrosiphum*), Puerto Rico (Smith *et al.*, 1963), North Carolina (Olive, 1963), the Rocky Mountain area (Palmer, 1952, in *Macrosiphum*), and Canadian *Solidago*-feeders (Richards, 1972).

Uroleucon ambrosiae (Thomas) Brown Ambrosia Aphid Plate 115

Medium-sized to rather large, broadly spindle-shaped aphids, red brown to dark brown or dull red with black siphunculi and a pale cauda. On numerous species of Compositae, usually colonizing the flower stems (*Achillea, Ambrosia, Aster, Cichorium, Coreopsis, Eupatorium, Lactuca, Rudbeckia, Senecio, Solidago, Taraxacum, Xanthium*, etc.) in North, Central, and South America. Monoecious holocyclic in temperate North America, with alate males. Probably the name is being applied to a complex of very similar species. Olive (1963) separated a *Lactuca*- (and *Cichorium*)-feeding form in North Carolina as *U. pseudambrosiae*, but samples from *Lactuca* and *Cichorium* in other parts of North and South America agree better with *U. ambrosiae*. Smith *et al.* (1964) give an account of *U. ambrosiae* in Puerto Rico. In Louisiana, where *U. ambrosiae* is common in sugar cane fields, it has been shown to transmit sugar cane mosaic and maize dwarf mosaic viruses (Koike, 1977). McLain (1980) studied interrelations of ants, coccinellids, and *U. ambrosiae* on wild *Lactuca*. $2n = 12$.

Uroleucon cichorii (Koch)

Large (2.7–4.0 mm), metallic brown, broadly spindle-shaped aphids with black antennae and siphunculi and a pale yellow cauda. On *Cichorium* and related genera of Compositae (*Crepis, Hieracium, Lactuca, Lampsana, Leontodon*, etc.) in Europe, and also recorded from Armenia, the Middle East (Iran, Iraq, Turkey), Mongolia, and Eritrea. Monoecious holocyclic, with alate males. *U. cichorii* is a member of a group of closely related species in Europe. Hille Ris Lambers (1939) discussed variation in the group as a whole, erected several subspecies of *cichorii* based on host-plant preferences observed in the field, and regarded *cichorii s. str.* in the Netherlands as a form specific to *Cichorium*, characterized by a long terminal process (more than 5.5 times longer than base

VI). This distinction does not apply generally, and the aphid on *Cichorium* should be regarded as potentially capable of living on other genera of Compositae. $2n = 12$ is recorded for a *Crepis*-feeding form of *U. cichorii* (subspecies *grossus* Hille Ris Lambers; Gut, 1976).

Uroleucon (Uromelan) compositae (Theobald)   Plate 116

Medium-sized to large (1.9–4.1 mm), broadly spindle-shaped, shiny very dark red to almost black aphids with black siphunculi and cauda, colonizing flower stems and in lower numbers along the mid-ribs of the leaves of Compositae. Widely distributed in Africa and on the Indian subcontinent; also recorded from Réunion, Mauritius, Taiwan, Surinam and Sicily. *Vernonia* spp. are favoured host plants after the rains in Africa (Eastop, 1958), but many other Composite species are colonized, particularly those growing in moist or shaded situations at the end of the dry season. In India U. compositae can be an important pest of safflower (Bindra and Rathore, 1967); field resistance of safflower varieties to *U. compositae* was studied by Bhumannavar and Thontadarya (1979). It is occasionally found on plants outside the Compositae, e.g. *Malva* (Müller and Schöll, 1958), *Morus* (Devaiah *et al.*, 1976), and is recorded as a vector of passion fruit woodiness virus. Apparently anholocyclic everywhere; no sexual morphs have been recorded. *U. compositae* is difficult to separate from the south-east Asian species *U. gobonis* (q.v.) and may be an anholocyclic race of that species. Early records of *U. jaceae* or *U. solidaginis* on safflower in Africa or India probably all refer to *U. compositae*. $2n = 12$ (Kurl and Misra, 1979).

Uroleucon formosanus (Takahashi)

Medium-sized to large, dark red aphids with black siphunculi and a yellow cauda, on flower stems and along the mid-ribs of leaves of Compositae in South East Asia (Japan, China, Taiwan, Korea). The preferred host plants are *Lactuca* spp. and *Sonchus* spp.; *Senecio, Taraxacum*, and *Youngia japonica* may also be colonized. Monoecious holocyclic, with alate males, in Japan, but anholocyclic in Taiwan (Takahashi, 1923). Takahashi (1925) recorded the natural enemies of *U. formosanus* in Taiwan, and reported brachypterization of the alate morph by a hymenopterous parasite. $2n = 14$ according to Shinji (1941).

Uroleucon (Uromelan) gobonis (Matsumura)

Medium-sized to large, broadly spindle-shaped, dark greenish-brown to black aphids with black siphunculi and a black cauda, on flower stems or under leaves of Compositae in South East Asia (Korea, Mongolia, China, Japan, Taiwan). It is very common on *Arctium lappa* in Japan, and also frequently colonizes species of *Cirsium, Saussurea*, and *Synurus*. Holocyclic and anholocyclic overwintering both occur in Japan, whereas in Taiwan reproduction is exclusively parthenogenetic (Takahashi, 1923). In Raychaudhuri (1980), *U. gobonis* is recorded from a species of *Carthamus* in West Bengal, but the distinction between this species and *U. compositae* is not as clear as indicated by the key provided, and this record should perhaps be referred to *compositae*, which is the common aphid on safflower in India. $2n = 14$ (Shinji) or 12^* (Blackman).

Uroleucon (Uromelan) helianthicola Olive

Medium-sized to large, broadly spindle-shaped, red to reddish-brown aphids with appendages, including siphunculi and cauda, black; living on *Helianthus* spp. including *H. tuberosus*. Widespread in the U.S.A. The biology and sexual morphs are not recorded. Differences from closely related species on other plants are discussed by Olive (1963). $2n = 12$.

Uroleucon (Uromelan) jaceae (L.)

Medium-sized to large (2.5–3.5 mm), broadly spindle-shaped, dark reddish brown to almost black, sometimes with a metallic lustre; antennae, legs (except for bases of femora), siphunculi, and cauda are jet black. On flower stems, rarely on lower sides of leaves along mid-ribs, of *Centaurea* spp.; in Europe, the Middle East, and Central Asia eastwards to Pakistan. Most earlier European records of *U. jaceae* from other genera of Compositae probably apply to other, closely related species of *Uromelan*. However, Ilharco (1979) found *U. jaceae* colonizing *Carthamus tinctorius* in Portugal. Monoecious holocyclic on Compositae in Europe, probably anholocyclic in the Middle East. $2n = 12$.

Uroleucon leonardi (Olive)

Medium-sized to large, broadly spindle-shaped, dark brown to almost black aphids with black siphunculi and a pale yellow cauda, described from *Rudbeckia* spp. in Massachusetts, U.S.A. (Olive, 1965), and also recorded from Delaware, New Jersey, New York, and Pennsylvania. Biology is not recorded.

Uroleucon pseudambrosiae (Olive)
see *U. ambrosiae* (Thomas)

Uroleucon rudbeckiae (Fitch)

Medium-sized to large, broadly spindle-shaped aphids, shiny, bright orange red in colour with blackish antennae, legs, and siphunculi (although the siphunculi are distinctly paler at their bases), and a pale to slightly dusky cauda. Often in large numbers on the stems of *Rudbeckia* spp., and occurring throughout North America. Monoecious holocyclic, with alate males. $2n = 12$.

Utamphorophora Knowlton   Aphidinae: Macrosiphini

A genus for about 10 species of *Myzus-* and *Hyalomyzus*-like aphids with swollen siphunculi, the apterae often having sensoria on antennal segment III. Six species are American and associated with Rosaceae and/or Gramineae or, in one case, *Commelina*. Two of the Far Eastern species assigned to *Utamphorophora* live on Saxifragaceae, one at least probably migrating to an unknown secondary host. The other Far Eastern species and the single European species live on ferns. The group is probably polyphyletic. Miyazaki (1971) revised the Japanese species (in *Utamphorophora* and *Taiwanomyzus*), and the American species are to be found in accounts of *Rhopalosiphum, Myzus*, and *Amphorophora*.

Utamphorophora humboldti (Essig)   Plate 127

Apterae on grasses are medium-sized, apple green with light brown head and pale siphunculi and cauda. Immature stages have a pair of distinct dorsal longitudinal stripes. Alatae have a green abdomen with variably developed dark intersegmental markings. This species lives in aggregations along the upper sides of the leaf blades, or hidden in the flower heads, of various Gramineae (*Dactylis, Festuca, Lolium, Poa*, etc.). Heteroecious holocyclic in North America, with *Physocarpus* spp. as primary hosts. Introduced into England, where it produces sexual morphs in autumn but is probably mainly anholocyclic on grasses. (Stroyan (1979a) gives a detailed account of this aphid. $2n = 20$.

Vesiculaphis del Guercio   Aphidinae: Macrosiphini

An oriental genus of about 12 species of mostly small aphids associated with Ericaceae and/or Cyperaceae. Four species occur on *Rhododendron*. The only widely distributed one, *V. caricis,* is a small oval, dorsally flattened yellow-brown aphid heteroecious between *Rhododendron* spp. and Cyperaceae (*Carex, Cyperus*); it occurs in Japan, Korea, Taiwan, India, Hawaii, California, and eastern U.S.A. Uye (1925) studied the life cycle in Japan. On the North American continent it is known only from *Rhododendron*. Two other species are recorded from *Rhododendron* in West Bengal, one of them, *V. grandis*, being unusually large (4.0–4.6 mm). Raychaudhuri (1980) provided a key to the Indian species of *Vesiculaphis*. The fourth species, *V. kongoensis*, was described from *R. reticulatum* in Japan, and only the alate morph is known. Miyazaki (1980) has fully revised the genus.

Viteus Shimer   Phylloxeridae

A genus for *vitifoliae*, the alatae of which have paler abdominal stigmal plates and a shorter distal sensorium on the third antennal segment than the common European *Quercus*-feeding *Phylloxera*. The generic classification of the Phylloxeridae is likely to remain unsatisfactory until the 34 species of *Phylloxera* described from *Carya* in North America are better known.

Viteus vitifoliae (Fitch) (= *Phylloxera vastatrix*)   Grape Phylloxera

Appearance in life: Very small, yellow insects forming hairy, scabrous galls on the undersides of vine leaves (but opening on the upper surface of the leaf); or on vine roots causing bird's head-like swelling and blackening of rootlets (see Cornu, 1879, for fine illustrations of the effects of *Viteus* on vine roots). Apterae and alatae 0.7–1.4 mm.

Host plants: Vitis spp., including *V. vinifera*, the roots of which are continuously attacked by parthenogenetic forms of *V. vitifoliae*, and usually killed.

Distribution: Of North American origin, now in Europe, the Mediterranean, the Middle East, Africa, Korea, Australia, New Zealand, and South America.

Biology: A two-year cycle involving a sexual phase and leaf-galling and root-feeding stages normally occurs on American vine species, but on European vine (*Vitis vinifera*), *V. vitifoliae* normally lives continuously on the roots, reproducing parthenogenetically. Leaf galls occur in Europe on cultivars derived from hybrids between *vinifera* and American vine species. More has probably been written about *V. vitifoliae* than about any other pest species. For a general account and history in relation to the wine-growing industry see Ordish (1972). Coombe (1963) reviewed the pest status, distribution, and control of *V. vitifoliae* in relation to its possible introduction into South Australia. Stevenson (1964) studied population ecology of root-living stages in Ontario, Canada.

Wahlgreniella Hille Ris Lambers   Aphidinae: Macrosiphini

A genus for about 6 species with elongate swollen siphunculi, a cauda with only 5 hairs, and apterae often without sensoria on antennal segment III. Most of the species are northerly in distribution and associated with *Rosa* and/or Ericaceae, but recently a species was described from Argentina on Cucurbitaceae (Delfino, 1982). Accounts are available for western Europe (Hille Ris Lambers, 1949) and Britain (Stroyan, 1979a).

Wahlgreniella nervata (Gillette)   Plate 128

Apterae are small to medium-sized (1.4–2.5 mm), spindle-shaped, pale green to dull mid-green; the long, slightly swollen siphunculi are mainly pale but with dark apices. Alatae have a green abdomen with variably developed dark dorsal markings, sometimes coalesced into an irregular-shaped and much perforated central patch. *W. nervata* colonizes the shoots or mature leaves of wild and cultivated roses in England, Mexico, and western North America. In North America it is apparently heteroecious, migrating from *Rosa* to Ericaceae (*Arbutus, Arctostaphylos, Pieris*), although the life cycle needs further confirmation. The recently introduced English population is anholocyclic on rose, and morphologically distinguishable from another anholocyclic form of *Wahlgreniella* on *Arbutus* in Europe (ssp. *arbuti* Davidson), which is apparently a separate introduction of North American *W. nervata s. lat.* The European forms on *Rosa* and *Arbutus* can both be reared on the other's host plant in the laboratory, but apparently maintain their respective host-plant preferences in the field. $2n = 12$.

Wahlgreniella vaccinii (Theobald)

Apterae are small to medium-sized (1.6–2.2 mm), rather slender oval, shining yellow green; the siphunculi are slightly swollen on the distal half and dusky towards their apices. Alatae have a yellow-green abdomen without any dark dorsal markings. *W. vaccinii* lives on undersides of leaves of *Vaccinium vitis-idaei*, and is also recorded from *Arctostaphylos uva-ursi*, in Europe (England, Scotland, Norway, Sweden, Germany) and North America (Colorado, Massachusetts, Oregon, Manitoba). However, North American aphids from *Vaccinium* have a much longer terminal process to the antenna, and should perhaps be regarded as a distinct species. The life cycle is unclear; a holocycle would be expected in northern Europe and oviparae occur in England, but no males have yet been recorded outside North America.

Watabura Matsumura Pemphiginae: Eriosomatini

A monotypic genus in Japan characterized by the single developed tarsal claw; the second claw is much reduced and inconspicuous, or sometimes completely absent.

Watabura nishiyae Matsumura (= *Aphidounguis mali* Takahashi)

Small, white, oval, dorsally convex aphids with white woolly wax secretion, on roots of apple or quince, as yet known only from Japan. Heteroecious holocyclic, forming clusters of leaf roll galls in spring on the primary host, *parvifolia* (Akimoto 1983). The synonymy of *Aphidounguis* Takahashi and *Watabura* Matsumura suggested by Hille Ris Lambers (1967b) is retained here despite the reservations of Akimoto (1983).

D. *Techniques*

Here we give only a brief account of methods of collecting, preserving, and making slide preparations of aphids, in connection with the identification of species colonizing crop plants. For more detailed techniques of aphid study consult van Emden (1972a).

COLLECTING

Aphids may be collected by sweeping or beating, or simply by careful searching of appropriate parts of the plant. Sweeping is *not* recommended here as crop plants which can be swept are unlikely to be weed-free, and it is impossible to be certain of the plant species which a swept aphid had been feeding on. Beating onto a white tray or card placed underneath the plant must also be done very carefully to ensure that the aphids falling onto the tray do actually originate from the plant in question. It is best, if at all possible, to examine plants and try to find the aphids *in situ*, so that no mistakes are made.

It is usually considered best to bring the aphids back to the laboratory alive on a part of the host plant, rather than to collect them directly into preservative. Polythene or glass tubes stoppered with cotton wool, or polythene bags with tissue paper inside to soak up excess moisture, are suitable containers. Aphid colonies often consist of mainly immature individuals, and the proportion of adults in the sample can be increased by keeping the aphids alive for a few days in a cool place before preserving them. Adults, especially alatae, should be left for 1–2 days after they have reached maturity in order to develop their full pigmentation. This also provides a method of rearing out any parasites. Predators must of course be first excluded.

It is preferable to note down fullest possible collection data at the time of collection, including host plant, locality, and date, and it is also very important to note biological information such as colour of the aphids in life, position of colony on host plant, and whether or not there is ant attendance. It is probably best to keep a field notebook for this purpose, and give each sample a collection number which is written on a slip of paper and inserted in the tube with the specimens.

PRESERVATION AND MOUNTING

Aphids for morphological examination should be preserved in tightly stoppered tubes filled with 80–95% ethyl alcohol. For prolonged storage one volume of 75% w/w lactic acid may be added after a few days to every two volumes of alcohol containing specimens, and the tubes plugged with cotton wool and kept either under alcohol or on a cushion of cotton wool soaked in alcohol, in an airtight glass jar.

Maceration to remove the soft tissues of the specimens prior to mounting is best carried out with the specimen tubes in a water bath kept near boiling point. The following stages are involved.

(1) Gently boil the specimens in 95% alcohol for 1–2 min.
(2) Decant or pipette off alcohol, add about 1 cm depth of 10% potassium hydroxide (KOH) solution, simmer for 3–5 min.
(3) Decant or pipette off KOH solution and wash the specimens free of all KOH using 2–3 changes of distilled water, leaving them to soak for at least 10 min each time.

The water-based Berlese mountant is frequently used by aphid workers, but balsam mounts are recommended here because of their proven permanence and resistance to a wide range of climatic conditions. The macerated specimens need to be totally dehydrated and cleared before mounting in Canada balsam. This can be done by taking the specimens through a series of progressively higher grades of alcohol, but the following procedure (developed by J.H. Martin) has also been found to give satisfactory results.

(4) Remove distilled water, add 1 cm depth of glacial acetic acid, and leave for 2–3 min. Pipette off and repeat with fresh glacial acetic acid. Pipette off.
(5) Add clove oil as clearing agent (specimens will float). Leave for 10–20 min until specimens are clear.
(6) Transfer 1–2 aphids to a drop of fairly thin Canada balsam on a clean microslide and rapidly arrange the appendages.
(7) Dip a clean coverslip in xylene and immediately lower it carefully onto the specimens displayed in the drop so as to spread the mountant evenly without trapping air bubbles.
(8) Dry the slide horizontally in an oven at 50°C for about 1 week.

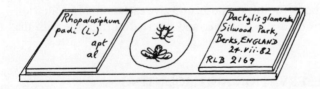

Fig. 13   Labelled slide preparation

LABELLING AND STORAGE OF SLIDE-MOUNTED PREPARATIONS

Aphid workers have found that thick card labels, gummed to the slide with an impact adhesive, help to protect the coverslip and enable slides to be stacked vertically while awaiting attention. The system of labelling most frequently used is shown in Fig. 13. Slides may be stored horizontally in trays, or vertically in slotted cabinet drawers or boxes. For larger collections a compact and versatile system involves the vertical storage of slides in individual envelopes made of paper or cellulose acetate, with interspersed tab cards providing the collection with an integral index (Eastop, in press).

E. *Sources of Information*

REGIONALLY CLASSIFIED FAUNAL WORKS

World

Averill (1945)	Index to host-plant genera and species mentioned by Patch (1938).
Baker (1920)	Keys to genera then recognized, now out of date.
Börner (1930)	Keys to genera, less reliable for extra-European groups.
Börner and Heinze (1957)	Accounts of economically important aphids with extensive bibliographies.
Eastop and Hille Ris Lambers (1976)	List of genera and species with synonyms.
Martin (1983)	Key to tropical pest aphids.
Patch (1938)	Aphid/host-plant records arranged under plant families, with references to sources.
Richards (1976)	Aphids listed under genera and species of host plants arranged alphabetically.

West Palaearctic

Heie (1980, 1982)	Keys to major groups and genera and species of Mindarinae, Thelaxinae, Anoeciinae and Pemphiginae (1980), and Drepanosiphinae including Chaitophorinae (1982).
Heinze (1960, 1961)	Keys to genera and species of Myzini together with generic diagnoses (many species illustrated).
Heinze (1962)	Lachninae, Adelgidae, Phylloxeridae; keys to genera and species, generic diagnoses, illustrated.
Hille Ris Lambers (1931 to date)	Revisions of many groups, particularly Macrosiphini in *Temminckia* **3–9** (1938–1953).

Müller (1969)	Keys to genera and species.
Müller (1975a)	Keys to commonly trapped alatae illustrated with photomicrographs.
Shaposhnikov (1964)	Keys to Aphidoidea of European U.S.S.R.
Stroyan (1952)	Keys, diagnoses, and line drawings of economically important aphids.
Stroyan (1977)	Keys, diagnoses, and line drawings of Chaitophorinae and Drepanosiphinae.
Szalay–Marzso (1969)	Keys, line drawings, and photographed colonies on horticultural crops.
Szelegiewicz (1977)	Keys and illustrations for genera and species of Lachninae, Chaitophorinae and Drepanosiphinae.
Szelegiewicz (1978)	Keys and illustrations for genera and species of Lachninae.

Middle East

Bodenheimer and Swirski (1957)	Keys, extensive review of literature.
Habib and El-Kady (1961)	Keys to Egyptian genera and species.

Eastern Palaearctic

Higuchi (1972)	Revision of Japanese Drepanosiphinae.
Higuchi and Miyazaki (1969)	Catalogue of Japanese aphids and their host plants.
Ivanovskaya (1977)	Keys, diagnoses for aphids of Western Siberia.
Miyazaki (1971)	Revision of Japanese Macrosiphini.
Narzikulov (1963)	Revision of Phloeomyzinae, Lachninae, Anoeciinae, Pemphiginae, Drepanosiphinae, Chaitophorinae, and Pterocommatinae of Tadzhikistan and adjacent republics.
Narzikulov and Umarov (1969)	Revision of Macrosiphini of Tadzhikistan and adjacent republics.
Nevskii (1929)	Revision of aphids of Central Asia.
Paik (1965)	Revision of aphids of Korea.
Paik (1972)	Revision of aphids of Korea (in Korean).
Szelegiewicz (1974)	Annotated list of aphids from northern Korea.
Tanaka (1976)	Well-illustrated account of aphids on vegetables in Japan.

Afrotropical

Eastop (1958)	Keys to East African aphids.
Eastop (1961a)	Keys to West African aphids.
Orian (1973)	Keys to aphids of Mauritius.
Potgieter and Dürr (1961)	South African aphids listed systematically and by host plants.
Quednau (1962, 1964)	Annotated additions to South African fauna.

Oriental

Behura (1963)	Check list and bibliography of Indian aphids.
Behura (1966)	Supplement to Behura (1963).
Calilung (1967, 1968)	Revision of Philippine aphids.
Das and Raychaudhuri (1983)	Account of aphids of Nepal.
Ghosh, A.K. (1974)	Keys, illustrations of economically important Indian aphids.
Ghosh, A.K. (1980, 1982a)	Keys to major groups and revision of Indian Chaitophorinae, Lachninae.
Raychaudhuri (1980)	Revision of aphids of north-eastern India.
Takahashi (1931)	Revision of Taiwan aphids.
Tao (1962–1970)	Keys to groups of Chinese aphids.
Zhang and Zhong (1983)	Keys to Chinese aphids of economic importance.

Australian

Eastop (1966)	Revision of Australian aphids.

Polynesian

Cottier (1953)	Revision of New Zealand aphids.
Palmer, J. (1974)	Annotated list of aphids from Subantarctic islands of New Zealand.
Zimmerman (1948)	Revision of Hawaiian aphids.

Nearctic

Hottes and Frison (1931)	Revision of aphids of Illinois.
Medlar and Ghosh (1969)	Keys to alatae from north central U.S.A.
Palmer, M. (1952)	Revision of Rocky Mountain aphids.
Robinson (1979)	Annotated list of aphids from north-west Canada, Yukon, and Alaska.

Smith, C.F. (1974)   Revision of North American Pemphiginae.
Smith and Parron (1978)   List of North American aphids annotated with synonyms, host plants, geographical distribution, plus host-plant catalogue.

Neotropical

Blanchard (1939)   Revision of Argentinian aphids.
Carrillo (1974, 1977)   Annotated lists of aphids from Chile.
Cermeli (1970a)   Key and illustrations to economically important aphids in Venezuela.
Cermeli (1970b, 1973)   Supplements to Cermeli (1970a).
Holman (1974)   Revision of Cuban aphids.
Smith and Cermeli (1979)   List of South American aphids, annotated with synonyms, host plants, and geographical distribution.
Smith, Martorell, and
  Pérez-Escolar (1963)   Keys to aphids of Puerto Rico.
Smith, Gauda, Martorell, and
  Pérez-Escolar (1971)   Keys to aphids of Puerto Rico, illustrations, and host-plant catalogue.

GENERAL BIOLOGY

Blackman (1975); Bodenheimer and Swirski (1957, extensive literature review until *c.* 1954); Dixon (1973, 1977); Heie (1980, pp. 44–64, concise summary); Kennedy and Stroyan (1959, useful discussion and ideas); Lowe (1973).

MORPHOLOGY, ANATOMY, AND PHYSIOLOGY

Morphology   Heie (1980, pp. 24–44, concise general summary)
                                       Weber (1928, detailed account of *Aphis fabae*)
Sensoria   Bromley, Dunn, and Anderson (1979); Shambough et al. (1978)
Anatomy   Ponsen in Harris and Maramorosch (1977); Ponsen (1982)
Mouthparts   Forbes in Harris and Maramorosch (1977); Kidd (1976)
Penetration of plant tissues   Pollard in Harris and Maramorosch (1977)
Reproductive system:
  female   Dixon and Dhama (1980)
  male   Glowacka, Klimaszewsky, Szelegiewicz, and Wojcierhanda (1974)

Honeydew, triglycerides, pigments, cardenolides	Brown (1975)
Pheromones	Nault and Montgomery (1979)

GENETICS AND DEVELOPMENT

Eggs, diapause	Behrendt (1963)
Eggs, embryology	Böhmel and Jancke (1942)
Cytogenetics and genetics	Blackman (in press)
Parthenogenetic development	Bruslé (1963); Blackman (1978)
Polymorphism	Lees (1966); Hille Ris Lambers (1966); Steel and Lees (1977)

MIGRATION AND DISPERSAL

Berry and Taylor (1968); Close and Tomlinson (1975); Johnson (1969, review of insect flight); Kring (1972, flight behaviour); Taylor and Taylor (1977); Taylor and Woiwood (1980); Taylor et al. (1980); Thresh (1983, long-range, in relation to virus transmission); Wiktelius (1981, *Rhopalosiphum*, long-range migration into Sweden).

RELATIONSHIPS WITH OTHER INSECTS

Ants	Kleinjan and Mittler (1975, chemistry); Kunkel (1973); Way (1963)
Natural enemies:	
General	Börner and Heinze (1957); Cavallaro (1983); Hagen and van den Bosch (1968); Hodek (1966)
Predators	
Neuroptera	New (1975)
Coccinellidae	Baumgaertner et al. (1981); Hodek (1973)
Cecidomyidae	Harris (1973)
Chamaemyidae	no review available
Syrphidae	Schneider (1969)
Parasites	
Aphidiidae	Mackauer and Starý (1967); Starý (1979); Raychaudhuri, Poddar, and Raychaudhuri (1982)
Aphelinidae	Kalina and Starý (1976, European); Graham (1976, European)
Fungal pathogens	I.M. Hall in Lowe (1973, pp. 30–39); R.A. Hall and Burges (1979); Milner (1981)

HOST-PLANT RELATIONSHIPS

General	Kennedy and Fosbrooke in van Emden (1972b, pp. 129–40); Eastop in van Emden (1972b, pp. 157–78); van Emden in Lowe (1973, pp. 54–64); Hille Ris Lambers in Hedberg (1980, pp. 114–19); Eastop in Hedberg (1980, pp. 120–34)
Virus transmission	Harris and Maramorosch (1977); Hille Ris Lambers (1972); Swenson in Lowe (1973, pp. 92–102).
Influence of host plant on polymorphism	Mittler in Lowe (1973, pp. 65–75)

BIBLIOGRAPHIES

Sharma (1969–77); C.F. Smith (1972); Szelegiewicz (1969a).

F. References

(N.B. *Some of the titles have been abbreviated.*)

Addicott, J.F. (1981) Synonymy of *Aphis heraclella* Davis 1919, with *Aphis helianthi* Monell 1879. *Can. Ent.* **113**: 167–69.
Akimoto, S. (1983) A revision of ... *Eriosoma* and its allied genera in Japan. *Insecta matsum. (N.S.)* **27**: 37–106.
Aliniazee, M.T. (1983) Carbaryl resistance in the Filbert aphid. *J. econ. Ent.* **76**: 1002–4.
Alma, A. and Arzone, A. (1983) Reperti inediti del ciclo eterogonico di *Myzus varians*. *Atti XIII Congresso naz. ital. ent.* **13**: 431–36.
Alston, F.H. and Briggs, J.B. (1977) Resistance genes in apple and biotypes of *Dysaphis devecta*. *Ann. appl. Biol.* **87**: 75–81.
Ankersmit, G.W. and Dijkman, H. (1983) Alatae production in the cereal aphid *Sitobion avenae*. *Neth. J. Pl. Path.* **89**: 105–12.
Aoki, S. (1975) Descriptions of the Japanese species of *Pemphigus* and its allied genera. *Insecta matsum.* N.S. **5**: 1–56, 3 plates.
Aoki, S. and Miyazaki, M. (1978) Notes on the pseudoscorpion-like larvae of *Pseudoregma alexanderi*. Kontyû **46**: 433–38.
Archangelsky, P.P. (1917) On the biology of *Pterochloroides persicae*. *Turkestan Ent. Stn Tashkent*, 70 pp.
Arzone, A. (1979) Afide di attualita su Pesco: *Myzus varians* Davidson. *Inftore fitopatol.* **1979(8)**: 3–6.
Averill, A.W. (1945) Supplement to food-plant catalogue of Aphids of the World. *Bull. Me agric. Expt. Sta.* **393S**, 50 pp.
Baker, A.C. (1917) Life history of *Macrosiphum illinoisensis*, the grape vine aphis. *J. agric. Res.* **11**: 83–89.
Baker, A.C. (1919) The houghton gooseberry aphis. *J. econ. Ent.* **12**: 433–40.
Baker, A.C. (1920) Generic classification of the hemipterous family Aphididae. *Bull. U.S. Dep. Agric.* **826**: 109 pp.
Baker, A.C. and Davidson, W.M. (1916) Woolly Pear Aphis. *J. agric. Res.* **6**: 351–60.
Baker, A.C. and Davidson, W.M. (1917) A further contribution to the study of *Eriosoma pyricola*. *J. agric. Res.* **10**: 65–74.
Baker, A.C. and Turner, W.F. (1916) Morphology and biology of the green apple aphis. *J. agric. Res.* **5**: 955–93.
Baker, H. (1935) *Phylloxera devastatrix* Perg. on Pecans. *J. econ. Ent.* **28**: 681–85.
Balachowsky, A. (1933) I. Sur la présence en France de *Capitophorus fragaefolii* ... II. Sur l'existence de nouveaux foyers d'*Aphis forbesi* Weed. *Revue Path. vég. Ent. agric. Fr.* **20**: 256–67.
Barbagallo, S. (1966a) Contributo alla conoscenza degli afidi degli agrumi. 1. *Aphis spiraecola* Patch. *Boll. Lab. Ent. agr. Filippo Silvestri* **24**: 49–83.

Barbagallo, S. (1966b) L'afidofauna degli agrumi in Sicilia. *Entomologica, Bari* **2**: 201–60.
Barbagallo, S. (1974) Osservazioni sugli afide (Homoptera, Aphidoidea) del carciofo (*Cynara scolymus* L.). *Boll. Lab. Ent. agr. Filippo Silvestri* **31**: 197–252.
Barbagallo, S. and Stroyan, H.L.G. (1982) Osservazione biologiche, ecologiche e tassonomiche sull'afidofauna della Sicilia. *Frustula ent (N.S.)* **3**: 1–182.
Barlow, C.A. (1962) Development, survival and fecundity of the potato aphid, *Macrosiphum euphorbiae* (Thomas), at constant temperatures. *Can. Ent.* **94**: 667–71.
Barlow, N.D. and Dixon, A.F.G. (1980) *Simulation of Lime Aphid Population Dynamics.* Simulation monographs, Wageningen, 164 pp.
Barnes, M.M. and Moffitt, H.R. (1978) A five year study of the effects of the walnut aphid ... in coastal orchards. *J. econ. Ent.* **71**: 71–74.
Baronio, P. (1971) Ricerche su ... *Dysaphis plantaginea* ... in un meloto. Nota preliminaire. *Boll. Oss. Mal. Piante* **2**: 1–13 (1967–70).
Barson, G. and Carter, C.I. (1972) A species of Phylloxeridae, *Moritziella corticalis* (Kalt.) new to Britain, and a key to the British oak-feeding Phylloxeridae. *Entomologist* **105**: 130–34.
Basu, A.N. (1961) Some aphids new to India, with description of a new subspecies. *Current Science* **30**: 390–91.
Basu, A.N. (1969) Four new genera and nine new species of aphids from West Bengal, India. *Orient. Insects* **3**: 169–86.
Basu, R.C., Chakrabarti, S., and Raychaudhuri, D.N. (1969) Record of the sexuales of *Aphis craccivora* Koch from India. *Orient. Insects* **2**: 349–51 (1968).
Basu, R.C. and Raychaudhuri, D.N. (1976a) Studies on the aphids from India. xxv. The genus *Myzus* with five new species from eastern India. *Orient. Insects* **10**: 93–112.
Basu, R.C. and Raychaudhuri, D.N. (1976b) Studies on the aphids from eastern India. xxix. Genus *Macrosiphoniella. Orient. Insects* **10**: 295–306.
Baumgaertner, J.U., Frazer, B.D., Gilbert, N, Gill, B., Gutierrez, A.P., Ives, P.M., Nealis, V., Raworth, D.A. and Summers, C.G. (1981) Coccinellids (Coleoptera) and aphids. *Can. Ent.* **113**: 975–1048.
Behrendt, K. (1963) Über die Eidiapause von *Aphis fabae* Scop. *Zool. Jb. Abt. Allg. Zool. Physiol. Tiere* **70**: 309–98.
Behura, B.K. (1963) Aphids of India: A survey of published information. *Recent advances in zoology in India* **1961**: 25–78.
Behura, B.K. (1966) Supplement to 'Aphids of India: A survey of published information'—1. *J. Utkal Univ.* **3**: 40–65.
Bell, A.C. (1983) The life-history of the leaf curling plum aphid *Brachycaudus helichrysi* in Northern Ireland and its ability to transmit potato virus Y[e] (AB). *Ann. appl. Biol.* **102**: 1–6.
Belli, G., Corbetta, G. and Ostler, R. (1975) Richerche e osservazione sull' epidemiologia e sulle posserverta di prevenzione del 'giallume' del riso. *Riso* **24**: 395–63.
Bem, F. and Murrant, A.F. (1979) Host range, purification and serological properties of heracleum latent virus. *Ann. appl. Biol.* **92**: 243–56.
Bennett, S.H. (1955) The biology, life history and methods of control of the leaf curling plum aphid *Brachycaudus helichrysi* (Kltb.). *J. Hort. Sci.* **30**: 252–59.
Berest, Z.L. (1980) [Parasites and predators of the aphids *Brachycolus noxius* and *Schizaphis graminum* in crops of barley and wheat in Nikalaev and Odessa regions] [In Russian]. *Vest. Zool.* **1980** (2): 80–81.
Berry, R.E. and Taylor, L.R. (1968) High-altitude migration of aphids in maritime and continental climates. *J. Anim. Ecol.* **37**: 713–22.
Bhanotar, R.K. and Ghosh, L.K. (1969) An oviparous morph of *Pentalonia nigronervosa* Coquerel (Aphididae: Homoptera) from West Bengal, India. *Bull. Ent. Soc. India* **10**: 97–99.

References

Bhargava, P.D., Mathur, S.C., Vyas, H.K., and Anwer, M. (1971) Note on screening fennel (*Foeniculum vulgare* L.) varieties against aphid (*Hyadaphis coriandri* (Das)) infestation. *Indian J. agric. Sci.* **41**: 90–92.

Bhattacharya, D.K. and Chakrabarti, S. (1982) The genus *Rhopalosiphoninus* from India. *Orient. Insects* **15**: 287–93 (1981).

Bhumannavar, B.S. and Thontadarya, T.S. (1979) Varietal response of safflower (*Carthamus tinctorius* L.) to the aphid *Dactynotus compositae* Theobald. *Curr. Res., Karnataka* **8**: 134–36.

de Biase, L.M. and Calambuca, E. (1979) Gli afidi (Homoptera-Aphidoidea) del noce (*Juglans regia* L.) in Campania. *Boll. Lab. Ent. agr. Filippo Silvestri* **36**: 81–107.

Bindra, O.S. and Rathore, Y.S. (1967) Investigation on the biology and control of the safflower aphid at Jabalpur. *JNKVV Res.* **1**: 60–63.

Bissell, T.L. (1978) *Aphids on Juglandaceae in North America*. Maryland Agric. Expt. Stn, iii + 78 pp.

Blackman, R.L. (1975) *Aphids* Ginn & Co., Aylesbury, 175 pp.

Blackman, R.L. (1976) Biological approaches to the control of aphids. *Phil. Trans. R. Soc. Lond. B.* **274**: 473–88.

Blackman, R.L. (1980a) Chromosome numbers in the Aphididae and their taxonomic significance. *Syst. Ent.* **5**: 7–25.

Blackman, R.L. (1980b) Chromosomes and parthenogenesis in aphids. *Symposia R. ent. Soc. Lond.* **10**: 133–48.

Blackman, R.L. (1981) Aphid genetics and host plant resistance. *Bull.SROP* **4**: 13–19.

Blackman, R.L. (in press) Aphid cytology and genetics, in Szelegiewicz, H. (Ed.) *Evolution and Biosystematics of Aphids*. Warsaw.

Blackman, R.L., Eastop, V.F., and Hills, M. (1977) Morphological and cytological separation of *Amphorophora* Buckton feeding on European raspberry and blackberry (*Rubus* spp.). *Bull. ent. Res.* **67**: 285–96, plate V.

Blackman, R.L., Takada, H., and Kawakami, K. (1978) Chromosomal rearrangement involved in insecticide reistance of *Myzus persicae*. *Nature, Lond.* **271**: 450–52.

Blanchard, E.E. (1939) Estudio sistemático de los afidoideos argentinos. *Physis* **17**: 857–1003.

Blanchard, E.E. (1944) Descriptiones y anotaciones de Afidoideos argentinos. *Acta. zool. lilloana* **2**: 15–62.

Blunck, H. (1948) Die grosse Pflaumenlaus *Brachycaudus cardui* L. im Schutz der Rasenameise *Lasius niger* L. *Z. Pflkrankh. PflPath. Pflschutz* **55**: 91–92.

Boakye, D.B. and Randles, J.W. (1974) Epidemiology of lettuce necrotic yellows virus in S. Australia III. *Aust. J. agric. Res.* **25**: 791–802.

Bochen, K. (1978) Morphology of the developmental stages of three species of the genus *Macrosiphoniella* del Guercio, 1911 (Homoptera, Aphididae). Bionomy of three species of the genus *Macrosiphoniella* del Guercio, 1911 (Homoptera, Aphididae). *Annls. Univ. Mariae Curie-Sklodowska* C (Biologie) **33**: 411–28, 429–42.

Bode, O. (1980) Untersuchungen zum Auftreten der Haferblattlaus *Rhopalosiphum padi* (L.) an ihren Winterwirt *Prunus padus* L. I. Biologie. *Z. angew. Ent.* **89**: 363–77.

Bodenheimer, F. and Swirski, E. (1957) *Aphidoidea of the Middle East*. Weizmann Science Press, Jerusalem, 378 pp.

Böhmel, W. and Jancke, O. (1942) Beitrag zur Embryona entwicklung der Wintereier von Aphiden. *Z. angew. Ent.* **29**: 636–58.

Börner, C. (1930) Beitrage zur einem neuen System der Blattlause. *Arch. klassif. phylogen. Ent.* **1**: 115–80.

Börner, C. (1949) Kleine Beiträge zur Monographie der europaischen Blattlause. *Beitr. tax. Zool.* **1**: 44–62.

Börner, C. (1952, 1953) Europae centralis Aphides. *Mitt. Thuring. bot. Ges.* **4**(3) 1–488, L.1, 1–259; L.2, 260–484 (1952); L.2, 485–488 (1953).

Börner, C. and Heinze, K. (1957) In Sorauer, P., *Handb. Pflkrankh.* 5th Ed. 5(4), (Aphidoidea, pp. 1–402).
Bondy, F.F. and Rainwater, C.F. (1939) Boll weevil and miscellaneous cotton insect investigations. *Rep. S. Carol. Exp. Stn* **52**: 124–34.
Bonnemaison, L. (1971) Action de la photopéroide sur la production des gynopares ailées de *Dysaphis plantaginea* Pass. *Annls Zool. Ecol. anim.* **2**: 523–54.
Bonner, A.B. and Ford, J.B. (1972) Some effects of crowding on ... *Megoura viciae*. *Ann. appl. Biol.* **71**: 91–98.
van den Bosch, R. (1977) Informe de la segunda visita a Chile como consultar para el programa de control biológico e integrader de los áfides de los cereales. *Investnes Progreso Agric.* **9**: 16–24.
van den Bosch, R., Hom, R., Matteson, P., Frazer, B.D., Messenger, P.S., and Davis, C.S. (1979) Biological control of the walnut aphid in California: Impact of the parasite *Trioxys pallidus*. *Hilgardia* **47**: 13 pp.
Bouchard, D., Tourneur, J.C., and Paradis, R.O. (1981) Bio-écologie d'*Aphidoletes aphidimyza* (Rondani) prédateur du puceron du pommier, *Aphis pomi* De Geer. *Ann. ent. Soc. Queb.* **26**: 119–30.
Boudreaux, H.B. and Tissot, A.N. (1962) The black-bordered species of *Myzocallis* of oaks. *Misc. Publs. ent. Soc. Am.* **3**: 122–44.
Briggs, J.P. (1965) The distribution, abundance and genetic relationships of four strains of the rubus aphid (*Amphorophora rubi* (Kalt.)) in relation to raspberry breeding. *J. Hort. Sci.* **40**: 109–17.
Brown, P.A. (1984) A note on *Myzus (Sciamyzus) cymbalariae* Stroyan, with a description of the male. *J. nat. Hist.* **17**: 875–80.
Brodel, C.F. and Schaefers, G.A. (1980) The influence of temperature in the production of sexuales by *Aphis rubicola* under short day conditions. *Entomologia exp. appl.* **27**: 127–32.
Bromley, A.K., Dunn, J.A., and Anderson, M. (1979) Ultrastructure of the antennal sensilla of aphids. 1. *Cell Tissue Res.* **203**: 427–42.
Broughton, W.B., and Harris, K.M. (1971) First recording of the sound produced by the black citrus aphid, *Toxoptera aurantii* (Boy.). *Bull. ent. Res.* **60**: 559–63.
Brown, K.S. (1975) The chemistry of aphids and scale insects. *Chem. Soc. Rev.* **4**: 263–88.
Bruslé, S. (1962) Chronologie du dévelopment embryonnaire des females parthénogénétique de *Brevicoryne brassicae*. *Bull. Soc. zool. Fr.* **87**: 396–410.
Burger, H.C. (1975) Key to the European species of *Brachycaudus*, subgenus *Acaudus* with redescriptions and a note on *B. persicae*. *Tijdschr. Ent.* **118**: 99–116.
Calilung, V.J. (1967, 1968) A faunistic study of Philippine Aphids. *Philipp. Agric.* **51**: 93–170 (1967); **52**: 393–429 (1968).
Calvert, D.J. and van den Bosch, R. (1972) Host range and specificity of *Monoctonus paulensis*, a parasite of certain Dactynotine aphids. *Ann. ent. Soc. Am.* **65**: 422–32.
Cammell, M.E. (1981) The black bean aphid, *Aphis fabae*. *Biologist, London* **28**: 247–58.
Carpenter, T.L., Neel, W.W., and Hedin, P.A. (1979) A review of host plant resistance of pecan, *Carya illinoiensis* to Insecta and Acarina. *Bull. ent. Soc. Am.* **25**: 251–57.
Carrillo, R. (1974, 1977) Aphidoidea de Chile I & II. *Agro. Sur.* **2**: 33–40 (1974); **5**: 109–14 (1977).
Carter, N. and Rabbinge, R. (1980) Simulation of models of the population development of *Sitobion avenae*. *Bull. SROP* **3**: 93–98.
Cartwright, B.O., Eikenbary, R.D., Honson, J.W., Farris, T.N., and Morrison, R.D. (1977) Field release and dispersal of *Menochilus sexmaculatus*, an imported predator of the greenbug, *Schizaphis graminum*. *Environmental Ent.* **6**: 699–704.

Carver, M. (1976) New and recent additions to the aphid fauna of Australia. *J. Aust. ent. Soc.* **15**: 461–65.
Carver, M. (1978) The black citrus aphids, *Toxoptera citricidus* (Kirkaldy) and *T. aurantii* (Boyer de Fonscolombe). *J. Aust. ent. Soc.* **17**: 263–70.
Carver, M. (1980) *Neotoxoptera* Theobald and *Pterocallis* Passerini in Australia. *J. Aust. ent. Soc.* **19**: 139–42.
Castanera, P., Loxdale, H.D. and Nowak, K. (1983) Electrophoretic study of enzymes from cereal aphid populations II ... identifying aphidiid parasitoids of *Sitobion avenae*. *Bull. ent. Res.* **73**: 659–65.
Cavallaro, R. (Ed.) (1983) *Aphid Antagonists*. A.A. Balkema, Rotterdam, 144 pp.
Cermeli, M. (1970a) Los afidos de importancia agricola en Venezuela y algunos observaciones sobre ellos. *Agronomia trop.* **20**: 15–61.
Cermeli, M. (1970b) Los afidos de Venezuela y sus plantas hospederas I. *Agronomia trop.* **20**: 249–56.
Cermeli, M. (1973) Los afidos de Venezuela y sus plantas hospederas II. *Agronomia trop.* **23**: 163–73.
Cermeli, M. and Smith, C.F. (1979) Keys to species of the genus *Picturaphis* with descriptions of new species. *Proc. ent. Soc. Wash.* **81**: 611–20.
Chakrabarti, S. (1976) A new synonymy and hitherto unknown morphs of *Chaetosiphon* Mordvilko described from India. *Sci. cult.* **42**: 335–36.
Chakrabarti, S. (1978) Sexual morphs of *Chromaphis hirsutustibis* Kumar and Lavigne from India. *Entomon (India)* **3**: 295–96.
Chakrabarti, S., Ghosh, A.K., and Chowdhuri, A.N. (1971) Aphids of Himachal Pradesh, India—III. *Orient. Insects* **4**: 447–52 (1970).
Chakrabarti, S. and Maity, S.P. (1978) Aphids of North West India: New subgenus, new species and new records of root-inhabiting aphids. *Entomon (India)* **3**: 265–67.
Chakrabarti, S., Maity, S.P., and Bhattacharya, D.K. (1982). New and little-known aphids infesting roots of Gramineae in North west Himalaya, India. *Orient. Insects* **16**: 99–111.
Chakrabarti, S., Mandal, A.K., and Raha, S. (1983) *Rhododendron*-infesting aphids of the Himalayas. *J. Linn. Soc. Zoology* **78**: 349–62.
Chakrabarti, S. and Raychaudhuri, D.N. (1975) Aphids from Sundardhunga Valley, Kumaon, Himalaya, India. *Orient. Insects* **9**: 195–211.
Chang, G.-s. see Zhang, G-x.
Chapman, R.F., Bernays, E.A. and Simpson, S.J. (1981) Attraction and repulsion of the aphid, *Cavariella aegopodii*, by plant odors. *J. chem. Ecol.* **7**: 881–88.
Ciampolini, M. and Martorelli, M. (1978) Comparsa in Italia dell' afide lignicola delle prunoidea, *Pterochloroides persicae* (Cholod.). *Boll. Zool. agr. Bacchic.* **14**: 189–96.
Close, R.C. and Tomlinson, A.I. (1975) Dispersal of the grain aphid *Macrosiphum miscanthi* from Australia to New Zealand. *N.Z. Ent.* **6**: 62–65.
Cockbain, A.J. and Costa, C.L. (1973) Comparative transmission of bean leaf roll and pea enation mosaic viruses by aphids. *Ann. appl. Biol.* **73**: 167–76.
Comstock, J.H. (1879) The Japan lily aphis. *Rep. U.S. Dep. Agric.* **1879**: 220–21.
Converse, R.H., Daubeny, H.A., Stace-Smith, R., Russell, L.M., Koch, E.J., and Wiggans, S.C. (1971) Search for biological races in *Amphorophora agathonica* Hottes on red raspberries. *Can. J. Pl. Sci.* **51**: 81–85.
Cook, W.C. (1963) Ecology of the pea aphid (*Acyrthosiphon pisum*) in the Blue Mountain area of Eastern Washington and Oregon. *Tech. Bull. U.S. Dep. Agric.* **1287**: 1–48.
Coombe, B.G. (1963) *Phylloxera* and its relation to South Australian Viticulture. *Tech. Bull. Dep. Agric. S. Aust.* **31**: 1–90.

Cornu, M.M. (1879) Etudes sur le *Phylloxera vastatrix. Mem. Acad. Sci. Inst. Fr.* **26**: 1–357 (1877/8).
Corpuz-Raros, L.A. and Cook, E.F. (1974) A Revision of North American *Capitophorus* van der Goot and *Pleotrichophorus* Börner. *Smithson. Contr. Zool.* **156**: 143 pp.
Costa, A.S. and Grant, A.J. (1951) Studies on transmission of the tristeza virus by the vector *Aphis citricidus. Phytopathology* **41**: 105–13.
Cottier, W. (1953) Aphids of New Zealand. *Bull. N.Z. Dep. Scient. Ind. Res.* **106**: 382 pp.
Cutright, C.R. (1925) Subterranean aphids of Ohio. *Bull. Ohio agric. Exp. Stn* **387**: 175–238.
Dahl, M.L. (1968) Biologische und morphologische Untersuchungen über den Formenkreis der Schwarzen Kirschenlaus *Myzus cerasi* (F.). *Dt. ent. Z.* (N.F.) **15**: 281–312.
Dani, R.C. and Majumdar, N. (1978) Preliminary observations on the incidence of root aphids on different rice cultivars. *Sci. cult.* **44**: 88–89.
Danielsson, R. (1982). The species of the genus *Eriosoma* Leach having *Ribes* L. as secondary host plant. *Entomologica scand.* **13**: 341–58.
Das, B. (1918) Aphididae of Lahore. *Mem. Indian Mus.* **6**: 135–274.
Das, B.C. and Raychaudhuri, D.N. (1983) Aphids of Nepal. *Misc. Publ. Occ. Pap. Rec. Zool. Surv. India* **51**: 79 pp.
Davatchi, A.G. (1958) Étude biologique de la faune entomologique des *Pistacia* sauvages et cultivés. *Rev. Path. Vég. Ent. Agr. Fr.* **37**: 3–166.
David, S.K. (1956) Notes on South Indian Aphids 1. Description of four new Species. *Indian J. Ent.* **18**: 1–9.
David, S.K. (1969) Some rare aphids in new Regions in India. *J. Bombay nat. Hist. Soc.* **66**: 323–26.
David, S.K. (1976) A taxonomic review of *Macrosiphum* in India. *Orient. Insects* **9**: 461–93 (1975).
David, S.K. (1977) Host-selection and speciation in some South Indian aphids. pp 19–21 in Ananthakrishnan, T.N. (ed.) *Insects and host-specificity*, Macmillan, Delhi.
David, S.K. and Hameed, S.F. (1975) One new species and two new records of aphids from Lahaul in N.W. Himalaya. *Orient. Insects* **9**: 213–19.
David, S.K., Narayanan, K. and Rajasingh, S.G. (1969) New records of aphids in India. *Bull. Ent. ent. Soc. India* **10**: 158–59.
David, S.K., Rajasingh, S.G. and Narayanan, K. (1971) The Myzaphidines of India with descriptions of three new species. *Orient. Insects* **4**: 395–406 (1970).
Davidson, W.M. (1914) Walnut aphides in California. *Bull. U.S. Dep. Agric.* **100**: 1–48.
Davidson, W.M. (1917) The reddish-brown plum aphis (*Rhopalosiphum nymphaeae* L.). *J. econ. Ent.* **10**: 350–53.
Davis, J.J. (1909) Biological studies on three species of Aphididae. *Tech. Ser. Bur. Ent. U.S.* **10**: 123–63.
Davis, J.J. (1914) New and little known species of Aphididae. *Can. Ent.* **46**: 41–51, 77–87, 121–34, 165–73, 226–36.
Davis, J.J. (1917) The Corn Root Aphid and methods of controlling it. *Fmrs Bull. U.S. Dep. Agric.* **881**: 1–12.
Delfino, M.A. (1982) Una nueva especie del genero *Wahlgreniella* Hille Ris Lambers, 1949. *Revta Soc. ent. argent.* **40**: 183–86.
Devaiah, M.C., Gubbaiah and Jayaramaiah, M. (1976) *Dactynotus compositae* (Theobald), a new aphid pest of mulberry (*Morus* spp.). *J. Bombay nat. Hist. Soc.* **73**: 234.
Dicker, G.H.L. (1940) The Biology of the *Rubus* Aphides. *J. Pomol.* **18**: 1–33.
Dicker, G.H.L. (1952) Studies in population fluctuations of the strawberry aphid *Pentatrichopus fragaefolii*. I. Enemies of the strawberry aphid. *Rep. E. Malling Res. Stn* **39**: 166–68.

Dieleman, F.L. and Eenink, A.H. (1977) Resistance in lettuce to *Nasonovia ribisnigri*. *Bull. SROP* **1977**(3): 67–8.
Dixon, A.F.G. (1971) The life cycle and host preferences of the bird cherry-oat aphid, *Rhopalosiphum padi* L. . . . *Ann. appl. Biol.* **68**: 135–47.
Dixon, A.F.G. (1973) *Biology of Aphids*. Edward Arnold Ltd., London, 58 pp.
Dixon, A.F.G. (1977) Aphid ecology: life cycles, polymorphism and population regulation. *A. Rev. Ecol. Syst.* **8**: 329–53.
Dixon, A.F.G. and Dharma, T.R. (1980) Number of ovarioles and fecundity in the black bean aphid, *Aphis fabae*. *Entomologia exp. appl.* **28**: 1–14.
Dixon, A.F.G. and Shearer, J.W. (1974) Factors determining the distribution of the aphid *Sipha kurdjumovi* on grasses. *Entomologia exp. appl.* **17**: 439–44.
Doncaster, J.P. (1954) Notes on the genus *Lipaphis* Mordv. 1928, and description of a new species. *Proc. R. ent. Soc. Lond.* (B) **23**: 83–88.
Doncaster, J.P. (1956) The Rice Root Aphid. *Bull. ent. Res.* **47**: 741–47.
Doncaster, J.P. (1966) Notes on some Indian aphids described by G.B. Buckton. *Entomologist* **99**: 157–60.
Duncan, C.D. (1922) The North American species of *Phylloxera* infesting Oak and Chestnut. *Can. Ent.* **54**: 267–76.
Dunn, J.A. (1959) The biology of the lettuce root aphid (and similar titles). *Ann. appl. Biol.* **47**: 475–91; 766–71; 772–77.
Dunn, J.A. (1960) The natural enemies of the lettuce root aphid *Pemphigus bursarius* (L.). *Bull. ent. Res.* **51**: 271–78.
Dunn, J.A. (1965) Studies on the aphid *Cavariella aegopodii* Scop. I. On willow and carrot. *Ann. appl. Biol.* **56**: 429–38.
Dunn, J.A. and Kempton, D.P.H. (1972) Resistance to attack by *Brevicoryne brassicae* among plants of Brussels sprouts. *Ann. appl. Biol.* **72**: 1–11.
Dunn, J.A. and Kirkley, J. (1966) Studies on the aphid *Cavariella aegopodii* Scop. II. On secondary hosts other than carrot. *Ann. appl. Biol.* **58**: 213–17.
Dürr, H.J.R. (1983) *Diuraphis noxia* (Mordvilko), a recent addition to the aphid fauna of South Africa. *Phytophalactica* **15**: 81–83.
Eastop, V.F. (1953) A study of the Tramini. *Trans. R. ent. Soc. Lond.* **104**: 385–413.
Eastop, V.F. (1958) *A Study of the Aphididae of East Africa*. Colonial Research Publication, H.M.S.O. London, 126 pp.
Eastop, V.F. (1961a) *A Study of the Aphididae of West Africa*. British Museum (Natural History), London, 93 pp.
Eastop, V.F. (1961b) A key for the determination of *Schizaphis* Börner. *Entomologist* **94**: 241–46.
Eastop, V.F. (1966) A taxonomic study of Australian Aphidoidea. *Aust. J. Zool.* **14**: 399–592.
Eastop, V.F. (1971) Keys for the identification of *Acyrthosiphon*. *Bull. Br. Mus. nat. Hist. (Ent.)* **26**: 1–115.
Eastop, V.F. (1981) The wild hosts of aphid pests, pp. 285–98 in Thresh, J.M., *Pests, Pathogens and Vegetation*, Pitman, London.
Eastop, V.F. (1983) The biology of the principal aphid virus vectors, pp. 115–32 in Thresh, J.M. and Plumb, R. (eds) *Plant Virus Epidemiology*, Blackwell, Oxford.
Eastop, V.F. (in press) The acquisition and processing of taxonomic data, in Szelegiewicz, H. (Ed.) *Evolution and Biosystematics of Aphids*, Warsaw.
Eastop, V.F. and Hille Ris Lambers, D. (1976) *Survey of the World's Aphids*, W. Junk, The Hague, 573 pp.
Edelson, J.V. and Estes, P.M. (1983) Intracanopy distribution and seasonal abundance of the yellow pecan aphids . . . . *Envir. Ent.* **12**: 862–67.
Ehrhardt, P. (1963) Untersuchungen über Bau Function des Verdauungstraktes von *Megoura viciae* Buckt. unter besonderer Berücksichtigung der Nahrungsaufnahme und der Hongigtau abgabe. *Z. Morph. Ökol. Tiere* **52**: 597–677.

Eikenbary, R.D. and Rogers, C.E. (1974) Importance of alternate hosts in establishment of introduced parasites. *Proc. Tall Timbers Conference on Ecological Animal Control by Habitat Management.* N–S. 1–2 March 1973, Tallahassee, Fla, pp. 119–33.
El-Kady, E.A. see Kady.
van Emden, H.F. (Ed.) (1972a) *Aphid Technology.* Academic Press, London & New York, xiv + 344 pp.
van Emden, H.F. (Ed.) (1972b) Insect/Plant Relationships. *Symp. R. ent. Soc. Lond.* **6**: 215 pp.
van Emden, H.F. (1977) Failure of the aphid, *Myzus persicae*, to compensate for poor diet during early growth. *Physiol. Ent.* **2**: 43–48.
van Emden, H.F. and Bashford, M.A. (1969) A comparison of the reproduction of *Brevicoryne brassicae* and *Myzus persicae* in relation to soluble nitrogen concentration and leaf age (leaf position) in the Brussels sprout plant. *Entomologia exp. appl.* **12**: 351–64.
van Emden, H.F., Eastop, V.F., Hughes, R.D. and Way, M.J. (1969) The ecology of *Myzus persicae*. *A. Rev. Ent.* **14**: 197–270.
Eskanderi, F., Sylvester, E.S. and Richardson, J. (1979) Evidence for lack of propagation of potato leaf roll virus in its aphid vector *Myzus persicae*. *Phytopathology* **69**: 45–47.
Essig, E.O. and Kuwana, S.I. (1918) Some Japanese Aphididae. *Proc. Calif. Acad. Sci.* (4) **8**(3): 35–112.
Evelyn, J. (1693) *The Compleat Gardener.* Matthew Gilliflower, London (2 vols).
Evenhuis, H.H. (1963) De groene appeltakluis, *Aphis pomi* Geer, en haar parasitencomplex. *Meded. LandbHogesch. Gent* **28**: 784–91.
Ewing, H.E. (1916) Eighty seven generations in a parthenogenetic pure line of *Aphis avenae* Fabr. *Biol. Bull. mar. biol. Lab., Woods Hole* **31**: 53–112.
Faith, D.P. (1979) Strategies of gall formation in *Pemphigus* aphids. *J. N.Y. ent. Soc.* **87**: 21–37.
Falk, U., (1960) Über das Auftreten von Intermediärformen zwischen oviparem und geflügeltem viviparem Weibchen bei *Aphis craccivora* Koch. *Zool. Anz.* **165**: 388–92.
Fidler, J.H. (1951 *Aphis vaccinii* (Börner) new to the British Isles. A redescription and comparison with *Aphis callunae* Theobald. *Proc. R. ent. Soc. Lond.* (B) **20**: 37–43.
Firempong, S. (1976) Changes in populations of *Toxoptera aurantii* (Boy.) on cocoa in Ghana. *J. appl. Ecol.* **13**: 793–99.
de Fluiter, H.J. (1933) Bijdrage tot kennis der oecologie en morphologie van *Eriosoma lanuginosum* (Hartig), de 'bloedluis' onzer pereboomen. *Tijdschr. Plziekt.* **39**: 45–72.
Forbes, A.R. (1963) An occurrence of the Bulb and Potato Aphid *Rhopalosiphoninus latysiphon* (Davidson), on potato in British Columbia. *Proc. Ent. Soc. Br. Columb.* **60**: 30–32.
Forbes, A.R. (1964) The morphology, history and fine structure of the gut of the Green Peach Aphid *Myzus persicae* (Sulzer). *Mem. ent. Soc. Can.* **36**: 1–74.
Forbes, A.R. (1981) *Brachycolus asparagi* Madrilko, a new aphid pest damaging asparagus in British Columbia. *J. ent. Soc. Br. Columb.* **78**: 13–16.
Forrest, J.M.S. (1970) The effect of maternal and larval experience on morph determination in *Dysaphis devecta*. *J. Insect Physiol.* **16**: 2281–92.
Forrest, J.M.S. and Dixon, A.F.G. (1975) The induction of leaf-roll galls by the apple aphids *Dysaphis devecta* and *D. plantaginea*. *Annls appl. Biol.* **81**: 281–88.
Frazer, B.D. and Gill, B. (1981) Age, fecundity, weight and the intrinsic rate of increase of the lupin aphid *Macrosiphum albifrons*. *Can. Ent.* **113**: 739–45.
Frazier, N.W. (1975) Strawberry mild yellow-edge in North Carolina. *Pl. Dis. Reptr* **59**: 41–42.
Fronk, W.D. (1955) An economically important aphid new to the United States. *Pan-Pacific Ent.* **31**: 190.
Furk, C. and Prior, R.N.B. (1975) On the life cycle of *Pemphigus (Pemphiginus)*

populi Courchet with a key to British species of *Pemphigus* Hartig. *J. Ent.* (B) **44:** 265–80.

Gadiyappanavar, R.D. and Channabasavanna. G.P. (1973) Bionomics of the ragi (*Eleusine coracava*) root aphid, *Tetraneura nigriabdominalis* (Sasaki). *Mysore J. agric. Sci.* **7:** 436–44.

Galecka, B. (1966) The rôle of predators in the reduction of two species of potato aphids, *Aphis nasturtii* Kalt. and *A. frangulae* Kalt. *Ecol. pol.* (A) **14:** 245–74.

Gaponova, A.F. (1970) [The biological characteristics of the beet root aphid *Pemphigus fuscicornis* Koch in the conditions of the USSR.] pp. 51–57 in *Proc. Intern. Symp. Sugar Beet Protection* (1970) **18** (11–12): 357 pp.

Garg, A.K. and Sethi, G.R. (1978) Biology and seasonal incidence of *Carolinaia (Hysteroneura) setariae* (Thomas) infesting paddy in Delhi. *Indian J. Ent.* **40:** 221–23.

Georghiou, G.P. (1982, and in press) The occurrence of resistance to pesticides in arthropods. 1. An index of cases reported through 1980. (1981); F.A.O., Rome, xv + 172 pp; 2: in press.

Ghosh, A.K. (1974) Some new and little-known species of aphids from India. *Orient. Insects* **8:** 425–32.

Ghosh, A.K. (1975) Aphids of economic importance in India. *Indian Agric.* **18:** 81–214 (1974) reprinted in hard cover and paginated 1–134 + 2 errata.

Ghosh, A.K. (1976) A list of aphids from India and adjacent countries. *J. Bombay Nat. Hist. Soc.* **71:** 201–20.

Ghosh, A.K. (1980) Homoptera, Aphidoidea, part 1. General introduction and subfamily Chaitophorinae, in *The Fauna of India and adjacent countries,* Zoological Survey of India, Calcutta, 124 pp.

Ghosh, A.K. (1981) Review of *Kaltenbachiella* Schouteden. *Orient. Insects* **15:** 127–38.

Ghosh, A.K. (1982a) Homoptera, Aphidoidea, 2. Lachninae, in *The Fauna of India and adjacent countries,* Zoological Survey of India, Calcutta, 167 pp.

Ghosh, A.K. (1982b) Cervaphidini of the world. *Orient. Insects* **16:** 77–97.

Ghosh, M.R., Ghosh, A.K. and Raychaudhuri, D.N. (1971) Studies in aphids from Eastern India III. *Orient. Insects* **4:** 377–93.

Ghosh, A.K. and Raychaudhuri (1972) Studies on the aphids from Eastern India. 16. Descriptions of ... little-known species. *Proc. zool. Soc., Calcutta* **25:** 93–107.

Ghosh, M.R., Pal, P.K., and Raychaudhuri, D.N. (1974) Studies on the aphids from Eastern India. 21. The genus *Astegopteryx* and other related genera .... *Proc. zool. Soc., Calcutta* **27:** 81–116.

Ghosh, M.R. and Raychaudhuri, D.N. (1973) A study of the genus *Aiceona* Takahashi .... *Orient. Insects* **7:** 547–55.

Ghosh, M.R. and Raychaudhuri, D.N. (1981) Biology of *Nippolachnus piri* (Matsumura) infesting pear in West Bengal. *Entomon (India)* **6:** 229–32.

Gibson, R.W. (1971) Climatic factors restricting the distribution of the aphid *Rhopalosiphoninus latysiphon* to the subterranean parts of field potato plants. *Ann. appl. Biol.* **69:** 89–96.

Gilbert, N. and Hughes, R.D. (1971) A model of an aphid population—three adventures. *J. Anim. Ecol.* **40:** 525–34.

Gilmore, J.E. (1960) Biology of the black cherry aphid in the Willamette Valley, Oregon. *J. econ. Ent.* **53:** 659–61.

Glowacka, E., Klimazewski, S.M., Szelegiewicz, H., and Wojciechowski, W. (1974) Über den Bau des männlichen Fortpflanzungssystems der Lachniden. *Annls zool.* **32:** 39–49.

Gonzales, D., Miyazaki, M., White, W., Takada, H., Dickson, R.C. and Hall, J.C. (1979) Geographical distribution of *Acyrthosiphon kondoi* Shinji and some of its parasites and hyperparasites in Japan. *Kontyû* **47:** 1–7.

van der Goot, P. (1917/18) Zur Kenntniss der Blattläuse Java's. *Contr. Faune Indes néerl.* **1(3):** 301 pp.

Graham, M.W.R. de V. (1976) The British species of *Aphelinus* with notes and descriptions of other European Aphelinidae (Hymenoptera). *Syst. Ent.* **1**: 123–46.

Granovsky, A.A. (1928) A review of *Myzocallis* species inhabiting *Alnus*, with description of a new species. *Ann. ent. Soc. Am.* **21**: 546–64.

Griffiths, D.C. (1961) The development of *Monoctonus paludum* Marshall in *Nasonovia ribisnigri* on lettuce, and immunity reactions in other lettuce aphids. *Bull. ent. Res.* **52**: 147–63.

Grigarick, A.A. and Lange, W.H. (1962) Host relationships of the sugar-beet root aphid in California. *J. econ. Ent.* **55**: 760–64.

Grigorov, S. (1977) [Study on aphids of the genus *Dysaphis* C.B. on pear.] *Grudinarska; Lozarska Nauk* **14**: 22–28.

Grossheim, N.A. (1914) [The barley aphid *Brachycolus noxius* Mordvilko.] *Trudy estest.-istor. Muz., Simferopol* **3**: 35–73.

del Guercio, G. (1914) Intomo a due nuovi Vacunini del Castagno. *Redia* **9**: 285–91 (1913).

Gut, J. (1976) Chromosome numbers of parthenogenetic females of fifty-five species of Aphididae new to cytology. *Genetica* **46**: 279–85.

Guttierrez, A.P., Morgan, D.J., and Havenstein, D.E. (1971) The ecology of *Aphis craccivora* Koch and subterranean clover stunt virus ... in south-east Australia. *J. appl. Ecol.* **8**: 699–721.

Gutierrez, A.P., Havenstein, D.E., Nix, H.A., and Moore, P.A. (1974a) The ecology of *Aphis craccivora* Koch ... in temperate pastures. *J. appl. Ecol.* **11**: 1–20.

Guttierrez, A.P., Nix, H.A., Havenstein, D.E., and Moore, P.A. (1974b) The ecology of *Aphis craccivora* Koch ... phenology and migration .... *J. appl. Ecol.* **11**: 21–35.

Habib, A. and El-Kady, E.A. (1961) The Aphididae of Egypt. *Bull. Soc. ent. Égypte* **45**: 1–137.

Hafez, M. (1978) Einfuhr und Ansiedlung der Blattlauszehrwespe, *Aphelinus mali* (Hald.) in Saudi Arabia. *Anz. schadlingsk. Pfl.-Umweltschjutz* **51**: 29–30.

Hagen, K.S. and van den Bosch, R. (1968) Impact of pathogens, parasites and predators on aphids. *A. Rev. Ent.* **12**: 325–84.

Haine, E. (1955) Biologisch-ökologische Studien an *Rhopalosiphoninus latysiphon*. *D. Landwirtschaft-angew. wiss.* **29**: 1–58.

Hall, R.A. and Burges, H.D. (1979) Control of aphids in glasshouses with the fungus *Verticillium lecanii*. *Ann. appl. Biol.* **93**: 235–46.

Hand, S.C. and Williams, C.T. (1981) The overwintering of the rose-grain aphid *Metopolophium dirhodum* on wild roses. In Thresh, J.M. (Ed.), *Pests, Pathogens and Vegetation*, pp. 285–98. Pitman, London.

Harper, A.M. (1959) Gall aphids on poplar in Alberta 1. Descriptions of galls and distributions of aphids. *Can. Ent.* **91**: 489–96.

Harper, A.M. (1963) Sugar-beet root aphid, *Pemphigus betae* Doane, in southern Alberta. *Can. Ent.* **95**: 863–73.

Harris, K.F. and Maramorosch, K. (Eds) (1977) *Aphids as Virus Vectors*. Academic Press, New York, 559 pp.

Harris, K.M. (1973) Aphidophagous Cecidomyidae (Diptera): Taxonomy, biology and assessments of field populations. *Bull. ent. Res.* **63**: 305–25.

Hassan, M.S. (1957) Studies on the morphology and biology of *Aphis maidis* Fitch in Egypt. *Bull. Soc. ent. Égypte* **41**: 199–211.

Hayward, K.J. (1944a) El pulgón amarillo de la cana de azúcar (*Sipha flava* Forbes) en Tucumán. *Circ. Estac. exp. agric. Tucumán* **125**: 8 pp.

Hayward, K.J. (1944b) Contribucion a la bibliografia del pulgón amarillo de la cana azúcar (*Sipha flava* Forbes). *Publ. misc. Estac. exp. agric.* Tucumán **3**: 13 pp.

Hedberg, I. (Ed.) (1980) Parasites as plant taxonomists. *Symb. bot. upsal.* **22(4)**: 1–221.

References

Heie, O. (1962) A List of Danish Aphids. 3. *Ent. Meddr* **31**: 205–24.
Heie, O. (1964) Aphids collected in Iceland in August, 1961. *Ent. Meddr* **32**: 220–35.
Heie, O. (1979) Revision of the aphid genus *Nasonovia* Mordvilko, including *Kakimia* Hottes & Frison, with keys and descriptions of the species of the world. *Ent. scand. suppl.* **9**: 105 pp.
Heie, O. (1980, 1982) The Aphidoidea of Fennoscandia and Denmark. I. *Fauna ent. scand.* **9**: 236 pp. (1980): **11**: 176 pp. (1982).
Heie, O.E. and Friedrich, W.L. (1971) A fossil specimen of the North American hickory aphid (*Longistigma caryae* Harris), found in tertiary deposits in Iceland. *Ent. Scand.* **2**: 74–80.
Heinze, K. (1960, 1961) Systematic der mittel-europäischen Myzinae. *Beitr. Z. Ent.* **10**: 744–842 (1960): **11**: 24–96 (1961).
Heinze, K. (1962) Pflanzenschädliche Blattlausarten der Familien Lachnidae, Adelgidae und Phylloxeridae, eine systematische-faunistische Studie. *Dt. ent. Z.* **9**: 143–227.
Heryford, N.W. and Sokal, R.R. (1971) Seasonal morphometric variations in *Pemphigus populitransversus. J. Kansas ent. Soc.* **44**: 384–90.
Higuchi, H. (1968) A revision of the genus *Takecallis* Matsumura. *Insecta Matsum.* **31**: 25–33.
Higuchi, H. (1972) A taxonomic study of the subfamily Callipterinae in Japan. *Insecta Matsum.* **35**: 19–126.
Higuchi, H. and Miyazaki, M. (1969) A tentative catalogue of host plants of Aphidoidea in Japan. *Insecta Matsum.* **suppl.** **5**: 66 pp.
Hill, T. (1568) *The Profitable Arte of Gardening*. Thomas Mashe, London.
Hille Ris Lambers, D. (1931) Notes on the Aphididae of Venezia Tridentina, with descriptions of new species. *Memorie Mus. Stor. nat. Venezia trident.* **1**: 15–24.
Hille Ris Lambers, D. (1933, 1934) Notes on Theobald's 'The Plantlice or Aphididae of Great Britain'. *Stylops* **2**: 169–76 (1933); **3**: 25–33 (1934).
Hille Ris Lambers, D. (1938) Contributions to a monograph of the Aphididae of Europe. 1. The genus *Macrosiphoniella* Del Guercio, 1911. *Temminckia* **3**: 1–44.
Hille Ris Lambers, D. (1939a) Contributions to a monograph of the Aphididae of Europe. II. *Temminckia* **4**: 1–134.
Hille Ris Lambers, D. (1939b) On some Western European aphids. *Zool. Meded.* **22**: 79–119.
Hille Ris Lambers, D. (1944) Overzicht bladluizen in de Fruitteelt. *Fruitteelt* **34(31)**: 1–26.
Hille Ris Lambers, D. (1945) De Bloedvlekkenluis van Appel, *Sappaphis devecta* (Walker). *Tijdschr. Plziekt.* **51**: 57–72.
Hille Ris Lambers, D. (1947a) Contributions to a monograph of the Aphididae of Europe. III. *Temminckia* **7**: 179–319.
Hille Ris Lambers, D. (1947b) On some mainly western European aphids. *Zool. Meded.* **28**: 291–333.
Hille Ris Lambers, D. (1948) On Palestine aphids, with descriptions of new subgenera and new species (Hom. Aphid.). *Trans. R. ent. Soc. Lond.* **99**: 269–89.
Hille Ris Lambers, D. (1949) Contributions to a monograph of the Aphididae of Europe. IV. *Temminckia* **8**: 182–329.
Hille Ris Lambers, D. (1950a) Host plants and aphid classification. *Proc. intern. Congr. Ent., Stockholm* **8**: 141–44.
Hille Ris Lambers, D. (1950b) De Nederlandse bladluizen van framboos en braam. *Tijdschr. PlZiekt.* **56**: 253–61.
Hille Ris Lambers, D. (1952) New Aphids from Sweden. *Opusc. Ent.* **17**: 51–58.
Hille Ris Lambers, D. (1953a) Contributions to a monograph of the Aphididae of Europe. V. *Temminckia* **9**: 1–176.
Hille Ris Lambers, D. (1953b) Notes on aphids from *Cocos nucifera*. *Agric. J. Fiji* **24**: 211–15.

Hille Ris Lambers, D. (1954a) On some Eritrean aphids with descriptions of a new genus and new species. (Hom. Aphididae). *Boll. Lab. gen. agr. zool. Portici* **33**: 171–86 (volume published in 1956, separates in 1954?).
Hille Ris Lambers, D. (1954b) New Israel Aphids. *Bull. Res. Coun. Israel* **4**: 276–83.
Hille Ris Lambers, D. (1956a) A revision of the genus *Cervaphis* v.d. Goot, 1917 (Homopt., Aphid.) *Ent. Ber., Amst.* **16**: 130–36.
Hille Ris Lambers, D. (1956b) Lachnids from Elaeagnaceae (Hom., Aph.). *Z. angew. Ent.* **39**: 467–73.
Hille Ris Lambers, D. (1959) *Myzus (Nectarosiphon) certus* (Wlk.) as a problem in studies on flights of *Myzus (Nectarosiphon) persicae* (Sulz.) (Homoptera, Aphididae). *Ent. Ber., Amst.* **19**: 17–19.
Hille Ris Lambers, D. (1961) Notes on three North American Aphididae. *Fla Ent.* **44**: 181–83.
Hille Ris Lambers, D. (1965) On some Japanese Aphididae (Homoptera). *Tijdschr. Ent.* **108**: 189–203.
Hille Ris Lambers, D. (1966a) On Alpine Rose- and grass-infesting species of *Metopolophium* Mordvilko, 1914, with descriptions of two new species (Homoptera, Aphididae). *Mitt. Schweiz. ent. Ges.* **39**: 103–17.
Hille Ris Lambers, D. (1966b) Polymorphism in Aphididae. *A Rev. Ent.* **11**: 47–78.
Hille Ris Lambers, D. (1966c) Notes on California aphids, with descriptions of new genera and new species. *Hilgardia* **37**: 569–623.
Hille Ris Lambers, D. (1967a) New and little-known members of the aphid fauna of Italy. *Boll. Lab. Zool. agr. Bichic.* (2)**8**: 1–32.
Hille Ris Lambers, D. (1967b) Notes on some Japanese Eriosomatinae and their life cycle. *Ent. Ber., Amst.* **27**: 86–89.
Hille Ris Lambers, D. (1969) A key to, notes on, and descriptions of American *Pleotrichophorus* Börner. *Ent. Ber., Amst.* **29**: 165–80.
Hille Ris Lambers, D. (1970a) The Genus *Nearctaphis* Shaposhnikov, 1950. I & II. *Proc. Koninkl. Nederl. Akad. van Wetenschappan* (series C) **73**: 48–62, 63–74.
Hille Ris Lambers, D. (1970b) A study of *Tetraneura* Hartig, 1841, with descriptions of a new subgenus and new species. *Boll. Lab. Zool. agr. Bachic.* (2) **9**: 21–101 (1968–69).
Hille Ris Lambers, D. (1972) Aphids: Their life cycles and their role as virus vectors. pp. 36–56 in *Viruses of potatoes and seed-potato production*, Wageningen.
Hille Ris Lambers, D. (1973) *Masonaphis lambersi* MacGillivray, 1960, a new pest of Rhododendron in Europe. *Tijdschr. Plziekt.* **79**: 159–61.
Hille Ris Lambers, D. (1974) On American aphids, with descriptions of a new genus and some new species. *Tidschr. Ent.* **117**: 103–55.
Hille Ris Lambers, D. (1980) Aphids as botanists? *Symb. Bot. Upsal.* **22**(4): 114–19 (1979).
Hille Ris Lambers and Basu, A.N. (1966) Some new and little-known genera, subgenera, species and subspecies of Aphididae from India. *Ent. Ber., Amst.* **26**: 12–20, 27–36.
Hille Ris Lambers, D. and Van den Bosch, R. (1964) On the genus *Therioaphis* Walker, 1870, with descriptions of new species. *Zool. verhand.* **68**: 1–47.
Hille Ris Lambers, D. and Dicker, G.H.L. (1965) *Aphis triglochinis* Theobald, 1926, as a pest of Red Currant (*Ribes rubrum* L.) and Black Currant (*Ribes nigrum* L.). *Ent. Ber., Amst.* **25**: 5–6.
Hille Ris Lambers, D. and Takahashi, R. (1959) Some species of *Thoracaphis* and of nearly related genera from Java. *Tidschr. Ent.* **102**: 1–16.

References

Hille Ris Lambers, D. and Wildbolz, T. (1958) *Allocotaphis quaestionis* Börner in der Schweiz. *Mitt. schweiz. ent. Ges.* **31**: 317–19.
Hodek, I. (Ed.) (1966) *Ecology of Aphidophagous Insects*. Academy of Sciences, Prague, 360 pp.
Hodek, I. (1973) *Biology of Coccinellidae*. Dr W. Junk, the Hague, 260 pp.
Hodjat, S.H. (1981) Two new aphids from Iran and keys to related Middle Eastern species. *J. nat. Hist.* **15**: 365–74.
Holdsworth, R.P. (1970) Aphids and aphid enemies: Effect of integrated control in an Ohio apple orchard. *J. econ. Ent.* **63**: 530–35.
Holman, J. (1961) Descriptions of two new Aphid species (*Sitobion equiseti* sp.n. and *Linosiphon asperulophagus* sp.n.) from Czechoslovakia. *Cas. csl. Spol. ent.* **58**: 324–32.
Holman, J. (1965) Description of *Pleotrichophorus achilleae* sp. n. with notes on related species. *Acta ent. bohemoslavaca* **62**: 277–82.
Holman, J. (1972) *Nasonovia (Kakimia) brachycyclica* sp.n. on *Ribes grossulariae* L. from Czechoslovakia. *Acta ent. bohemoslavaca* **69**: 317–23.
Holman, J. (1974) *Los áfidos de Cuba*. Instituto del Libro, La Habana, 304 pp.
Holman, J. (1981) A review of the *Uroleucon* species confined to Asteraceae, Inuleae. *Acta ent. bohemoslavaca* **78**: 162–76.
Holman, J. and Szelegiewicz, H. (1974, 1978) Aphids of the genus *Macrosiphoniella* from Mongolia. *Acta ent. bohemoslavaca* **71**: 161–77 (1974); **75**: 178–93 (1978).
Hottes, F.C. (1949) Descriptions of some undescribed forms belonging to two little-known species of the family Aphididae. *Proc. biol. Soc. Wash.* **62**: 45–52.
Hottes, F.C. and Frison, H.F. (1931) The plant lice, or Aphididae, of Illinois. *Bull. Ill. Nat. Hist. Surv.* **19**: 121–447.
Hoyt, S.C. and Madsen, H.F (1960) Dispersal behaviour of the first instar nymphs of the woolly apple aphid. *Hilgardia* **30**: 267–99.
Hrdy, I. and Hrdlickova, H. (Eds) 1981. Integrated pest and disease control in hops. *Bull. SROP* **1981/IV/3**: 1–179.
Hsieh, C.Y. (1970) The aphids attacking rice plants in Taiwan (II). Studies on the biology of the red rice root aphid *Rhopalosiphum rufiabdominalis* (Sasaki). *Pl. Prot. Bull. Taiwan* **12**: 68–78.
Iglisch, I. (1966) Untersuchungen über die Biologie und phytopathologische Bedeutung der Holunderblattlaus, *Aphis sambuci* L. *Mitt. Biol. Bund. Land. Forstw. Berlin–Dahlem* **119**: 1–32.
Ilharco, F.A. (1966) On the identity of *Tavaresiella suberis* del Guercio. *Agron. lusitana* **26**: 83–89 (1964).
Ilharco, F.A. (1967) *Thelaxes suberi* (del Guercio) in Madeira Island. *Agron. lusitana* **27**: 97–101 (1965).
Ilharco, F.A. (1968) *Pentalonia nigronervosa* Coquerel na ilha da Madeira. Contribuicao para o conhecimento da sua bibliografia. *Bocagiana* **17**: 1–25.
Ilharco, F.A. (1979) 1. Aditamento ao catalogo dos afideos de Portugal continental. *Agron. lusitana* **39**: 253–94.
Inouye, M. (1956) Beiträge zur kenntnis der koniferen-läuse, verkommend im nördlichen teil Japans. *Spec. Rep. Hokkaido Govt Forest Res. Stn* **5**: 204–38.
I.O.B.C. (1980) Écologie des pucerons des céréales. Rev. due sous-groupe du Groupe de Travail 'Lutte intégrée en Céréales'. Colmar (France). *Bull. SROP* **1980/3**: 119 pp.
Ivanovskaya, O.I. (1977) [*Aphids of Western Siberia.*] Vol. 1, Adelgidae–Chaitophoridae, 272 pp.; vol. 2, Aphididae, 328 pp. Academy of Sciences, Novosibirsk.
Ivanovskaya, O.I. (1980) Aphids of the genus *Cavariella* del Guercio, 1911 in northern Asia. pp. 79–87 in Cherapanov, A.I. (Ed.). *Systematics and Ecology of Animals*.
Iwaki, M. and Auzay, H. (1978) Virus diseases of mungbean in Indonesia, pp. 169–72 in Cowell, R. (Ed.). *The 1st International Mungbean Symposium*. Office of Information Services, Vegetable Research and Development Centre, Taiwan.

Jacob, F.H. (1946) A new species of *Myzaphis* van der Goot associated with wild roses, *Myzaphis bucktoni* sp.n., and a comparison with *M. rosarum* (Kaltenbach). *Proc. R. ent. Soc. Lond.* (B) **15**: 110–17.

Jacob, F.H. (1949) A study of *Aphis sambuci* L. and a discussion of its bearing upon the study of the 'black aphids'. Parts 1 & 2. *Proc. R. ent. Soc. Lond.* (A) **24**: 90–98; 99–109.

Janiszewska-Cichocka, E. (1969) Zur Morphologie und Biologie de Ulmblattlaus, *Eriosoma ulmi* (L.). *Annls zool., Warsz.* **27**: 205–23.

Janiszewska-Cichocka, E. (1971) [Life-cycle of *Eriosoma ulmi* (L.).] *Reczn. Nauk roln.*, **E1**: 25–55.

Johnson, C.G. (1969) *The Migration and Dispersal of Insects by Flight.* Methuen, London, 763 pp.

Jones, T.H. and Gillette, C.P. (1918) Life history of *Pemphigus transversus. J. agric. Res.* **14**: 577–94.

Judge, F.D. (1968) Polymorphism in a subterranean aphid, *Pemphigus bursarius.* I. Factors affecting the development of sexuparae. *Ann. ent. Soc. Am.* **61**: 819–27.

El-Kady, E.A., Hassan, M.S., and Attia, A.A. (1971) The biology of the mealy plum aphid, *Hyalopterus pruni* (Geoffroy) in Egypt. Studies on the life cycle . . . . *Bull. Soc. ent. Egypte* **54**: 529–40; 579–82.

Kalina, V. and Starý, P. (1976) A review of the aphidophagous Aphelinidae (Hym., Chalcidoidea), their distribution and host range in Europe. *Studia ent. for.* **2**: 143–70.

Karczewska, M. (1975) [On the biology of the small plum aphid—*Brachycaudus helichrysi* (Kalt.).] *Polskie Pismo ent.* **45**: 583–96

Karczewska, M. and Stasiak, A. (1973) [On the seasonal development of the gooseberry aphid—*Hyperomyzus pallidus* H.R.L.]. *Polskie Pismo ent.* **43**: 827–35.

Kassanis, B. (1947) Studies on dandelion yellow mosaic and other virus diseases of lettuce. *Ann. appl. Biol.* **34**: 412–21.

Kawada, K. and Murai, T. (1979) Apterous males and holocyclic reproduction of *Lipaphis erysimi* in Japan. *Entomologia exp. appl.* **26**: 343–45.

Keep, E. and Briggs, J.B. (1971) A survey of *Ribes* spp. for aphid resistance. *Ann. appl. Biol.* **68**: 23–30.

Kennedy, G.G. and Schaefers, G.A. (1973) Evidence for nonpreference and anti-biosis in aphid-resistant Red Raspberry cultivars. *Envir. Ent.* **3**: 773–77.

Kennedy, G.G. and Schaefers, G.A. (1974) The distribution and seasonal history of *Amphorophora agathonica* Hottes in 'Latham' Red Raspberry. *Ann. ent. Soc. Am.* **67**: 356–58.

Kennedy, J.S., Day, M.F., and Eastop, V.F. (1962) *A Conspectus of Aphids as Vectors of Plant Viruses.* Commonwealth Institute of Entomology, London, 114 pp.

Kennedy, J.S. and Stroyan, H.L.G. (1959) Biology of aphids. *A. Rev. Ent.* **4**: 139–60.

Kidd, N.A.C. (1976) Factors influencing leaf vein selection in the lime aphid (*Eucallipterus tiliae* L.). *Oecologia* **23**: 247–54.

Kleinjan, J.E. and Mittler, T. (1975) A chemical influence of ants on wing development in aphids. *Entomologia exp. appl.* **18**: 384–88.

Knechtel, W.K. and Manolache, C.I. (1943) Observatii sapura sistematica unor specii de Aphide din România . . . . *Anal. Inst. cerc. agron.* **13**: 217–67 (1941).

Koike, H. (1977) Transmission by *Dactynotus ambrosiae* from mixed infections with sugarcane mosaic and maize dwarf mosaic virus strains. *Pl. Dis. Rptr* **61**: 724–77.

Kolesova, D.A. (1970) [The autumn forms of pear aphids.] *Zashch. Rast.* **15**: 33–35.

Kolesova, D.A. (1972) [The ecology of pear aphids of the genus *Anuraphis* Guercio in the Crimea.] *Vestnik Zoologii* No. **3**: 11–15.

Kolesova, D.A. (1974) [Pear aphids of the genus *Dysaphis* C.B.—two host species.] *Ent. Obozr.* **53**: 38–53.

Kolkaila, A.M. and Soliman, A.A. (1954) A study of the banana aphid, *Pentalonia nigronervosa* Coq. *Bull. Soc. Fouad 1 Ent.* **38**: 231–50.
Komakazi, S., Sakagami, Y. and Korenaga, R. (1979) [Overwintering of aphids on Citrus trees.] *Jap. J. appl. Ent. Zool.* **23**: 246–50.
Koronéos, J. (1939) *Les Insectes de l'Olivier dans le Pélion.* S.N. Taroussopoulos, Athens, 71 pp.
Kranz, J., Schmutterer, H. and Koch, W. (1977) *Diseases, Pests and Weeds in Tropical Crops.* Paul Parey, Berlin and Hamburg, 666 pp.
Kring, J.B. (1959) The life cycle of the melon aphid, *Aphis gossypii* Glover, an example of facultative migration. *Ann. ent. Soc. Am.* **52**: 284–86.
Kring, J.B. (1972) Flight behaviour of aphids. *A. Rev. Ent.* **17**: 461–92.
Kulkarni, P.P. and Kacker, R.K. (1979) Chromosomes of four species of aphids. *Bull. zool. Surv. India* **2**: 1–2.
Kunkel, H. (1973) Die Kotabagabe der Aphiden unter Einfluss von Ameisen. *Bonn. zool. Beitr.* **24**: 105–21.
Kurl, S.P. (1980a) Cytotaxonomy of the genus *Toxoptera*. *Entomon. (India)* **5**: 251–55.
Kurl, S.P. (1980b) Chromosome numbers of four species of Indian aphids. *Natl Acad. Sci. Left.* **3**: 185–86.
Kurl, S.P. and Misra, S.D. (1979) Karyological studies in two species of aphids. In Behura, B.K. (Ed.), *Recent Trends in Aphidological Studies*, Bhubaneswar. p. 46.
Kuznetsova, V.G. (1968) [Karyotypes of the aphids of subtribe Anuraphidina and possible ways of their evolution.] *Ent. obozr.* **47**: 767–81.
Kuznetsova, V.G. (1974) [The characteristics of the chromosomes of Aphidoidea.] *Tsitologiya* **16**: 803–9.
Kuznetsova, V.G. and Shaposhnikov, G. Kh. (1973) [Chromosome numbers of the aphids of the world fauna.] *Ent. obozr.* **52**: 116–35. [In Russian, English translation *Ent. Rev. Wash.* **52**: 78–96.]
Lal, O.P. (1969) Field studies for varietal resistance in rape and mustard against mustard aphid *Lipaphis erysimi* Kalt. *Z. angew. Ent.* **64**: 394–400.
Lance, R. (1980) Breeding lucerns that resist aphids. *Rur. Res.* **106**: 22–27.
Lange, W.H. (1965) Biosystematics of American *Pemphigus*. *Int. Congr. Ent. London* **12**: 102–4.
Leather, S.R. and Dixon, A.F.G. (1981) Growth, survival and reproduction of the bird-cherry aphid, *Rhopalosiphum padi*, on its primary host. *Ann. appl. Biol.* **99**: 115–18.
Leclant, F. (1963) Le Phylloxera du poirier. *Phytoma* **15**: 13–16.
Leclant, F. (1966) Contribution à l'étude des Aphidoidea du Languedoc meridional. *Annls Soc. hort. Hist. nat. Hérault* **106**: 119–134.
Leclant, F. (1967a) Un Aphididae américain nouveau pour la faune européene: *Nearctaphis bakeri* Cowen (Hom.). *Bull. Soc. ent. Fr.* **72**: 25–6.
Leclant, F. (1967b) Un *Schizaphis* nouveau du sud de la France. *Annls Soc. ent. Fr.* (N.S.) **3**: 451–57.
Leclant, F. (1968) Un *Pleotrichophorus* nouveau vivant sur *Erigeron*. *Annls. Soc. ent. Fr.* (N.S.). **4**: 365–70.
Leclant, F. and Remaudière, G. (1974) Un *Acyrthosiphon* nouveau vivant sur *Glaucium*. *Annls Soc. ent. Fr.* (N.S.) **10**: 875–83.
Lees, A.D. (1966) The control of polymorphism in aphids. *Advan. Insect Physiol.* **3**: 207–77.
Lees, A.D. (1973) Photoperiodic time measurement in the aphid *Megoura viciae*. *J. Insect Physiol.* **19**: 2279–316.
Lehane, L. (1982) Biological control of lucerne aphids. *Rur. Res.* **114**: 4–10.
Leonard, M.D. (1963) The distribution and habits of the mint aphid *Ovatus crataegarius* (Walker). *Proc. ent. Soc. Wash.* **65**: 55–62.
Liao, H.-T. (1976) Bamboo aphids of Taiwan. *Q. Jl Taiwan Mus.* **29**: 499–586.
Liao, H.-T. (1978) The *Greenidea* aphids of Taiwan. *J. agric. Res. China* **27**: 345–54.

Linné, Karl von (1758) *Systema naturae.* 10th edn. **1**: 823 pp. Laurentii, Salvii, Stockholm.
Lorriman, F. and Llewellyn, M. (1983) The growth and reproduction of hop aphid. ... biotypes resistant and susceptible to insecticides. *Acta ent. bohemoslov.* **80**: 87–95.
Lowe, A.D. (1966) Some records of aphids including six species not previously recorded from New Zealand. *N.Z. Jl Sci.* **9**: 357–60.
Lowe, A.D. (Ed.) (1973) *Perspectives in Aphid Biology Bull. ent. Soc. N.Z.* **2**: 123 pp.
Lowe, H.J.B. (1967) Interspecific differences in the biology of aphids on leaves of *Vicia faba.* I. Feeding behaviour. II. Growth and excretion. *Entomologica exp. appl.* **10**: 347–57; 413–20.
Lowe, H.J.B. (1981) Resistance and susceptibility to colour forms of the aphid *Sitobion avenae* in spring and winter wheats (*Triticum aestivum*). *Ann. appl. Biol.* **99**: 87–98.
Luk'yanchikov, V.P. and Sorokina, Z.A. (1979) [Virus disease of rise] *Zashch. Rast.* **1979 (12)**: 35–36.
MacGillivray, M.E. (1958) A study of the genus *Masonaphis* Hille Ris Lambers, 1939. *Temminckia* **10**: 1–131.
MacGillivray, M.E. (1963) The Yellow Rose Aphid, *Rhodobium porosum* (Sanderson), on strawberry. *Can. Ent.* **95**: 892–96.
MacGillivray, M.E. (1966) A new *Macrosiphum* species from New York State. *Can. J. Zool.* **44**: 1085–88.
MacGillivray, M.E. (1968) A review of twelve aphid species described as new by Edith M. Patch. *Ann. ent. Soc. Am.* **61**: 338–62.
MacGillivray, M.E. and Anderson, G.B. (1964) The effect of photoperiod and temperature on the production of gamic and agamic forms in *Macrosiphum euphorbiae* (Thomas). *Can. J. Zool.* **42**: 491–510.
Mackauer, M. and Starý, P. (1967) *The Aphidiidae of the World.* Le Francois, Paris, 175 pp.
Mackauer, M. and Way, M.Y. (1976) *Myzus persicae* Sulz., an aphid of world importance, chapter 4, pp. 51–119 in Ed. Delucchi, V.F., *Studies in Biological Control.* Cambridge University Press.
MacKay, P.A., Reeleder, D.J. and Lamb, R.J. (1983) Sexual morph production by ... *Acyrthosiphon pisum. Can. J. Zool.* **61**: 952–54.
Malenotti, E. (1923) Le stazione invernali dell'*Anuraphis persicae-niger*, Smith. *R. osserv. Fitopat. Verona e Province limitrofe*, 3 pp.
Manglitz, G.R., Calkins, C.O., Walstrom, R.J., Hintz, S.D., Kinder, S.D., and Peters, L.L. (1966) Holocyclic strain of the spotted alfalfa aphid in Nebraska and adjacent states. *J. econ. Ent.* **59**: 636–39.
Manglitz, G.R. & Russell, L.M. (1974) Cross matings between *Therioaphis maculata* (Buckton) and *T. trifolii* (Monell) and their implications in regard to the taxonomic status of the insects. *Proc. biol. Soc. Wash.* **76**: 290–96.
Mansour, F. (1981) The black-margined yellow pecan aphid (*Monellia caryella*). *Spec. Publs. Inst. Pl. Prot. Israel* **209**: 50–51.
Marboutie, G. (1976) Premieres résultats d'essais de lutte biologique en vergers de pêchers et de pommiers. *Revue Zool. Agric. Path. veg.* **75**: 23–30.
Marcovitch, S. (1925) The strawberry root louse in Tennessee. *J. agric. Res.* **30**: 441–49.
Markkula, M. (1953) Biologische-Ökologische Untersuchungen über die Kohlblattlaus, *Brevicoryne brassicae* (L.). *Suomal. eläin-ja kasrit. Seur. van eläin Julk* **15(5)**: 113 pp.
Markkula, M. and Rautapää, J. (1963) Three Aphidiids and one Pteromalid parasitising *Rubus* aphids. *Annls agric. fenn.* **2**: 103–4.
Martin, J.H. (1983) The identification of common aphid pests of tropical agriculture. *Trop. Pest Management* **29**: 395–411.
Mason, P.W. (1940) A revision of the North American aphids of the genus *Myzus. Misc. Publs U.S. Dep. Agric.* **371**: 1–30.

References

Mason, P.W. (1947) A new pear aphid. *Proc. ent. Soc. Wash.* **49**: 252–54.
Massonie, G., Maison, P., Monet, R. and Grassely, C. (1979) Mise en évidence de caractères de résistance à *Myzus varians* Davids. chez deux variéties de *Prunus persica* également résistantes à *Myzus persicae* (Sulzer). *Rev. Zool. Agric. Pat. Veg.* **78**: 1–5.
May, B. and Holbrook, F.R. (1978) Absence of genetic variability in the Green Peach Aphid, *Myzus persicae*. *Ann. ent. Soc. Am.* **71**: 809–12.
McLain, D.K. (1980) Relationships among ants, aphids and coccinellids on wild lettuce. *J. Georgia ent. Soc.* **15**: 417–18.
Medler, J.T. and Ghosh, A.K. (1970) Keys to species of alate aphids collected by suction, wind and yellow pan water traps in the North Central States, Oklahoma and Texas. *Res. Bull. Exp. Stn Univ. Wis.* **277**: 99 pp.
Meier, W. (1954) Über *Myzus varians* Davidson und einige weitere *Myzus*-Arten aus der Schweiz. *Mitt. Schweiz. ent. Ges.* **27**: 321–409.
Meier, W. (1958) Beiträge zur Kenntnis der auf Papilionaceen lebender *Acyrthosiphon*-Arten. *Mitt. Schweiz. ent. Ges.* **31**: 291–312.
Meier, W. (1961) Beiträge zur Kenntnis grünsteifigen Kartoffelblattlaus, *Macrosiphum euphorbiae* Thomas 1870, und verwandter Arten. *Mitt. Schweiz. ent. Ges.* **34**: 127–86.
Meier, W. and Kolar, O. (1970) Über eine für die Schweiz neue Blattlausart auf Rotklee. *Mitt. schweiz. Landwirts* **5**: 85–96.
Memmi, M., Pasqualini, E., and Briolini, G. (1979) Prospettive di lotta contro l'afide grigion del melo. *Inftore Agraria* **35**: 6227–33.
de Menezes, M. (1970) Reproducao sexuada de *Aphis spiraecola* Patch no Estado de Sao Paulo. *Biológico* **36**: 53–57.
Michel, E. (1942) Beiträge zur Kenntnis von *Lachnus (Pterochlorus) roboris* L., einer wichtigen Honigtauerzeugerin an der Eiche. *Z. angew Ent.* **29**: 243–81.
Micinski, B. and Ruskiewicz, M. (1975) [The biology of the hop aphid *Phorodon humuli* Schr.] *Prace Naukowe Instytuto Ochrony Róslin* **16**: 79–101 (1974).
Miller, R.L. (1929) A contribution to the biology of ... *Aphid spiraecola* Patch. *Bull. Fla agric. Exp. Stn* **203**: 431–76.
Milner, R.J. (1981) Recent changes in the nomenclature of entomophagous pathogens (fungi: Phycomycetes) of aphids. *J. Aust. ent. Soc.* **1981**: 20: 154.
Milner, R.J., Teakle, R.E., Lutton, G.G., and Dave, F.M. (1980) Pathogens of the blue green aphid *Acyrthosiphon kondoi* and other aphids in Australia. *Aust. J. Bot.* **28**: 601–19.
Mimeur, J.-M. (1936) Aphididae du Maroc (Huitième note). *Bull. Soc. Sci. nat. Maroc* **16**: 252–55.
Minoiu, N. (1973) [Vectors of the pox virus (Prunus virus 7 Christ) to plum.] *Anal. Inst. Cerc. pentru Protectia Plantelov* **9**: 49–56.
Mittler, T.E. (1976) Ascorbic acid and other chelating agents in the trace mineral nutrition of the aphid *Myzus persicae* on artificial diets. *Entomologia exp. appl.* **20**: 81–98.
Miyazaki, M. (1968a) On three aphids of *Rosa rugosa* occurring in Japan with description of a new species. *Kontyû* **36**: 274–84.
Miyazaki, M. (1968b) A new species of the genus *Moritziella* Börner from Japan. *Kontyû* **36**: 400–2.
Miyazaki, M. (1971) A revision of the tribe Macrosiphini of Japan. *Insecta matsum. (N.S.)* **20**: 43–83.
Miyazaki, M. (1980) A revision of the genus *Vesiculaphis* del Guercio, with descriptions of four new species. *Insecta matsum.* **20**: 43–83.
Möller, F.W. (1971a) *Macrosiphum stellariae* (Theobald)—eine bisher nicht von der grünstreifigen Kartoffelblattlaus *Macrosiphum euphorbiae* (Thomas) abgegrenste Art. *Dt. ent. Z.* **18**: 207–15.
Möller, F.W. (1971b) Bastardierung innerhalb der Artenkomplexes um die Grünstreifige Kartoffelblattlaus *Macrosiphum euphorbiae* (Thomas). *Beitr. Ent.* **21**: 531–37.

Mondal, P.K., Basu, R.C., and Raychaudhuri, D.N. (1976) Studies on the aphids from Eastern India XXX. The genus *Toxoptera*. *Orient. Insects* **10**: 533–40.
Mordvilko, A. (1921) [Aphids of Gramineae.] *Izv. sev. oblast. sta. Zashch. Rast. Vredit.* **3**: 1–72.
Mordvilko, A. (1935) Die Blattläuse mit unvollständigen Generationszyklus und ihre Entstehung. *Ergebn. Fortschr. Zool.* **8**: 36–328.
Moritsu, M. (1947) Four gall-forming aphids on cherry trees in Japan. *Mushi* **18**: 39–48.
Mostafawy, M. (1967) Morphologie, Biologie und phytopathologische Bedeutung der Gattung *Appelia* C.B. *Z. angew. Zool.* **54**: 373–432.
Muddathir, K. (1976) Studies on the biology of wheat aphids in the Gezira (D.R. Sudan). *Beitr. Ent.* **26**: 465–70.
Müller, F.P. (1958) Zwei weitere neue Blattlausarten aus Norddeutschland und ein neues Genus. *Beitr. Ent.* **8**: 84–98.
Müller, F.P. (1969) Aphidina—Blattläuse, Aphiden. *Exkursionsfauna von Deutschland* **II/2**: 51–141. Berlin.
Müller, F.P. (1970) Zucht- und Ubertragungsversuche mit Populationen un Klonen der Grünfleckigen Kartoffelblattlaus *Aulacorthum solani* (Kalt.). *Dt. ent. Z. (N.F.)* **17**: 259–70.
Müller, F.P. (1972) Beziehungen der Blattläuse des Formenkreises *Acyrthosiphon pelargonii* (Kaltenbach) zur Erdbeere. *TagBer. dt. Akad. LandwWiss. Berl.* **121**: 41–47.
Müller, F.P. (1973) Aphiden an Moosen. *Ent. Abhandl. Mus. Tierk. Dresden* **39**: 205–42.
Müller, F.P. (1975a) Bestimmungsschlüssel für geflügelte Blattlause in Gelbschalen. *Arch. Phytopath. Pflanzenschutz Berlin* **1**: 49–77.
Müller, F.P. (1975b) Incidence of the aphid *Acyrthosiphon gossypii* Mordvilko on legumes and cotton. *Beitr. Ent.* **25**: 257–60.
Müller, F.P. (1976) Hosts and non-hosts in subspecies of *Aulacorthum solani* (Kaltenbach) and intraspecific hybridizations. *Symp. Biol. Hung.* **16**: 187–90.
Müller, F.P. (1977a) Vergleich einer tropischen mit einer mitteleuropäischen Population von *Aphis craccivora* Koch. *Dt. ent. Z. (N.F.)* **24**: 251–60.
Müller, F.P. (1977b) Überwinterung und Fundatrix der Getreideblattlaus *Macrosiphum (Sitobion) avenae* (F.). *Arch. Phytopath. Pflanzenschutz Berlin* **13**: 347–53.
Müller, F.P. (1979) Morphologie und Biologie von *Aulacorthum speyeri* Börner vergleich mit *Aulacorthum watanabei* (Miyazaki). *Reichenbachia* **17**: 129–41.
Müller, F.P. (1980) Wirstpflanzen, Generationenfolge und reproduktive Isolation infraspezifischer Formen von *Acyrthosiphon pisum*. *Ent. exp. appl.* **28**: 145–57.
Müller, F.P. (1982) Das problem *Aphis fabae*. *Z. angew. Ent.* **94**: 432–46.
Müller, F.P., Hinz, B., and Möller, F.W. (1973) Übertragung des Enationenvirus der Erbse durch verischeiden Unterarten und Biotypen der Grünfleckig Kartoffelblattlaus *Aulacorthum solani* (Kaltenbach). *Zbl. Bakt.* (II) **128**: 72–80.
Müller, F.P. and Hubert-Dahl, M.L. (1979) Generationenfolge und reproduktive Isolation von *Ovatus crataegarius* (Walker) und *O. insitus* (Walker). *Dt. ent. Z.* **26**: 241–53.
Müller, F.P. and Karl, E. (1976) Beitrage zur Kenntnis der Bionomie und Morphologie der ... *Phorodon cannabis* .... *Beitr. Ent.* **2**: 455–63.
Müller, F.P. and Möller, F.W. (1968) Ein bermerkenswertes Massenauftreten von *Myzus ascalonicus* Doncaster in Freiland. *Arch. Freunde NatGesch. Mecklenb.* **14**: 44–55.
Müller, F.P. and Schöll, S.E. (1958) Some notes on the aphid fauna of South Africa. *J. ent. Soc. sth. Afr.* **21**: 382–414.
Murant, A.F. (1978) Recent studies on association of two plant virus complexes with aphid vectors. pp. 243–49 in Scott, P.R. and Bainbridge, A. (Eds) *Plant Disease Epidemiology*. Blackwell, Oxford, 329 pp.

Narzikulov, M.N. (1963) [Aphids of Tadzhikistan and the adjoining republics of Central Asia.] *Fauna Tadzhikistan* **9(1)**: 1–272 (1962). (In Russian.)
Narzikulov, M.N. and Umarov Sh.A. (1969) [Aphids of Tadzhikistan and contiguous regions of Central Asia.] *Fauna Tadzhikistan* **9(2)**: 1–229. (In Russian.)
Nault, L.R. and Montgomery, M.E. (1979) Aphid alarm pheromones. *Misc. Publs ent. Soc. Am.* **11(3)**: 23–31.
Neubauer, I., Raccah, B., Ishaaya, I., Aharonsen, N., and Swirski, E. (1981) The effect of hosts exchange on the population dynamics of the spiraea aphid *Aphis citricola* v.d.G. *Z. angew. Ent.* **91**: 231–36.
Nevskii, V.P. (1929) [The plant lice of Central Asia.] *Trudŷ uzbekist. opyt. Sta. Zaschch. Rast.* **16**: 1–425. (In Russian.)
New, T.R. (1975) The biology of Chrysopidae and Hemerobiidae (Nevroptera), with reference to their usage as biocontrol agents, a review. *Trans. R. ent. Soc. Lond.* **127**: 115–40.
Nielson, M.W. and Don, H. (1974) Probing behaviour of biotypes of the spotted alfalfa aphid on resistant and susceptible alfalfa clones. *Entomologia exp. appl.* **17**: 477–86.
Nielson, M.W., Lehman, W.F., and Kodet, R.T. (1976) Resistance in alfalfa to *Acyrthosiphon kondoi*. *J. econ. Ent.* **69**: 471–72.
Odebiyi, J.B. (1979) Field populations of *Sitobion nigrinectaria* Theo. attacking pigeon pea in Kenya. *Nigerian J. agric. Sci.* **1**: 135–39.
Okamoto, D. (1974) Varietal differences of alfalfa on the occurrence of an aphid, *Acyrthosiphon kondoi* Shinji. *Kinki Chugoku agric. Res.* **48**: 39–40.
Olive, A.T. (1963) The genus *Dactynotus* Rafinesque in North Carolina. *Misc. Publ. ent. Soc. Am.* **4**: 31–66.
Olive, A.T. (1965) Two new species of *Dactynotus* Rafinesque from the Eastern United States. *Proc. ent. Soc. Wash.* **67**: 41–45.
Olson, W.H. (1974) Dusky-veined walnut aphid studies. *Calif. Agric.* **28**: 18–19.
Ordish, G. (1972) *The Great Vine Blight*. J.M. Dent & Sons Ltd., London, 237 pp.
Orian, A. (1973) Studies on the Aphididae of Mauritius with additional notes on some forms from Madagascar. *Fauna of Mauritius, Insecta-Homoptera* **1(2)**: 30 pp.
Orlob, G.B. (1961) Biology and taxonomy of cereal and grass aphids in New Brunswick. *Can. J. Zool.* **39**: 495–503.
Osler, R. and Longoni, C.E. (1975) Mancata transmissione del giallume del riso mediante l'afide *Sipha glyceriae*. *Riso* **25**: 73–76.
Ossiannilsson, F. (1959) Contributions to the knowledge of Swedish aphids. *K. LantbrHögsk. Annlr* **25**: 375–527.
Paddock, F.B. (1915) The turnip louse. *Bull. Texas Agr. Exp. Stn.* **180**: 7–77.
Paetzold, D. & Vater, G. (1967) Populationsdynamische Untersuchungen an den Parasiten und Hyperparasiten von *Brevicoryne brassicae* (L.). *Acta ent. bohemoslovaca* **64**: 83–90.
Paik, W.M. (1965) *Aphids of Korea*. Seoul National University, 160 pp.
Paik, W.H. (1972) Aphidoidea. *Illustrated Encyclopaedia of Fauna and Flora of Korea*, Volume 13, Insecta 5, 751 pp. (In Korean.)
Painter, R.H. and Pathak, M.D. (1962) The distinguishing features and significance of the four biotypes of the corn leaf aphid, *Rhopalosiphum maidis* (Fitch). *Int. Congr. Ent. Vienna* **11** (2): 110–15.
Pal, N.B. and Khuda-Bukhsh, A.R. (1980) Chromosome numbers of fifteen species of aphids from the north-west Himalayas. *Chromosome Information Service* **29**: 14–15.
Paliwal, Y.C. (1980) Transmission of barley yellow dwarf isolates by the cereal root aphid *Rhopalosiphum rufiabdominalis*. *Can. J. Pl. Path.* **2**: 90–92.
Palmer, J.M. (1974) Arthropoda of the subantarctic islands of New Zealand (2) Hemiptera: Aphididae. *J. R. Soc. N.Z.* **4**: 303–6.
Palmer, M.A. (1952) *Aphids of the Rocky Mountain Region*. Thomas Say Foundation, Denver, 452 pp.

Parker, J.R. (1916) The western wheat aphis (*Brachycolus tritici* Gill.) *J. econ. Ent.* **9:** 182–87.
Patch, E.M. (1910) Gall aphids of the elm. *Bull. Me agric. Exp. Stn* **181:** 193–240.
Patch, E.M. (1912a) Elm leaf curl and woolly apple aphid. *Bull. Me agric. Exp. Stn* **203:** 235–58.
Patch, E.M. (1912b) Woolly aphid migration from elm to mountain ash. *J. econ. Ent.* **5:** 395–98.
Patch, E.M. (1913) Woolly aphids of the elm. *Bull. Me agric. Exp. Stn* **220:** 259–73.
Patch, E.M. (1915) Two clover aphids. *J. agric. Res.* **3:** 431–33.
Patch, E.M. (1919) Key to the Eastern species of *Macrosiphum*. *Bull. Me agric. Exp. Stn* **282:** 216–18.
Patch, E.M. (1923) Hemiptera of Connecticut: Aphididae. *Bull. Conn. St. geol. nat. Hist. Surv.* **34:** 250–329.
Patch, E.M. (1927) Two currant aphids that migrate to willow-herbs. *Bull. Me agric. Exp. Stn* **336:** 1–8.
Patch, E.M. (1938) Food-plant Catalogue of the Aphids of the World. *Bull. Me agric. Exp. Stn* **393:** 1–431.
Pergande, T. (1904) North American Phylloxerinae affecting *Hicoria (Carya)* and other trees. *Proc. Davenport Acad. Sci.* **9:** 185–273.
Petterssen, J. (1971) Studies on four grass inhabiting species of *Schizaphis* III (a) Host plants. *Swedish J. agric. Res.* **1:** 133–38.
Pelvat, B., Aubert, J.J., and Baggiolini, M. (1970) Nouvelles observations sur les pucerons nuisibles au prunier et au pêcher en Suisse romande. *Rev. Suisse Vitic. Arboric.* **2:** 69–74.
Pintera, A. (1956) Aphids, sg. *Tuberculaphis* Börner new for Slovakia. *Sb. faun. Praći ent. Odd. nar. Mus. Praze* **1:** 39–42.
Ponsen, M.B. (1972) The site of potato leaf roll virus multiplication in its vector, *Myzus persicae*. An anatomical study. *Meded. LandbHoogesch. Wageningen* **72–16:** 1–147.
Ponsen, M.B. (1982) The digestive system of *Glyphina* and *Thelaxes*. *Meded. LandbHoogesch. Wageningen* **82–9:** 1–10.
Potgieter, J.T. and Durr, H.R.L. (1961) A host plant index of South African plant lice with a list of species found on each plant recorded. *Annale Univ. Stellenbosch, Ser. A* **36:** 219–38.
Prasad, Y.K. and Phadke, K.G. (1980) Population dynamics of *Lipaphis erysimi* (Kalt.) on different varieties of *Brassica* species. *Indian J. Ent.* **42:** 54–63.
Prior, R.N.B. (1971) Some notes on new or uncommon aphids recently found in Britain. *Zool. J. Linn. Soc.* **50:** 397–430.
Prior, R.N.B. (1976) Keys to the British species of *Metopolophium* with one new species. *Syst. Ent.* **1:** 271–79.
Prior, R.N.B. and Stroyan, H.L.G. (1964) A new subspecies of *Acyrthosiphon malvae* (Mosley) from *Poterium sanguisorba* L. *Proc. R. ent. Soc. Lond.* (B) **33:** 47–49.
Pussard, R. (1932) Sur la présence en France d'*Anuraphis persicae-niger* Smith. *Bull. Soc. ent. Fr.* **37:** 110–13.
Quednau, F.W. (1962) A list of aphids so far unrecorded from South Africa, with descriptions of two new species. *S. Afr. J. agric, Sci.* **5:** 253–64.
Quednau, F.W. (1964) Further notes on the aphid fauna of South Africa. *S. Afr. J. agric Sci.* **7:** 659–72.
Quednau, F.W. (1979) A list of Drepanosiphine aphids from the Democratic Peoples Republic of Korea with taxonomic notes and descriptions of new species. *Ann. zool.* **34:** 501–25.
Rämert, B. (1977) Sallatsrotlusen (*Pemphigus bursarius*). Biologi. Observationer i Skane. *Växtskyddsnotiser* **41:** 83–87.
Rakauskas, R.P. (1980) [Aphids of fruit trees and berry-bearing bushes of south-east Lithuania.] *Akademii Nauk Litovskoï SSR, B* (1980) **2(90):** 33–43.

Randles, J.W. and Crowley, N.C. (1970) Epidemiology of lettuce necrotic yellows virus in South Australia I. Relationship between disease incidence and activity of *Hyperomyzus lactucae* (L.). *Aust. J. agric. Res.* **21**: 447–53.
Rankin, W.H. (1927) Mosaic of raspberries. *Bull. N.Y. agric. Exp. Stn* **543**: 60 pp.
Rautapää, J. (1967) The bionomics of raspberry aphids. *Suom. hyönt. Aikak.* **6**: 127–44.
Rautapää, J. (1970) Preference of cereal aphids for various cereal varieties and species of Gramineae, Juncaceae and Cyperaceae. *Annls agric. fenn.* **9**: 267–77.
Raychaudhuri, D.N. (1956) Revision of *Greenidea* and related genera. *Zool. Verh., Leiden* **31**: 106 pp.
Raychaudhuri, D.N. (1980) [Ed.] *Aphids of North-East India and Bhutan.* Zoological Society, Calcutta. 521 pp. + 6 pp. errata.
Raychaudhuri, D.N. and Banerjee, C. (1974) A study of the genus *Melanaphis* with descriptions of new taxa from India. *Orient. Insects* **8**: 365–89.
Raychaudhuri, D.N., Ghosh, M.R., and Basu, R.C. (1978) Studies on the aphids from eastern India 27. The genera *Acyrthosiphon* Mordvilko, *Metopolophium* Mordvilko and *Rhodobium* Hille Ris Lambers. *Proc. zool. Soc. Calcutta* **28**: 115–131 (1975).
Raychaudhuri, D.N., Ghosh, M.R., and Basu, R.C. (1980a) Subfamily Aphidinae, in Raychaudhuri, D.N. (Ed.) *Aphids of North-East India and Bhutan.* Zoological Society, Calcutta. 521 pp. + 6 pp errata.
Raychaudhuri, D.N., Pal, P.K., and Ghosh, A.K. (1980b) Anoeciinae, Pemphiginae, in Raychaudhuri, D.N. (Ed.) *Aphids of North-East India and Bhutan.* Zoological Society, Calcutta, 521 pp. + 6 pp. errata.
Raychaudhuri, D.N., Pal, P.K., and Ghosh, M.R. (1978) Root-infesting aphids of North East India. *Entomon, India* **3**: 239–64.
Raychaudhuri, D.N., Poddar, S.C., and Raychaudhuri, D. (1982) Study of the genus *Aphidius* (Hymenoptera: Aphidiidae) of India. *Entomon (India)* **7**: 11–22.
Réal, P. (1955) Le cycle annuel du puceron de l'arachide (*Aphis leguminosae* Theob.) en Afrique noire francaise et son determinisme. *Rev. Path. veg.* **34**: 3–122.
Remaudière, G. (1952) Contribution à l'étude des Aphidoidea de la fauna Francaise ... Myzinae et Dactynotinae. *Rev. Pat. veg. ent. agric. Fr.* **31**: 232–63.
Remaudière, G. and Tao, C. Chia Chu (1957) Les Fordinae du *Pistacia sinensis*. *Rev. Pat. veg. ent. agric. Fr.* **36**: 223–36.
Richards, W.R. (1960) A synopsis of the genus *Rhopalosiphum* Koch in Canada. *Can. Ent.* **92**: supplement 13, 51 pp.
Richards, W.R. (1962) A new species of *Rhopalosiphum* Koch. *Can. Ent.* **94**: 969–72.
Richards, W.R. (1963) The Myzaphidines of Canada. *Can. Ent.* **95**: 680–704.
Richards, W.R. (1965) The Callaphidini of Canada. *Mem. ent. Soc. Can.* **44**: 1–49.
Richards, W.R. (1966) A new species of *Myzocallis* Passerini, with a key to the North American species. *Can. Ent.* **98**: 876–80.
Richards, W.R. (1968) New combinations and redescription of *Monellia californica* with a key to the species of *Monelliopsis*. *Can. Ent.* **100**: 437–40.
Richards, W.R. (1972a) Review of the *Solidago*-inhabiting aphids in Canada with descriptions of three new species. *Can. Ent.* **104**: 1–34.
Richards, W.R. (1972b) Three new species of *Aulacorthum* ... key to Canadian species. *Can. Ent.* **104**: 1017–23.
Richards, W.R. (1976) A host index for species of Aphidoidea described during 1935 to 1969. *Can. Ent.* **108**: 499–550.
Rivnay, E. (1938) Factors affecting the fluctuation in the population of *Toxoptera aurantii* Boy. in Palestine. *Ann. appl. Biol.* **25**: 143–54.
Robert, Y. (1969a) Premières observations sur la biologie de *Capitophorus horni* Börner dans l'ouest de la France. *Annls Zool. Ecol. anim.* **1**: 39–54.
Robert, Y. (1969b) Les pucerons de l'artichaut dans l'ouest de la France. Sur une

espèce particulièrement nuisible, *Capitophorus horni* Börner. *C.R. hebd. Séanc. Acad. Agric. Fr.* **1969**: 410–20.

Robert, Y. (1971) Epidémiologie de l'enroulement de la pomme de terre; capacité vectrice de stades et de formes des pucerons *Aulacorthum solani* Kltb., *Macrosiphum euphorbiae* Thomas et *Myzus persicae* Sulz. *Potato Res.* **14**: 130–39.

Robert, Y. and Rabasse, J.M. (1977) Role écologique de *Digitalis purpurea* dans la limitation naturelle des populations du puceron strié de la pomme de terre *Aulacorthum solani* par *Aphidius urticae* dans l'ouest de la France. *Entomophaga* **22**: 373–82.

Roberti, D. (1939a) Contributi alla conoscenza degli afidi d'Italia II .... *Boll. Lab. Zool. Portici* **31**: 137–57.

Roberti, D. (1939b) Contributi alla conoscenza degli afidi d'Italia III. Fordini. *Boll. Lab. Ent. agr. Filippo Silvestri* **3**: 34–105.

Roberti, D. (1972) Contributi alla conoscenza degli afidi d'Italia VIII. *Tetraneura (Tetraneurella) akinere* Sasaki. *Entomologica, Bari* **8**: 141–205.

Robinson, A.G. (1963) Previously undescribed forms of *Pseudocercidis rosae* Richards with biological notes on the species. *Can. Ent.* **95**: 636–39.

Robinson, A.G. (1964) *Asiphonaphis* Wilson & Davis in North America. *Can. Ent.* **96**: 1093–97.

Robinson, A.G. (1965) A new genus, new species and previously undescribed morphs of aphids. *Can. Ent.* **97**: 1009–15.

Robinson, A.G. (1968) Two new species of aphids from Canada. *Can. Ent.* **100**: 275–79.

Robinson, A.G. (1979) Annotated list of aphids of Northwest Canada, Yukon and Alaska. *Manitoba Ent.* **13**: 23–29.

Robinson, A.G. and Hsu, S.J. (1963) Host plant records and biology of aphids on cereal grains and grasses in Manitoba. *Can. Ent.* **95**: 134–37.

Robinson, A.G. and Rojanavongse, V. (1976) A key to the Nearctic species of *Aphis* L. on *Ribes* spp., with descriptions of two new species from Manitoba. *Can. Ent.* **108**: 159–64.

Rochow, W.F. (1960) Specialization among greenbugs in the transmission of barley yellow dwarf virus. *Phytopathology* **50**: 881–84.

Rogerson, J.P. (1947) The oat bird-cherry aphid, *Rhopalosiphum padi* L. and comparison with *R. crataegellum* Theob. *Bull. ent. Res.* **38**: 157–76.

Rothschild, M., Euw, J. van, and Reichstein, T. (1970) Cardiac glycosides in the oleander aphid, *Aphis nerii*. *J. Insect Physiol.* **16**: 1141–45.

Rueda, L.M. and Calilung, V.J. (1975) Biological study of the sugar cane woolly aphid, *Ceratovacuna lanigera* Zehntner on five varieties of sugar cane. *Philipp. Ent.* **3**: 129–47.

Russell, L.M. (1966) *Aphis craccae* L., an aphid newly discovered in the United States. *USDA Coop. Econ. Ins. Rpt* **16**: 1021–23.

Sagar, P. and Singh, D.P. (1981) Chemical control of the aphid *Aphis affinis* del Guercio —a pest of Japanese mint, *Mentha arvensis* L. in the Punjab. *Entomon, India* **6**: 73–79.

Savary, A. (1953) Le puceron cendré du poirier en Suisse Romande. *Landw. Jb. Schweiz* **67**: 247–314.

Schaefers, G.A. (1960) A systematic study of the strawberry aphid complex (*Pentatrichopus* spp.). *Ann. ent. Soc. Am.* **53**: 783–93.

Schaefers, G.A. and Allen, W.A. (1962) Biology of the strawberry aphids, *Pentatrichopus fragaefolii* (Cockerell) and *P. thomasi* Hille Ris Lambers, in California. *Hilgardia* **32**: 393–431.

Schaefers, G.A. and Judge, F.D. (1972) Relationship between parent size and the production of winged forms in the strawberry aphid, *Chaetosiphon fragaefolii*. *J. Insect Physiol.* **18**: 1049–60.

Schmütterer, H. (1969) Der Einfluss einiger Faktoren auf die Leistung der Wickenlaus

Megoura viciae (Buckt.) bei der Übteragung des Enationenvirus der Erbse. *Z. angew. Ent.* **64**: 24–50.

Schneider, F. (1969) Bionomics and physiology of aphidophagous Syrphidae. *A. Rev. Ent.* **14**: 103–24.

Schoene, W.J. and Underhill, J. (1935) Life history and migration of the apple woolly aphis. *Tech. Bull. Va agric. Exp. Stn* **57**: 1–31.

Schwartz, R.E. (1965) Jahreszeitliche schwankungen in Verhalten von *Toxoptera citricidus* Kirk., gegenüber gelden Anlockungsmitteln. *Z. PflKrankh. PflPath. PflSchutz.* **72**: 84–89.

Sedlag, U. (1953) Wurzelläuse an Futter- und Zuckerrüben. *Anz. Schädlingsk.* **26**: 51–52.

Semal, J. (1956) Transmission of beet mosaic virus from *Stellaria media* and *Capsella bursa-pastoris* by *Myzus ascalonicus* Doncaster. *Nature* **178**: 501–2.

Setokuchi, O. (1976) [Ecology of *Longiunguis sacchari* (Zehntner) infesting sorghum IV. Varietal difference of sorghums in the aphid occurrence.] *Proc. Assoc. Pl. Prot. Kyushu* **22**: 139–41. (Japanese, Engl. Summary.)

Setzer, R.W. (1980) Intergall migration in the aphid genus *Pemphigus*. *Ann. ent. Soc. Am.* **73**: 327–31.

Shambaugh, G.F., Frazier, J.L., Castell, A.E.M., and Coons, L.B. (1978) Antennal sensilla of seventeen aphid species. *Int. J. Insect Morph. Embryol.* **7**: 389–404.

Shands, W.A. and Simpson, G.W. (1971) Seasonal history of the buckthorn aphid and suitability of alder-leaved buckthorn as a primary host in north-eastern Maine. *Tech. Bull. Univ. Me* **51**: 24 pp.

Shands, W.A., Simpson, G.W., Muesebeck, C.F.W., and Wave, H.E. (1965) Parasites of potato-infesting aphids in north-eastern Maine. *Tech. Bull. Me agric. Exp. Stn* **719**: 77 pp.

Shands, W.A., Simpson, G.W., and Wave, H.E. (1972) Seasonal population trends and productiveness of the potato aphid on swamp rose in north-eastern Maine. *Tech. Bull. Univ. Me* **52**: 35 pp.

Shaposhnikov, G. Kh. (1956) [The phylogenetic basis of the systematics of the short-tailed aphids (Anuraphidina).] *Trudy zool. Inst., Leningrad* **23**: 215–322. (In Russian.)

Shaposhnikov, G. Kh. (1964) in Bey-Bienko, G.Y., *Classification Keys to the insects of the European part of the U.S.S.R., Aphidinea*, **1**: 489–616. (In Russian, English translation in 1967 by Israel Program for Scientific Translations Ltd, Jerusalem. Aphidinea pp. 616–799.)

Sharma, M.L. (1969–1977) *Bibliography of Aphidoidea* (4 vols.) Editions Paulines, Sherbrooke, Quebec.

Shiga, M. (1975) Populations dynamics of *Myzus persicae* (Sulzer) and *Hyalopterus arundinis* (F.) in peach orchards. Pp. 9–14 in Yasumatsu, K. and Mari, H. (Eds), Approaches to Biological Control. University of Tokyo Press, 142 pp.

Shinji, O. (1931) The evolutional significance of the chromosomes of the Aphididae. *J. Morph.* **51**: 374–433.

Shinji, O. (1941) *Monograph of Japanese Aphids.* Tokyo, 1215 pp. (In Japanese.)

Singer, M.C. and Smith, B.D. (1976) Use of the plant growth regulator chlormequat chloride to control the aphid *Hyperomyzus lactucae* on black currants. *Ann. appl. Biol.* **82**: 407–14.

Sluss, R.R. (1967) Population dynamics of the walnut aphid, *Chromaphis juglandicola* (Kalt.) in northern California. *Ecology* **48**: 41–58.

Smith, C.F. (1960a) Aphids on 'Cacao' in the Dominican Republic. *J. Agric. Univ. P. Rico* **44**: 154–56.

Smith, C.F. (1960b) New species of Aphidae From Puerto Rico. *J. Agric. Univ. P. Rico* **44**: 157–62.

Smith, C.F. (1965) *Tiliphagus lycoposugus* new genus, new species from *Tilia americana* and *Lycopus virginicus*. *Ann. ent. Soc. Am.* **58**: 781–86.

Smith, C.F. (1972) Bibliography of the Aphididae of the World. *Tech. Bull. N. Carol. agric. Exp. Stn* **216**: 717 pp.
Smith, C.F. (1974) Keys to and descriptions of the genera of Pemphiginae in North America. *Tech. Bull. N. Carol. agric. Exp. Stn* **226**: 61 pp.
Smith, C.F. and Cermeli, M.M. (1979) An annotated list of Aphididae of the Caribbean Islands and South and Central America. *Tech. Bull. N. Carol. agric. Exp. Stn* **259**: 131 pp.
Smith, C.F., Gauda, S.M., Martorell, L.F., and Pérez-Escolar, M.E. (1971) Additions and corrections to the Aphididae of Puerto Rico. *J. Agric. Univ. P. Rico* **55**: 192–258.
Smith, C.F. and Knowlton, G.F. (1977) The genus *Rhopalosiphoninus* Baker in North America. *U.S. Dept. agric. Coop. Pl. Pest Rep.* **2(8)**: 75–80.
Smith, C.F., Martorell, L.F., and Pérez-Escolar, M.E. (1963) Aphididae of Puerto Rico. *Tech. Pap. Univ. agric. Exp. Stn P. Rico* **37**: 121 pp. (published 1964?)
Smith, C.F. and Parron, C.S. (1978) An annotated list of Aphididae of North America. *Tech. Bull. N. Carol. agric. Exp. Stn* **255**: 428 pp.
Smith, L.M. (1936) Biology of the mealy plum aphid, *Hyalopterus pruni* (Geoffroy). *Hilgardia* **10**: 167–209.
Smith, R.H. (1923) The clover aphis, biology, economic relationship and control. *Res. Bull. Idaho agric. Exp. Stn* **3**: 1–75.
Sokal, R.R., Bird, J., and Riska, B. (1980) Geographic variation in *Pemphigus populicaulis* in eastern North America. *Biol. J. Linn. Soc.* **14**: 163–200.
Sokal, R.R. and Riska, B. (1981) Geographic variation in *Pemphigus populitransversus*. *Biol. J. Linn. Soc.* **15**: 201–33.
Sokolova, N.P. and Ponomareva, M.S. (1968) [Characteristics of the damage caused to red currant by the currant leaf-gall aphid (*Cryptomyzus ribis* L.).] *Dokl. mosk. sel. 'khoz. Akad. K.A. Timiryazeva* **143**: 159–62.
Soliman, L.B. (1927) A comparative study of ... the genus *Macrosiphum* ... in California. *Univ. Calif. Publs Ent.* **4**: 89–158.
Sorin, M. (1962) The life history of *Melanaphis bambusae* (Fullaway), and the penetration of stylets into host plants. *Kontyû* **30**: 221–29.
Sorin, M. (1970) *Longiunguis* of Japan. *Insecta matsum.* Suppl. **8**: 5–17.
Sorin, M. (1971) Two new species of Aphididae from Japan. *Mushi* **45**: 59–63.
Sorin, M. (1980) Two new species of the genus *Lachnus* Burmeister from Japan. *Bull. Kôgakkau Univ.* **18**: 1–10 (reprint).
Stacherska, B. (1975) [*Diuraphis mühlei* Börn—a new pest of seed cultivation of timothy.] *Pr. nauk. Inst. Ochr. Rośl.* **15**: 187–90.
Starks, K.J. and Burton, R.L. (1977) Greenbugs: Determining biotypes, culturing and screening for plant resistance, with notes on rearing parasites. *Tech. Bull. U.S. Dept. Agric.* **1556**: 18 pp.
Starks, K.J. and Mirkes, K.A. (1979) Yellow sugarcane aphid: Plant resistance in cereal crops. *J. econ. Ent.* **72**: 486–88.
Starý, P. (1970) Aphid migration and impact of an indigenous parasite, *Aphidius transcaspicus* Telenga, on populations of *Hyalopterus pruni* (Geoffr.) in Iraq. *Bull. Soc. ent. Égypte* **53**: 185–98.
Starý, P. (1975) Parasites (Hymenoptera, Aphidiidae) of leaf-curling apple aphids in Czechoslovakia. *Acta ent. bohemoslov.* **72**: 99–114.
Starý, P. (1979) Aphid parasites (Hymenoptera, Aphidiidae) of the Central Asian area. *Rozpr. csl. Acad. Ved.* **89**(3): 1–116.
Starý, P. and Gonzales, D. (1978) Parasitoid spectrum of *Acyrthosiphon* aphids in Central Asia. *Entomologica scand.* **9**: 140–45.
Steel, C.G.M. .and Lees, A.D. (1977) The role of neurosecretion in the photo-periodic control of polymorphism in *Megoura viciae*. *J. exp. Biol.* **67**: 117–35.
Stenseth, C. (1970) Undersokelser over bladluis pa plommer. *Meld. Norg. LandbrHoisk.* **49**(18): 21 pp.

References

Stevenson, A.B. (1964) Seasonal history of root-infesting Phylloxera vitifoliae (Fitch) in Ontario. *Can. Ent.* **96:** 979–87.
Stoetzel, M.B. (1981) Two new species of *Phylloxera* on pecan. *J. Georgia ent. Soc.* **16:** 127–44.
Stoetzel, M.B. and Tedders, W.L. (1981) Investigation of two species of *Phylloxera* on pecan in Georgia. *J. Georgia ent. Soc.* **16:** 144–50.
Stroyan, H.L.G. (1950) *Jacksonia papillata* Theobald: A redescription with biological and taxonomic notes. *Proc. R. ent. Soc. Lond.* (B) **19:** 90–95.
Stroyan, H.L.G. (1952) The identification of aphids of economic importance. *Pl. Path.* **1:** 9–14, 42–48, 92–99, 123–29.
Stroyan, H.L.G. (1955) Recent additions to the British aphid fauna II. *Trans. R. ent. Soc. Lond.* **106:** 283–340.
Stroyan, H.L.G. (1957) *The British Species of Sappaphis Matsumura.* Part 1. H.M.S.O., London, 59 pp.
Stroyan, H.L.G. (1960) Three new subspecies of aphids from Iceland. *Ent. Medd.* **29:** 250–65.
Stroyan, H.L.G. (1963) *The British Species of Dysaphis Börner (Sappaphis auctt. nec Mats.).* Part 2. H.M.S.O., London, 119 pp.
Stroyan, H.L.G. (1964) Notes on hitherto unrecorded or overlooked British aphid species. *Trans. R. ent. Soc. Lond.* **116:** 29–72.
Stroyan, H.L.G. (1969a) Notes on some species of *Cavariella* Del Guercio 1911. *Proc. R. ent. Soc. Lond.* (B) **38:** 7–19.
Stroyan, H.L.G. (1969b) On a collection of aphids from Inverness-shire, with the description of a new species. *Trans. Soc. Brit. Ent.* **18:** 227–46.
Stroyan, H.L.G. (1972) Additions and amendments to the check list of British aphids. *Trans. R. ent. Soc. Lond.* **124:** 37–79.
Stroyan, H.L.G. (1977) Homoptera Aphidoidea (Part): Chaitophoridae and Callaphididae. *Handbk Ident. Br. Insects* II 4(a): 130 pp.
Stroyan, H.L.G. (1979a) Additions to the British aphid fauna. *Zool. J. Linn. Soc.* **65:** 1–54.
Stroyan, H.L.G. (1979b) An account of the alienicolous morphs of the aphid *Patchiella reaumuri* (Kalt.). *Zool. J. Linn. Soc.* **67:** 259–67.
Stroyan, H.L.G. (1981) A North American lupin aphid found in Britain. *Pl. Path.* **30:** 253.
Stroyan, H.L.G. (1982) Revisionary notes on the genus *Metopolophium* Mordvilko 1914, with keys to European species and descriptions of two new taxa. *Zool. J. Linn. Soc.* **75:** 91–140.
Sullivan, D.J. and van den Bosch, R. (1971) Field ecology of the primary parasites and hyperparasites of the potato aphid, *Macrosiphum euphorbiae*, in the East San Francisco Bay area. *Ann. ent. Soc. Am.* **64:** 389–94.
Swenson, K.G. (1971) Relation of sexupara production in the woolly pear aphid, *Eriosoma pyricola*, to tree growth in the field. *Can. Ent.* **103:** 256–60.
Swirski, E. (1954a) Fruit tree aphids of Israel. *Bull. ent. Res.* **45:** 623–38.
Swirski, E. (1954b) *Aphis punicae* Pass. in Israel. *Bull. Res. Coun. Israel* **4:** 2 pp. (reprint).
Swirski, E. (1963) Notes on the plant lice of Israel. *Israel J. agric. Res.* **13:** 9–23.
Swirski, E., Wyoski, M., Greenberg, S., and Cohen, M. (1969) Varietal susceptibility of pear trees in Israel to attack by *Aphanostigma piri* Chol. *Israel J. Ent.* **4:** 243–50.
Sylvester, E.S. and McClain, E. (1978) Rate of transovarial passage of sowthistle yellow vein virus in selected subclones of the aphid *Hyperomyzus lactucae*. *J. econ. Ent.* **71:** 17–20.
Sylvester, E.S., Richardson, J., and Frazier, N.W. (1974) Serial passage of strawberry crinkle virus in the aphid *Chaetosiphon jacobi*. *Virology* **59:** 301–6.
Szalay-Marzso, L. (1969) [Aphids in Horticulture.] *Mezőgazdaśagi Kiado,* Budapest, 187 pp. (In Hungarian.)

Szelegiewicz, H. (1962) Materialy do poznania mszyc Polski I. Podrodzina Lachninae. *Fragm. fauna* **10:** 63–98.
Szelegiewicz, H. (1969a) [Aphids—pests of plants, vectors of virus diseases and the cause of production losses. Bibliography.] *Kom. ochr. Rosl. Polskiej Akad. Nauk*, 250 pp.
Szelegiewicz, H. (1969b) Zwei neue arten der Gattung *Therioaphis* Walk. aus Ungarn. *Acta zool. hung.* **15:** 455–62.
Szelegiewicz, H. (1974) A list of aphids from the Democratic People's Republic of Korea I. Adelgidae to Chaitophoridae. *Fragm. faun.* **19:** 455–66.
Szelegiewicz, H. (1977) Levéltetvek I.—Aphidinea I. *Fauna Hung.* **17**(18): 1–175. (In Hungarian.)
Szelegiewicz, H. (1978) Aphidoidea Lachnidae. *Klucze Oznacz. Owad. Pol.* **17**(5a)101: 107 pp. (In Polish.)
Szelegiewicz, H. (1980) Aphids of the genus *Macrosiphoniella* del Guercio from the Democratic People's Republic of Korea. *Annls zool.* **35:** 419–73.
Taimr, L., Kudelová, A., and Kriz, J. (1978) Diurnal periodicity of the flight activity of migrantes alatae of *Phorodon humuli* Schrank. *Z. angew. Ent.* **86:** 373–80.
Takada, H. (1979) Esterase variation in Japanese populations of *Myzus persicae* (Sulzer), with special reference to resistance to organophosphorus insecticides. *Appl. Ent. Zool. Tokyo* **14:** 245–55.
Takada, H. (1982) Influence of photoperiod and temperature on the production of sexual morphs in a green and a red form of *Myzus persicae*. I. In the laboratory. II. Under natural conditions. *Kontyû* **50:** 233–45; 353–64.
Takada, H. and Shiga, M. (1974) Description of a new species and notes on ... the genus *Aclitus* (Hymenoptera, Aphidiidae). *Kontyû* **42:** 283–92.
Takahashi, R. (1923) Aphididae of Formosa, part 2. *Rep. Govt Res. Inst. Dep. Agric. Formosa* **4:** 1–173.
Takahashi, R. (1924) Aphididae of Formosa, part 3. *Rep. Govt Res. Inst. Dep. Agric. Formosa* **10:** 1–121.
Takahashi, R. (1925) Aphididae of Formosa, part 4. *Rep. Govt Res. Inst. Dep Agric. Formosa* **16:** 1–65.
Takahashi, R. (1931) Aphididae of Formosa, part 6. *Rep. Govt Res. Inst. Dep. Agric. Formosa* **53:** 1–127.
Takahashi, R. (1936) Some Aphididae from South China and Hainan I. *Lingnan Sci. J.* **15:** 595–606.
Takahashi, R. (1950) List of the Aphididae of the Malay Peninsula, with descriptions of new species. *Ann. ent. Soc. Am.* **43:** 587–607.
Takahashi, R. (1958) On the aphids of *Ceratovacuna* in Japan. *Kontyû* **26:** 187–90.
Takahashi, R. (1960) *Kurisakia* and *Aiceona* of Japan. *Insecta matsum.* **23:** 1–10.
Takahashi, R. (1962) Key to the genera and species of Greenideini of Japan, with descriptions of a new genus and three new species. *Trans. Shikoka ent. Soc.* **7:** 65–73.
Takahashi, R. (1964) *Macrosiphum* of Japan. *Kontyû* **32:** 353–59.
Takahashi, R. (1965a) Some species of *Aulacorthum* of Japan. *Insecta matsum.* **27:** 99–113.
Takahashi, R. (1965b) *Myzus* of Japan. *Mushi* **38:** 43–78.
Takahashi, R. (1965c) Some new and little-known Aphididae of Japan. *Insecta matsum.* **28:** 19–61.
Takahashi, R. (1966) Descriptions of some new and little-known species of *Aphis* of Japan, with a key to species. *Trans. Am. ent. Soc.* **92:** 519–56.
Tahon, J. (1964) Note sur *Hyperomyzus staphyleae* Koch, un aphide vecteur important des virus de la jaunisse de la betterave. *Parasitica* **20:** 17–22.
Takano, S. (1934) On the morphology and biology of *Ceratovacuna lanigera* Zehnt., and the relation of its outbreaks to environmental conditions in Formosa. *J. Formosan Sug. Plrs' Ass.* **11:** 481–528.

Takeda, S. (1979) Spatial distribution of the apple leaf-curling aphid *Myzus malisuctus* Matsumura and the spiraea aphid, *Aphis spiraecola*, on apple seedlings. Movement of adults and larvae ... on the apple tree. *Appl. Ent. Zool., Tokyo* **14:** 356–69; 487–89.

Talhouk, A.S. (1977) Contribution to the knowledge of almond pests in East Mediterranean countries VI. The sap-sucking pests. *Z. angew. Ent.* **83:** 248–57.

Tamaki, G. and Allen, W.W. (1969) Competition and other factors influencing the population dynamics of *Aphis gossypii* and *Macrosiphoniella sanborni* on greenhouse chrysanthemums. *Hilgardia* **39:** 447–505.

Tambs-Lyche, H. (1959) A new species of *Schizaphis* Börner attacking *Phleum pratense* in Norway. *Norsk. ent. Tidsskr.* **11:** 1–2.

Tanabe, C. and Mishima, R. (1929) [Studies concerning the aphid *Cinacium iakusuiensis* Kishida (Phylloxeridae).] *Nogaku kenkyu* **13:** 255–89. (In Japanese)

Tanabe, C. and Mishima, R. (1930) [*Results of the studies on Cinaciun iakusuiensis* Kishida.] Nara agric. Exp. Stn, Nara 183 pp. (In Japanese)

Tanaka, T. (1961) [The rice root aphids, their ecology and control.] *Spec. Bull. Coll. Agric. Utsunomiya* **10:** 83 pp. (In Japanese)

Tanaka, T. (1976) [*Aphids on vegetables in Japan.*] Japanese Plant Protection Association, 220 pp. (In Japanese.)

Tao, C. Chia-Chu (1958) Corrections and additions to the aphid fauna of China. *J. agric. Res., Taipei* **8:** 1–10.

Tao, C. Chia-Chu (1961) Revision of the genus *Toxoptera* Koch, 1856. *Q. Jl Taiwan Mus.* **14:** 257–60.

Tao, C. Chia-Chu (1962a) Revision of the Chinese Aphinae. *Pl. Prot. Bull., Taiwan* **4:** 95–110.

Tao, C. Chia-Chu (1962b) [Aphid fauna of China: Greenideinae.] *Sci. Yearbk Taiwan Mus.* **5:** 62–75. (In Chinese.)

Tao, C. Chia-Chu (1963) Revision of Chinese Macrosiphinae. *Pl. Prot. Bull., Taiwan* **5:** 162–205.

Tao, C. Chia-Chu (1966) Revision of Chinese Hormaphinae. *Q. Jl Taiwan Mus.* **19:** 165–79.

Tao, C. Chia-Chu (1970) Revision of Chinese Eriosomatinae. *Q. Jl Taiwan Mus.* **23:** 135–49.

Tao, C. Chia-Chu and Tan, M.F. (1961) Identification, seasonal population and chemical control of citrus aphids of Taiwan. *J. agric. Res., Taipei* **10:** 41–53.

Taylor, L.R. (1977) Migration and the spatial dynamics of an aphid, *Myzus persicae*. *J. Anim. Ecol.* **46:** 411–23.

Taylor, L.R. and Taylor, R.A.J. (1977) Aggregation, migration and population mechanics. *Nature, Lond.* **265:** 415–21.

Taylor, L.R. and Woiwood, I.P. (1980) Temporal stability as a density-dependent species characteristic. *J. Anim. Ecol.* **49:** 209–24.

Taylor, L.R., Woiwood, I.P., Tatchell, G.M., Dupuch, M.J., and Nicklen, J. (1980) Synoptic monitoring for migrant insect pests in Great Britain and Western Europe III. *Rep. Rothamsted exp. Stn* **1980**(2): 23–151.

Tedders, W.L. (1977) *Trioxys pallidus* and *Trioxys complanatus* as parasites of *Monellia costalis, Monelliopsis nigropunctata* and *Tinocallis caryaefoliae*. *Ann. ent. Soc. Am.* **70:** 687–90.

Tedders, W.L. and Osburn, M. (1970) Tests with aldecarb, disulfoton and phorate for aphid control on pecans. *J. Georgia ent. Soc.* **5:** 58–60.

Thakur, J.R. and Dogra, G.S. (1980) Woolly apple aphid, *Eriosoma lanigerum*, research in India. *Trop. Pest Management* **26:** 8–12.

Theobald, F.V. (1926, 1927, 1929) *The Plant Lice or Aphididae of Great Britain*. **1:** 372 pp. (1926); **2:** 411 pp. (1927); **3:** 364 pp. (1929). Headley Bros., Ashford, Kent.

Thomas, K.H. (1962) Die Blattläuse des Formenkreises *Brachycaudus prunicola* (Kalt.). *Wissensch Zeits. Univ. Rostock Math.-Naturwiss.* **11:** 325–42.

Thomas, K.H. (1968) Die Blattläuse aus der engeren Verwandtschaft von *Aphis gossypii* Glover und *A. frangulae* Kaltenbach unter besonderer Berücksichtigung ihres vorkommens an Kartoffel. *Ent. Abh. Mus. Tierk. Dresden* **26**: 337–89.

Thresh, J.M. (1983) The long-range dispersal of plant viruses by arthropod vectors. *Phil. Trans. R. Soc. Lond.* **B302**, 497–528.

Tomiuk, J. and Wöhrmann, K. (1980) Population growth and population structure of natural populations of *Macrosiphum rosae* (L.). *Z. angew. Ent.* **90**: 464–73.

Tremblay, E. (1963) Notulae aphidologicae I. Notizie su alcuni afidi dannosi. *Boll. Lab. ent. agr. Filippo Silvestri* **20**: 11–30.

Tseng, S. and Tao, C. Chia-Chu (1938) New and unrecorded aphids of China. *J. W. China Border Res. Soc.* **10**: 195–224.

Tsinovskii, Ya. and Egina, K.Ya (1977) [Entomophagous fungi in the control of apple aphids.] *Zashch. Rast* **1977**(7); 26–7. (In Russian.)

Tuatay, N. and Remaudière, G. (1964) Première contribution au catalogue des Aphididae de la Turquie. *Revue Path. vég. Ent. agric. Fr.* **43**: 243–78.

Uye, T. (1925) *Vesiculaphis caricis* (Fullaway). *Insect Wld* **29**: 218–23, 254–60. (In Japanese.)

Varma, A., Somadder, K., and Kishore, R. (1978) Biology, bionomics and control of *Melanaphis indosacchari* David, a vector of sugarcane grassy shoot disease. *Indian J. agric. Res.* **12**: 65–72.

Velimirović, V. (1977) [Contribution to the study of the peach aphid *Pterochloroides persicae* Cholodk.] *Zastita Bilja* **28**: 3–7. (In Croat.)

Verma, K.D. (1969) A new subspecies of *Impatientinum impatiensae* (Shinji) and the male of *Protrama penecaeca* Stroyan from N.W. India. *Bull. Ent. ent. Soc. India* **10**: 102–3.

Verma, K.D. (1972) A new genus and two new species of aphids from N.W. India. *Bull. Ent. ent. Soc. India* **12**: 97–99.

Vickerman, G.P. and Wratten, S.D. (1979) The biology and pest status of cereal aphids in Europe: A review. *Bull. ent. Res.* **69**: 1–32.

Walker, A.L., Bottrell, D.G., and Cate, J.R. (1972) Bibliography on the greenbug, *Schizaphis graminum* (Rondani). *Bull. ent. Soc. Am.* **18**: 161–73.

Wang, C.L., Siang, N.I., Chang, G.S. and Chu, H.F. (1962) Studies on the soybean aphid *Aphis glycines* Matsumura. *Acta ent. sin.* **11**: 31–44. (In Chinese with English summary.)

Wave, H.E., Shands, W.A., and Simpson, G.W. (1965) Biology of the foxglove aphid in the north-eastern United States. *Tech. Bull. U.S. Dep. Agric.* **1338**: 40 pp.

Way, M.J. (1963) Mutualism between ants and honeydew-producing Homoptera. *A. Rev. Ent.* **8**: 307–44.

Weber, H. (1928) Skelett, Muskulatur und Darm der Schwarz-Blattlaus *Aphis fabae* Scop. *Zoologica* **28**: 1–120.

Wene, G.P. and White, A.N. (1953) The cabbage root aphid. *Ohio J. Sci.* **53**: 332–34.

Wertheim, G. (1954) Studies on the biology and ecology of the gall-producing aphids of the tribe Fordini in Israel. *Trans. R. ent. Soc. Lond.* **105**: 79–96.

Whitham, T.G. (1978) Habitat selection by *Pemphigus* aphids in response to resource limitation and competition. *Ecology* **59**: 1164–76.

Wiktelius, S. (1981) *Studies on Aphid Migration*. Sveriges Lantbruksuniversitet, Uppsala, 88 pp.

Wildermuth, V.L. and Walter, E.V. (1932) Biology and control of the corn leaf aphid with special reference to the southwestern States. *Tech. Bull. U.S. Dep. Agric.* **306**: 21 pp.

Willcocks, F.C. (1925) *The Insect and Related Pests of Egypt.* Volume 2. Sultanic Agricultural Society, Cairo, 418 pp.

Wilson, F. (1960) A review of the biological control of insects and weeds in Australia and Australian New Guinea. *Tech. Comm. Commonw. Inst. biol. Control* **1**: 102 pp.

Wilson, H.F. (1909) Notes on *Lachnus caryae* Harris, under a new name. *Can. Ent.* **41**: 385–87.
Wood-Baker, C.S. (1970) Records of *Rhopalomyzus poae* Gillette in Britain with a note on its morphology. *Entomologist's mon. Mag.* **106**: 79–81.
Wool, D. and Koach, J. (1976) Morphological variation of the gall-forming aphid, *Geoica utricularia*, in relation to environmental variation. Pp. 239–72 in Karlin, S., and Nevo, E. (Eds) *Population Genetics and Ecology*, Academic Press, New York.
Worlidge, J. (1669) *Systema Agriculturae. The Mystery of Husbandry Discovered.* J.W. Gent, London.
Yano, K., Miyake, T. and Eastop, V.F. (1983) The biology and economic importance of rice aphids . . . a review. *Bull. ent. Res.* **73**: 539–66.
Yano, K., Miyake, T., and Hamasaki, S. (1982) [Discovery of a spotted alfalfa-like aphid (*Therioaphis trifolii* s. lat.) in Japan.] *Jap. J. appl. Ent. Zool.* **26**: 35–40.
Zavattari, E. (1921) Richerche sylla biologia dell' *Aploneura lentisci* Passerini. *Acta Zool. Stockh.* **2**: 241–92.
Zhang, G. (then Chang, G.) and Zhong, T. (1976) New species and new record of *Tuberocephalus* Shinji from China. *Acta ent. Sin.* **19**: 72–76. (In Chinese with English summary.)
Zhang, G. (then Chang, G.) and Zhong, T. (1979a) Three new species of *Greenidea* Schouteden from China. *Entomotaxon.* **1**: 115–20. (In Chinese with English summary.)
Zhang, G. (then Chang, G.) and Zhong, T. (1979b) Five new species of *Pemphigus* . . . From China. *Act ent. Sin.* **22**: 324–32. (In Chinese with English summary.)
Zhang, G. and Zhong, T. (1980) New species and new subspecies of Chinese Macrosiphinae. *Entomotaxon.* **2**: 53–64.
Zhang, G. and Zhong, T. (1982) New genera and new species of Chinese Callaphididae and Chaitophoridae. *Acta zoo-taxon. sin.* **7**: 67–77. (In Chinese with English summary.)
Zhang, G. and Zhong, T. (1983) [*Economic Insect Fauna of China, 25. Homoptera: Aphidinea I.*] Academia Sinica, Beijing, 387 pp. (In Chi ese.)
Zimmerman, E.C. (1948) *Insects of Hawaii*, Volume 5. *Homoptera: Sternorrhyncha.* University of Hawaii Press, Honolulu, 464 pp.
Zuniga, E. (1967) Los pulgones del duraznero en Chile Central. *Agricultura téc.* **27**: 32–39.
Zwölfer, H. (1957, 1958) Zur Systematik, Biologie und Ökologie unterirdisch lebender Aphiden, I–IV. *Z. angew. Ent.* **40**: 182–221, 528–75 (1957); **42**: 129–72, **43**: 1–52 (1958).

G. *Photographic Guide*

This series of photographs is the work of Mr P.V. York and staff of the Photographic Unit of the British Museum (Natural History).

LIST OF PHOTOGRAPHS OF SLIDE-MOUNTED APHIDS
(L = left, R = right)

Except where otherwise indicated, each species is represented by an apterous viviparous female (L) and an alate viviparous female (R). The scale on each photograph indicates the actual body length of the specimen.

Lachninae

1 *Longistigma caryae* (L = ovipara)
2 *Pterochloroides persicae*
3 L—*Lachnus roboris*, R—*L. tropicalis*
4 *Maculolachnus submacula*
5 *Trama troglodytes*
6 *Protrama radicis* (L = alatiform)

Chaitophorinae

7 *Atheroides serrulatus*
8 *Sipha flava*
9 *Sipha (Rungsia) maydis*

Drepanosiphinae

10 L—*Callaphis juglandis*, R—*Chromaphis juglandicola*
11 L—*Eucallipterus tiliae*, R—*Melanocallis caryaefoliae*
12 L—*Monellia caryella*, R—*Monelliopsis pecanis*
13 *Therioaphis trifolii* forma *maculata*

Aphidinae—Aphidini

14 *Hyalopterus pruni*
15 *Rhopalosiphum maidis*
16 *Rhopalosiphum insertum*
17 *Rhopalosiphum padi*
18 *Rhopalosiphum rufiabdominalis*
19 *Schizaphis graminum*
20 *Hysteroneura setariae*
21 *Aphis idaei*
22 *Melanaphis bambusae*
23 *Melanaphis sacchari*
24 *Melanaphis pyraria*
25 *Aphis (Protaphis) anuraphoides*
26 *Aphis armoraciae*
27 *Aphis rubicola*
28 *Aphis craccivora*
29 *Aphis fabae*
30 *Aphis glycines*
31 *Aphis gossypii*
32 *Aphis nasturtii*
33 *Aphis punicae*
34 *Aphis citricola*
35 *Aphis pomi*
36 *Aphis nerii*
37 *Aphis grossulariae*
38 *Aphis schneideri*
39 *Aphis forbesi*
40 *Aphis helianthi*
41 *Aphis illinoisensis*
42 *Aphis sambuci*
43 *Toxoptera aurantii*
44 *Toxoptera citricidus*
45 *Toxoptera odinae*
46 *Asiphonaphis pruni*

Aphidinae—Macrosiphini

47 Nearctaphis bakeri
48 Anuraphis subterranea
49 Dysaphis foeniculi
50 Dysaphis tulipae
51 Dysaphis devecta
52 Dysaphis apiifolia
53 Dysaphis pyri
54 Dysaphis plantaginea
55 Brachycaudus amygdalinus
56 Brachycaudus helichrysi
57 Brachycaudus schwartzi
58 Brachycaudus cardui
59 Brachycaudus persicae
60 Brevicoryne brassicae
61 Hyadaphis foeniculi
62 Hyadaphis coriandri
63 Lipaphis erysimi
64 Brachycorynella asparagi
65 Diuraphis noxia
66 Semiaphis dauci
67 Coloradoa rufomaculata
68 Capitophorus eleagni
69 Pleotrichophorus chrysanthemi
70 Cavariella aegopodii
71 Myzaphis rosarum
72 Longicaudus trirhodus
73 Chaetosiphon fragaefolii
74 Chaetosiphon tetrarhodum
75 Tuberocephalus momonis
76 Ovatus crataegarius
77 Phorodon humuli
78 Phorodon cannabis
79 Myzus cerasi
80 Myzus varians
81 Myzus persicae
82 Myzus ornatus
83 Myzus ascalonicus
84 Neotoxoptera oliveri
85 Picturaphis brasiliensis
86 Pentalonia nigronervosa
87 Eucarrazzia elegans
88 Hyperomyzus lactucae
89 Hyperomyzus rhinanthi
90 Nasonovia ribisnigri
91 Nasonovia (Kakimia) cynosbati
92 Cryptomyzus galeopsidis (L = fundatrix)
93 Cryptomyzus ribis (L = fundatrix)
94 Rhopalosiphoninus latysiphon
95 Rhopalosiphoninus staphyleae
96 Aulacorthum magnoliae
97 Aulacorthum (Neomyzus) circumflexum
98 Aulacorthum solani
99 Acyrthosiphon rogersii
100 Acyrthosiphon gossypii
101 Acyrthosiphon pisum
102 Acyrthosiphon kondoi
103 Cryptaphis poae
104 Metopolophium dirhodum
105 Metopolophium festucae
106 Rhodobium porosum
107 Fimbriaphis scammelli
108 Corylobium avellanae
109 Macrosiphum euphorbiae
110 Macrosiphum pallidum
111 Macrosiphum rosae
112 Sitobion africanum
113 Sitobion avenae
114 Sitobion fragariae
115 Uroleucon ambrosiae
116 Uroleucon compositae
117 Macrosiphoniella sanborni
118 Illinoia azaleae
119 Megoura lespedezae
120 Megoura viciae
121 Sinomegoura citricola
122 Glabromyzus rhois
123 Amphorophora sensoriata
124 Amphorophora agathonica
125 Amphorophora idaei
126 Amphorophora rubitoxica
127 Utamphorophora humboldti
128 Wahlgreniellla nervata

Anoeciinae

129 Anoecia corni

Greenideinae

130 Cervaphis schouteniae (apt.), rappardi (al.)
131 Greenidea ficicola

Hormaphidinae

132 Reticulaphis distylii
133 Astegopteryx nipae
134 Ceratovacuna lanigera
135 Cerataphis palmae

Pemphiginae

136 Pemphigus bursarius
137 Eriosoma lanigerum
138 Eriosoma (Schizoneura) lanuginosum
139 Tetraneura nigriabdominalis
140 Tetraneura ulmi
141 Smynthurodes betae
142 Forda formicaria
143 Forda marginata
144 Paracletus cimiciformis
145 Geoica utricularia
146 Baizongia pistaciae
147 Asiphonella dactylonii
148 Aploneura lentisci
149 Geopemphigus flocculosus
150 Aloephagus myersi

417

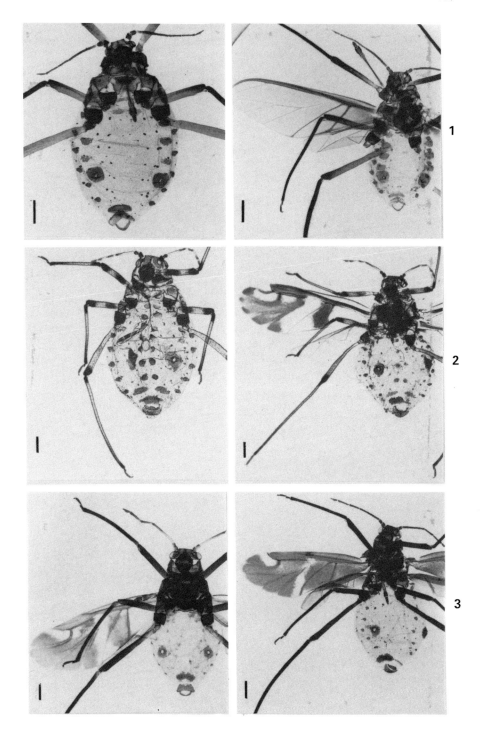

418

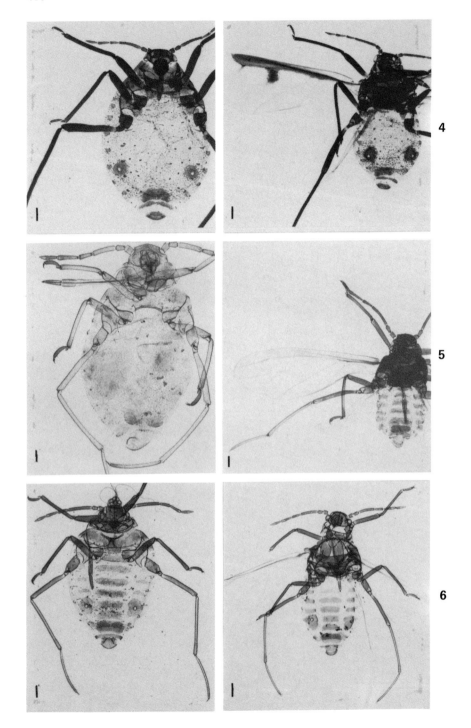

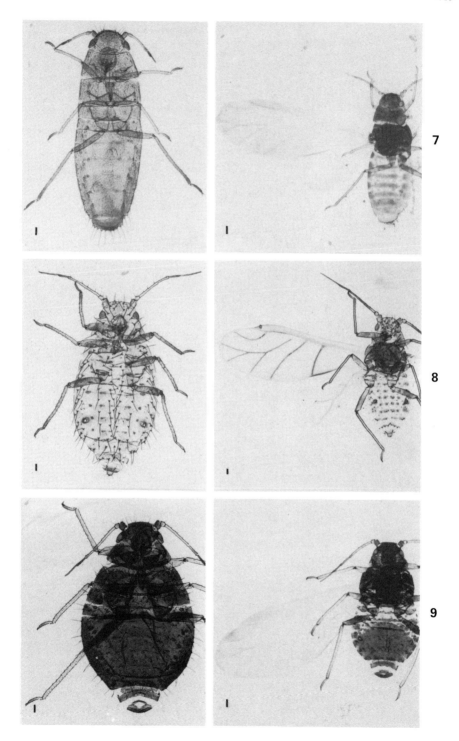

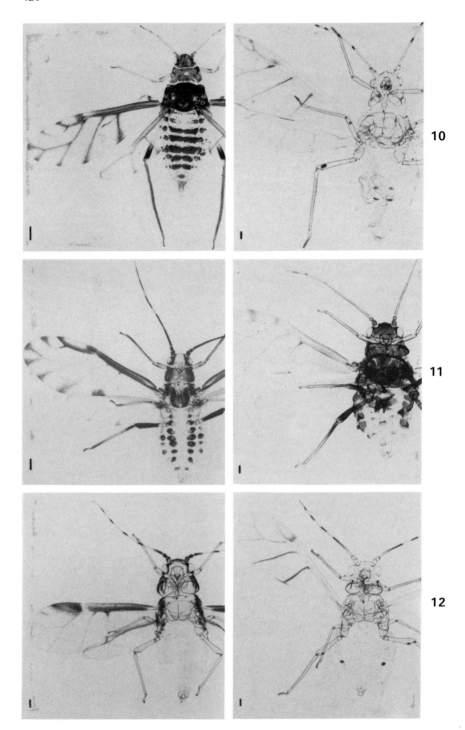

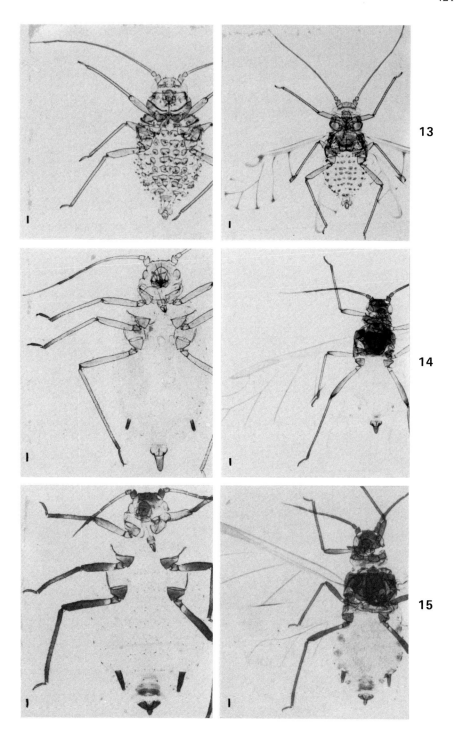

422

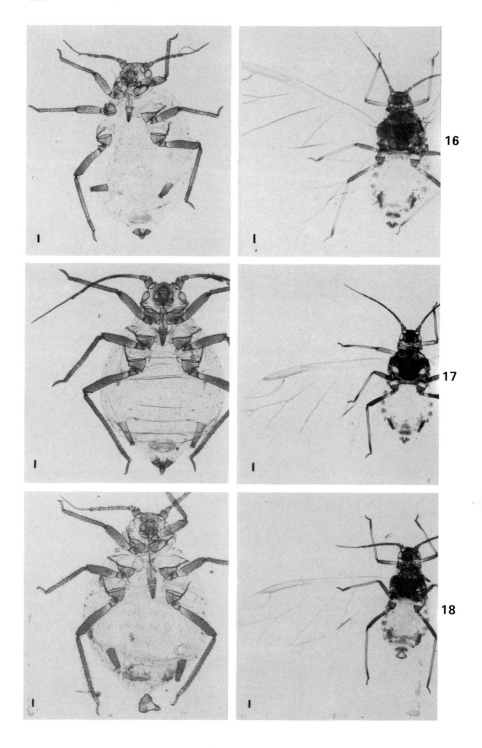

423

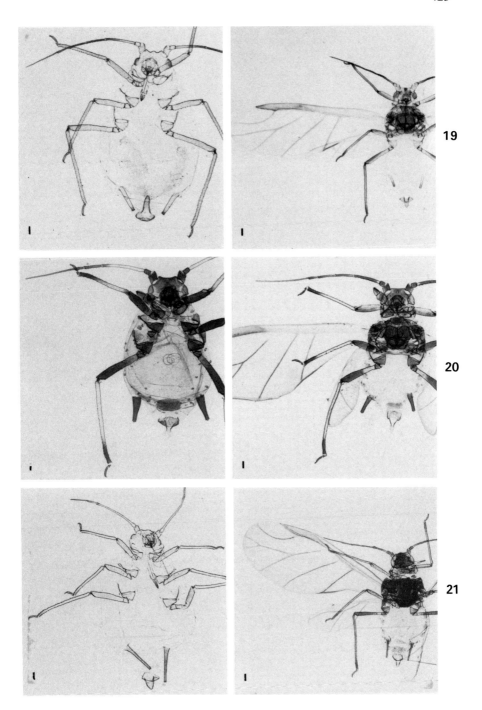

424

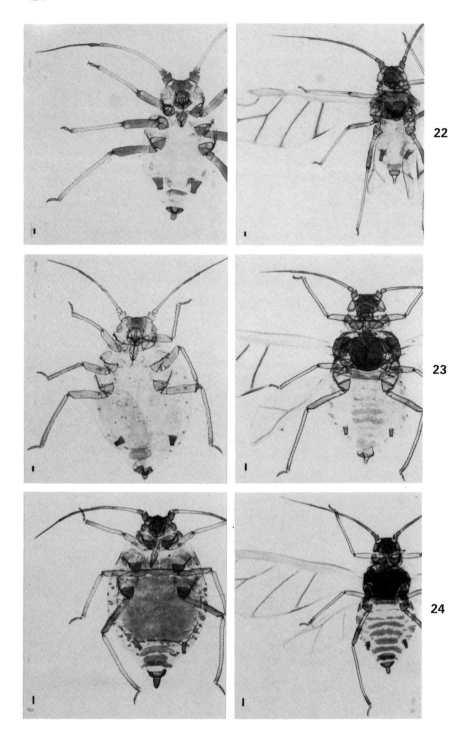

425

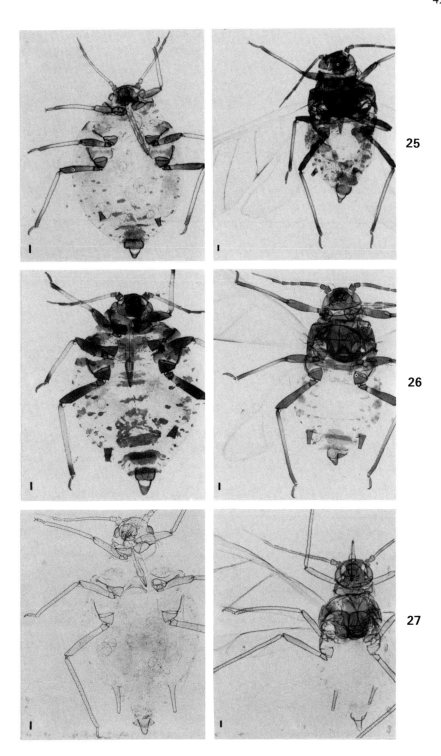

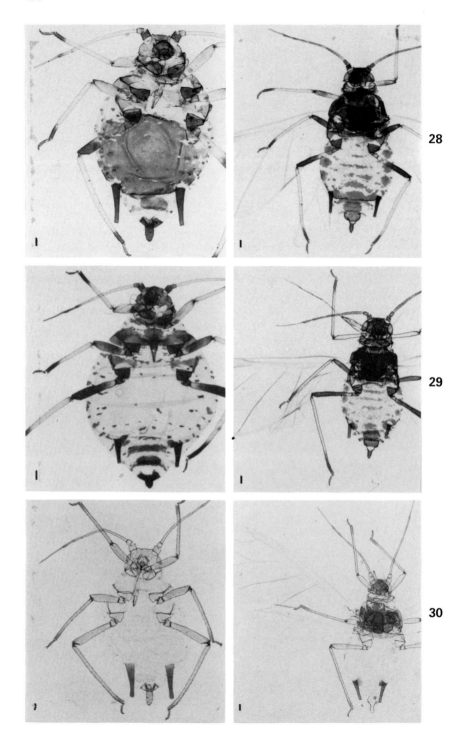

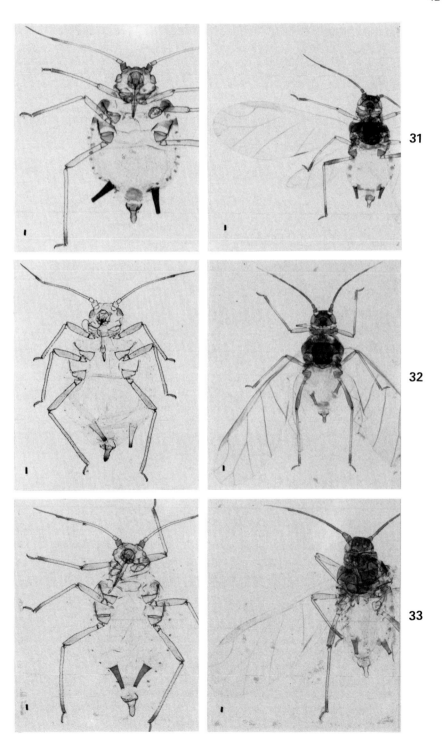

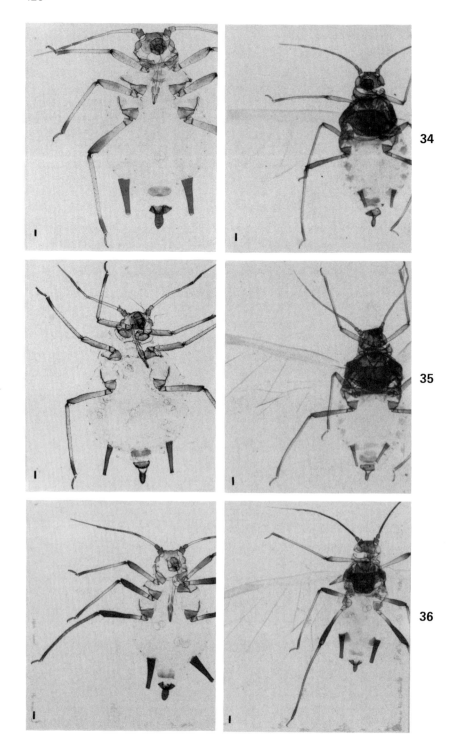

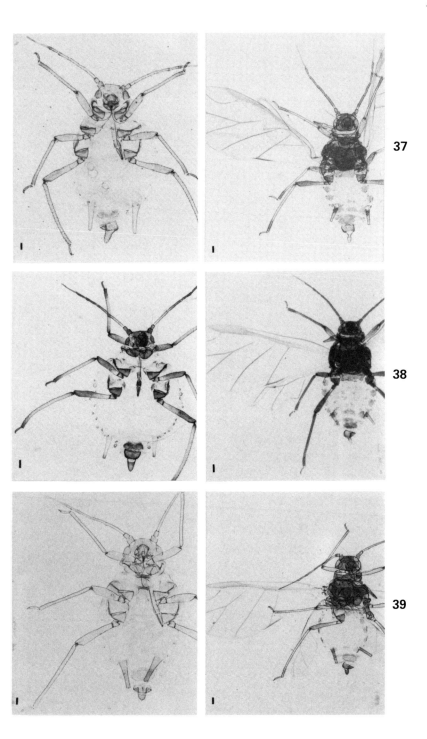

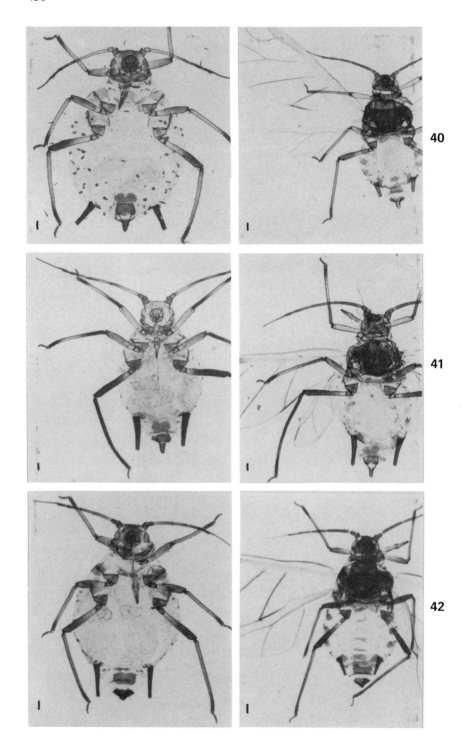

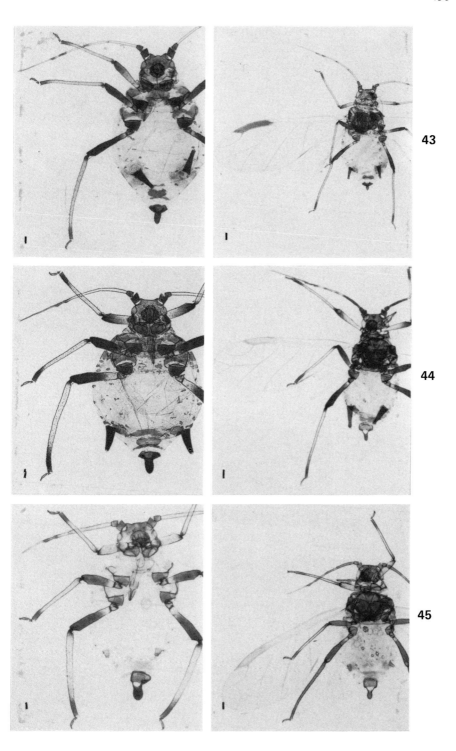

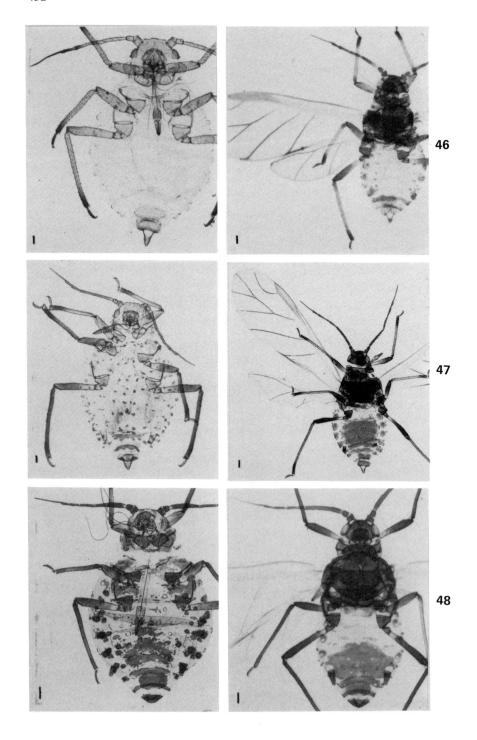

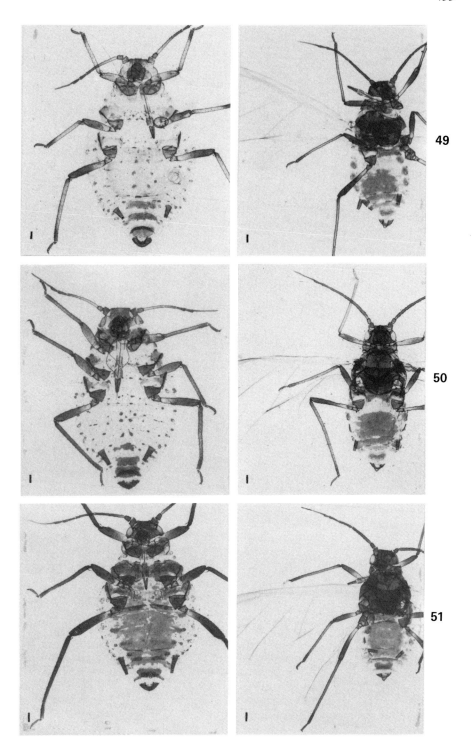

434

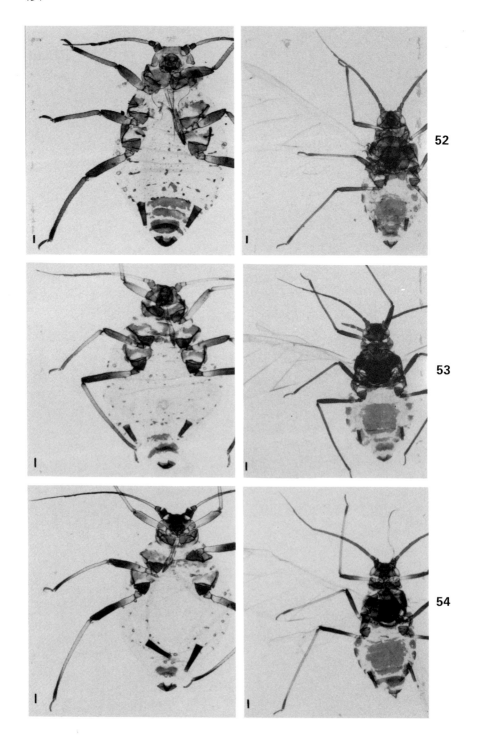

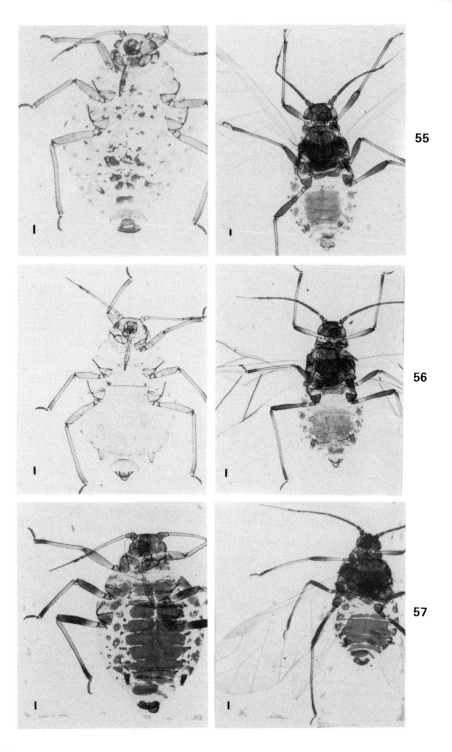

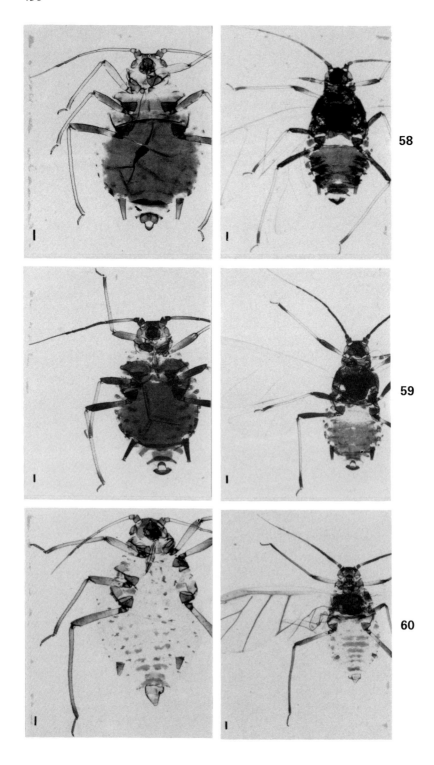

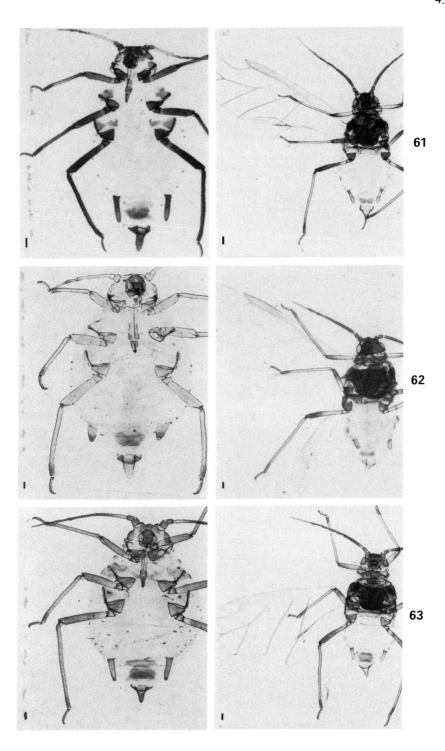

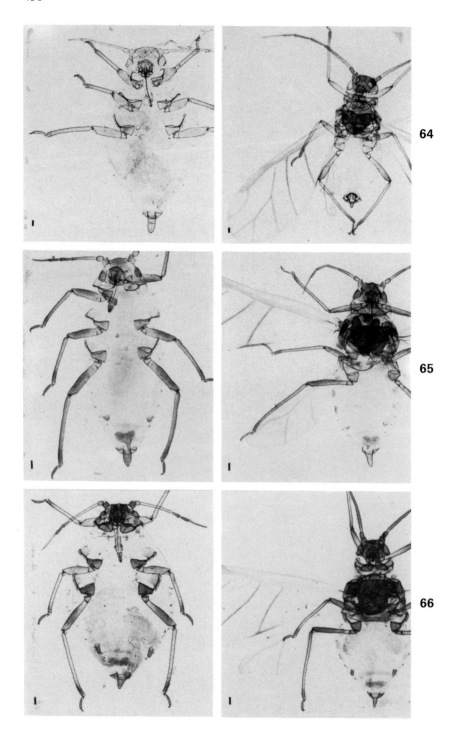

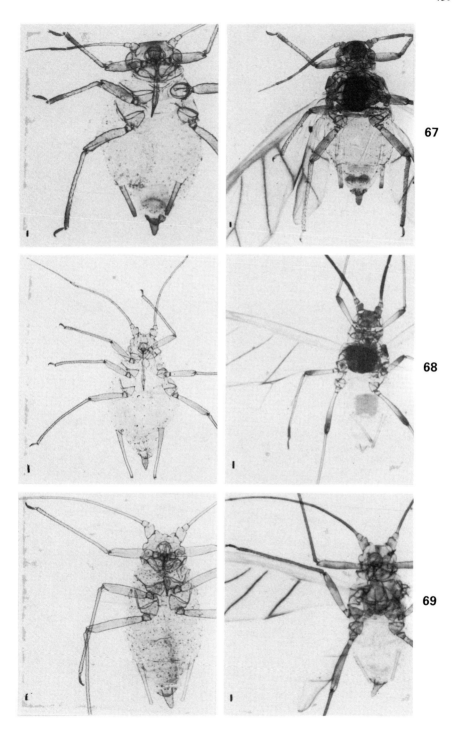

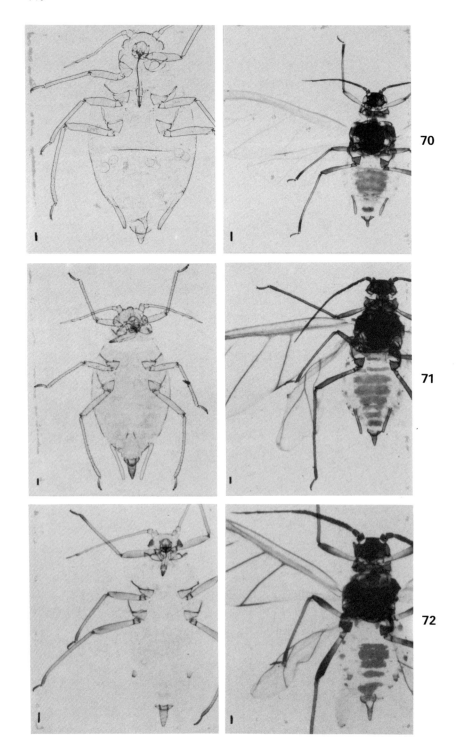

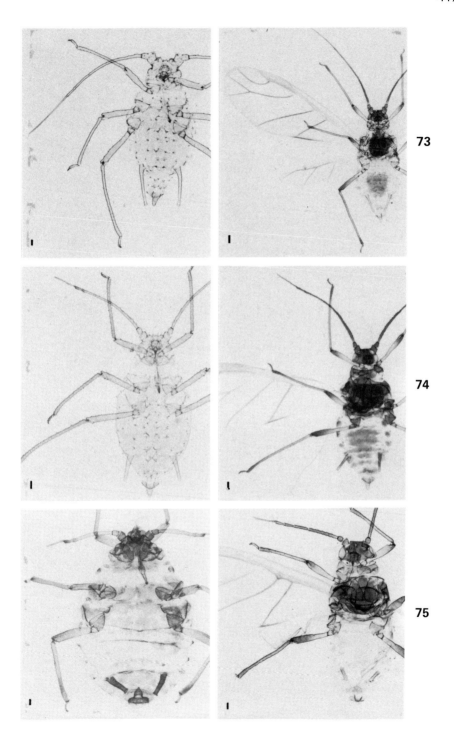

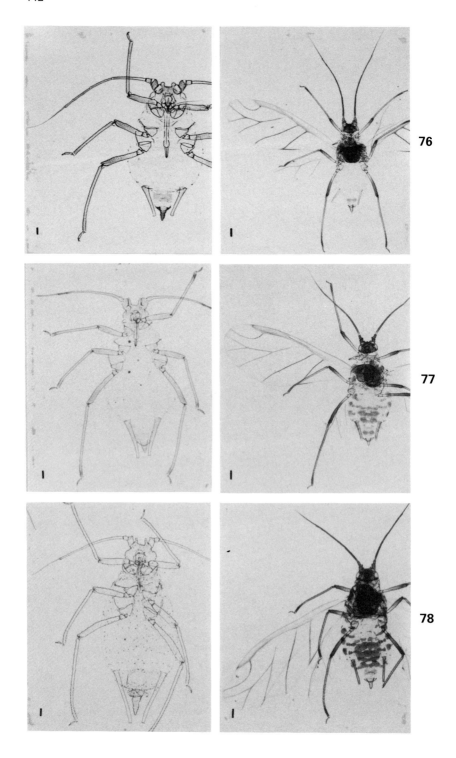

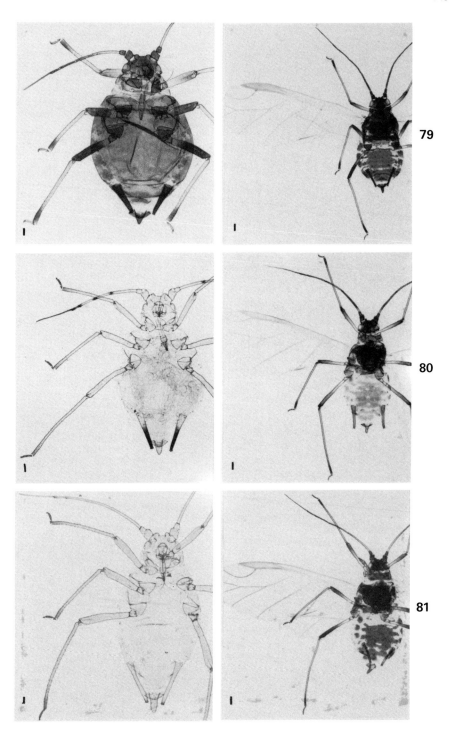

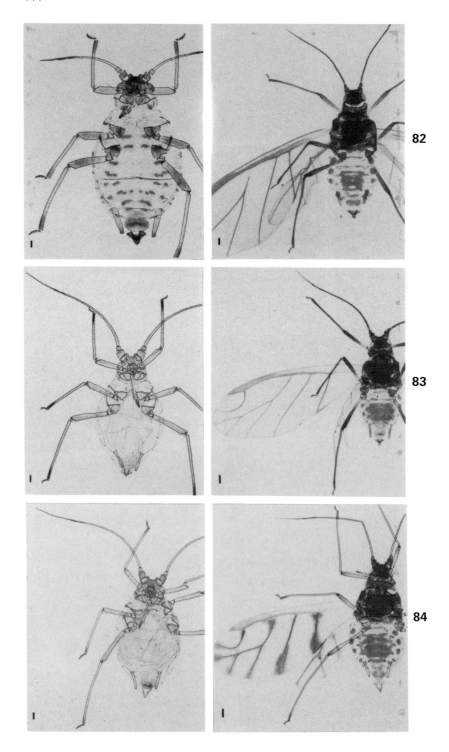

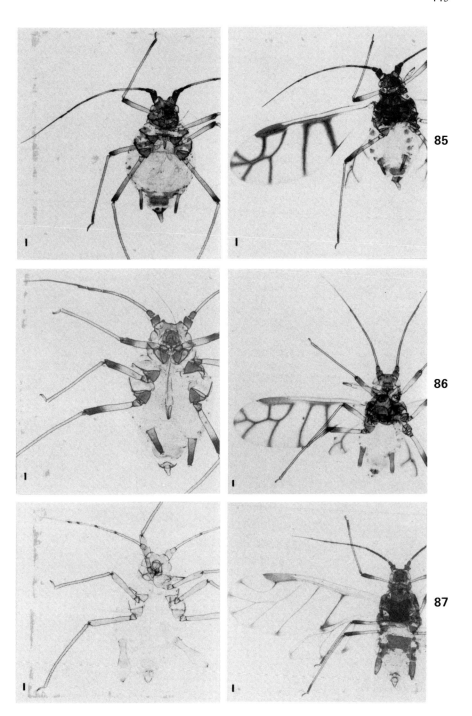

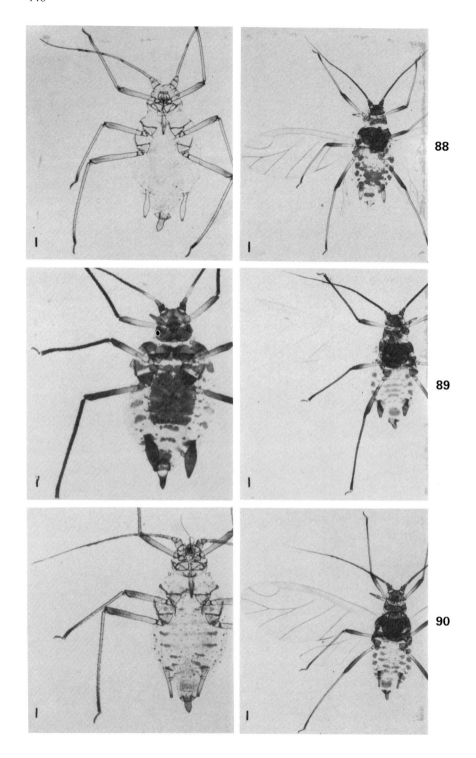

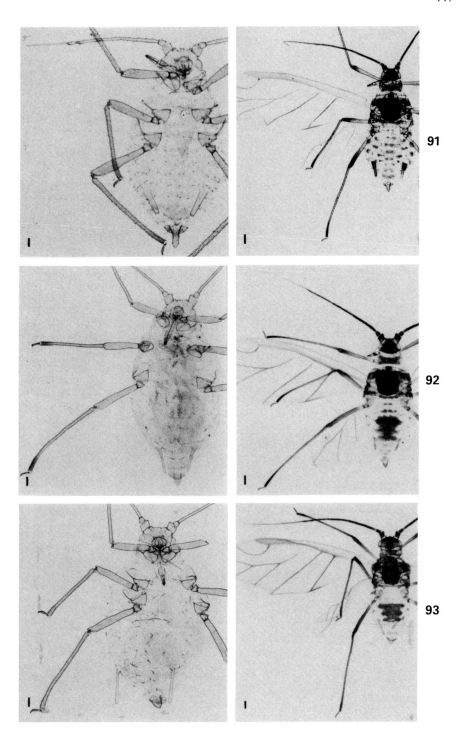

448

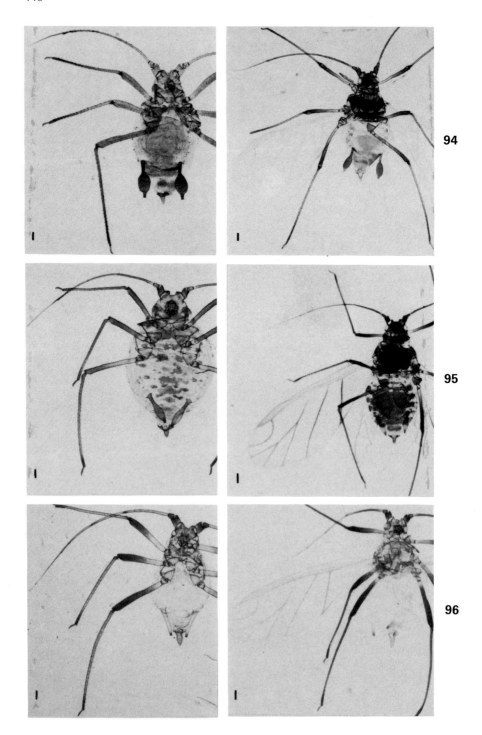

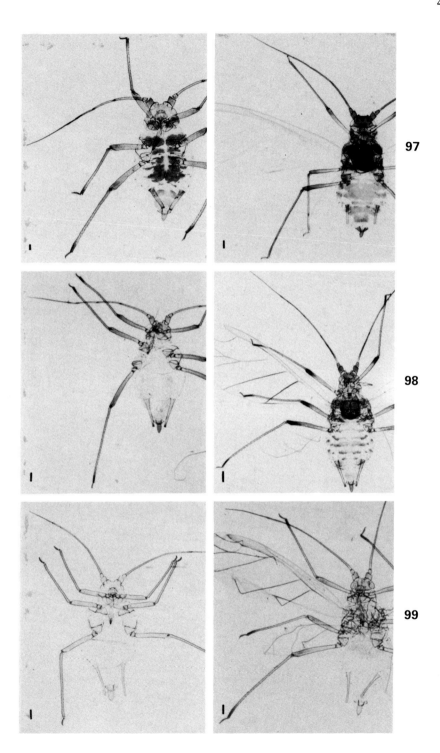

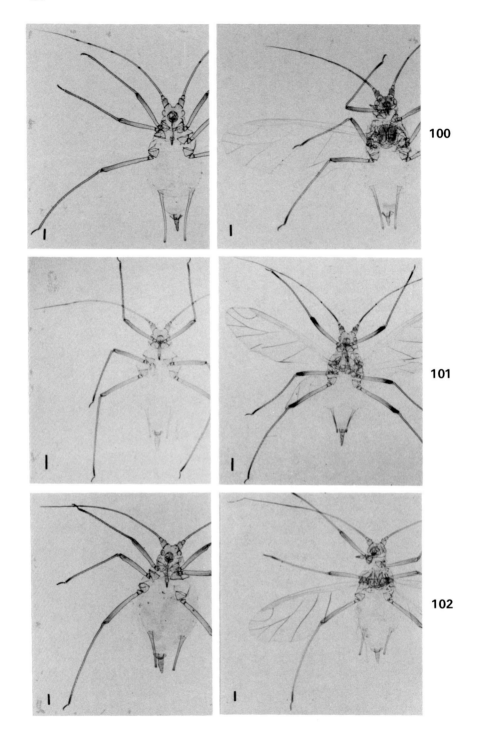

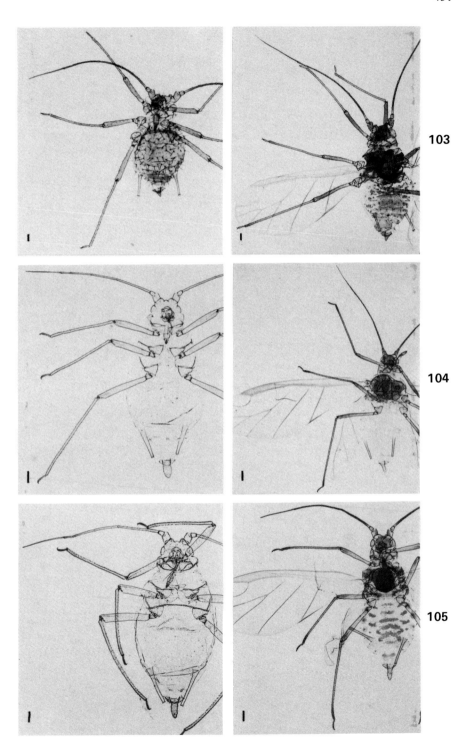

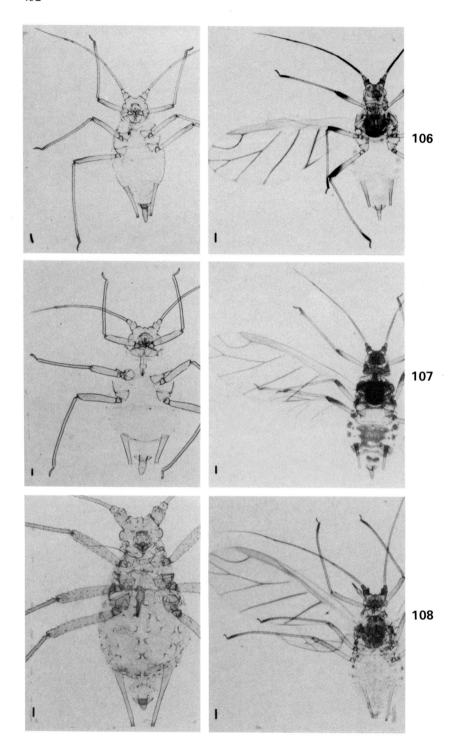

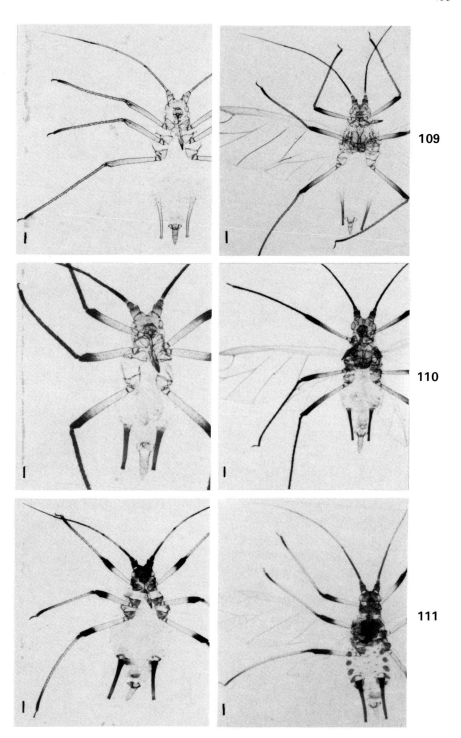

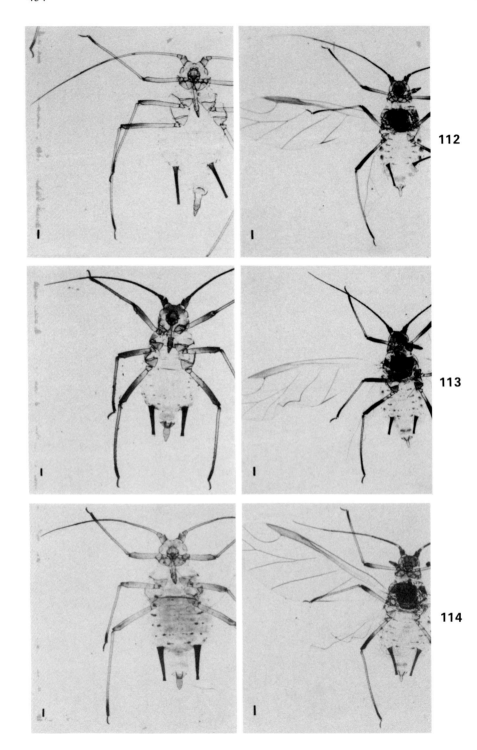

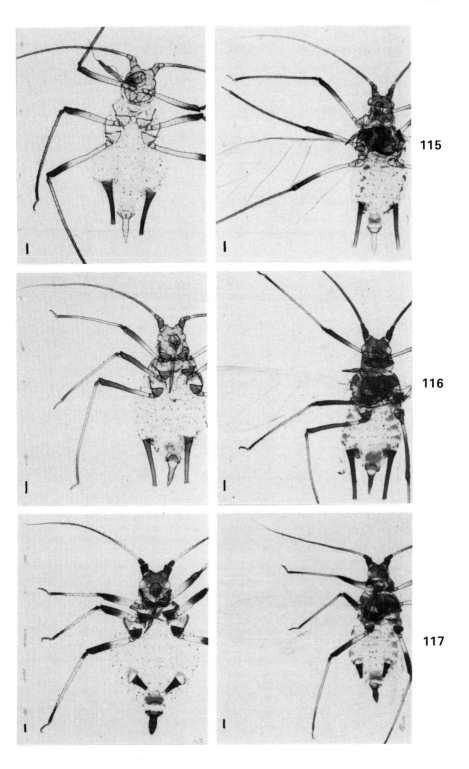

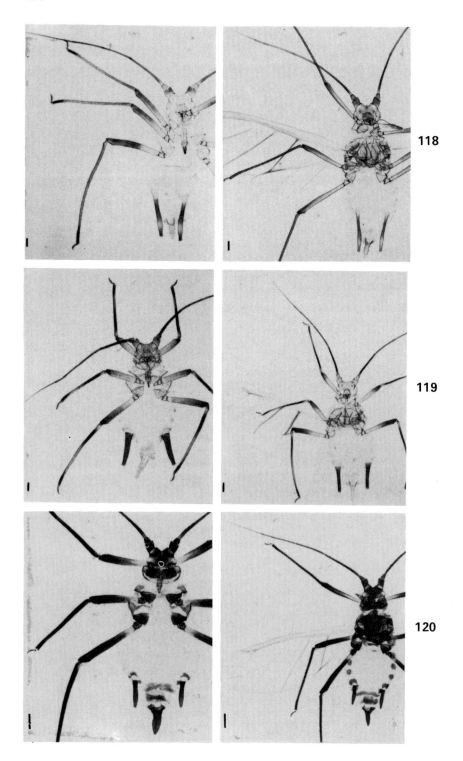

457

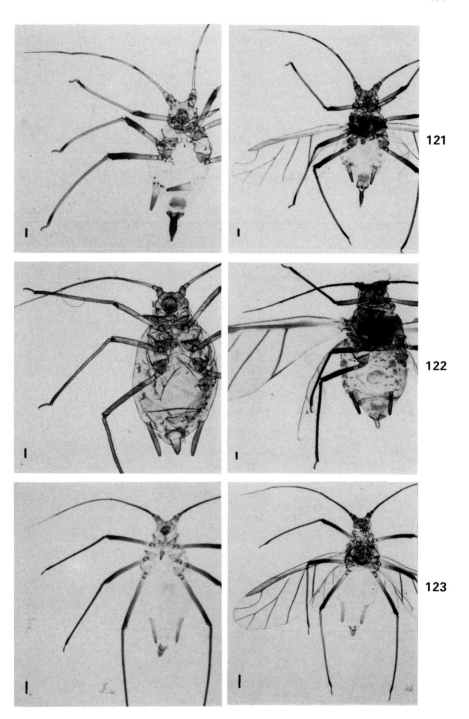

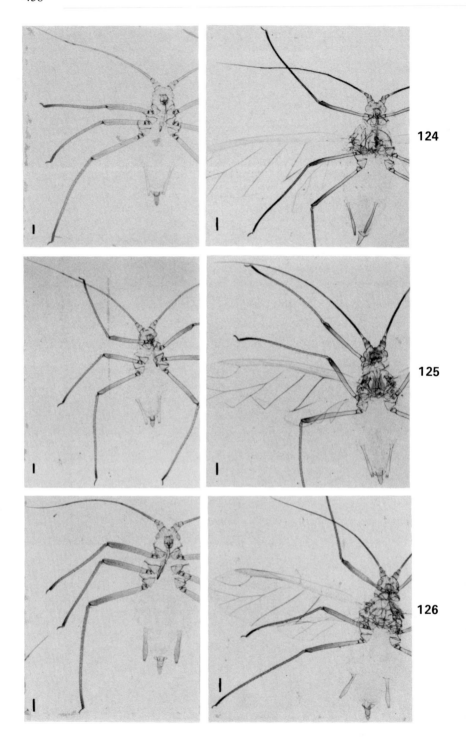

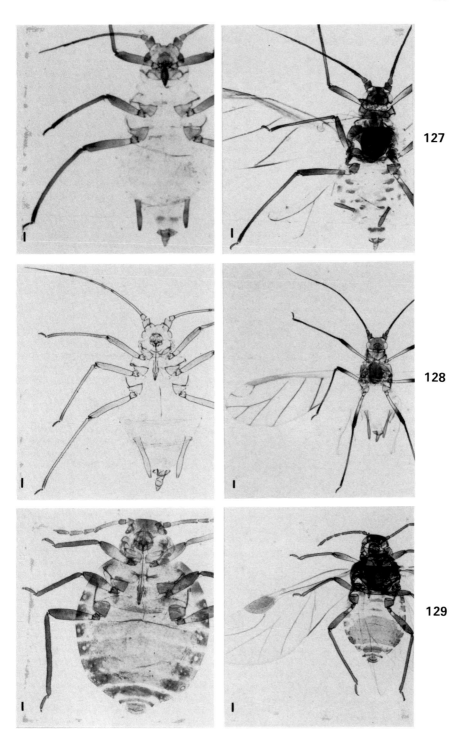

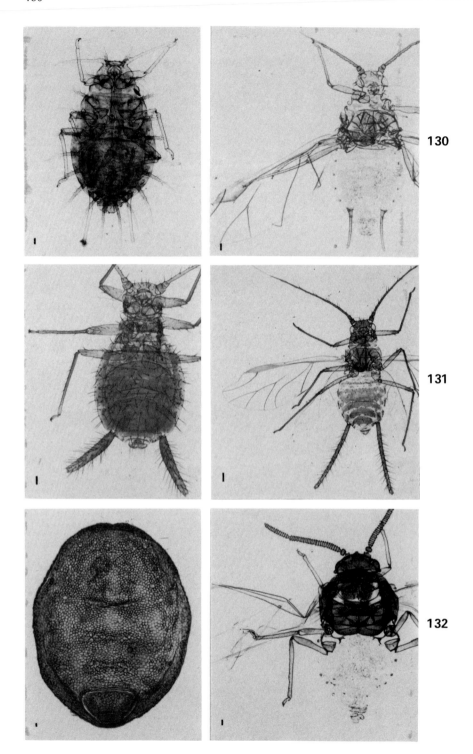

130

131

132

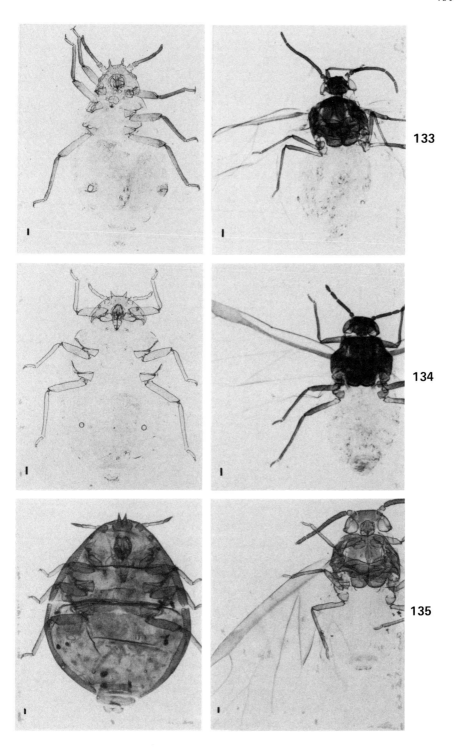

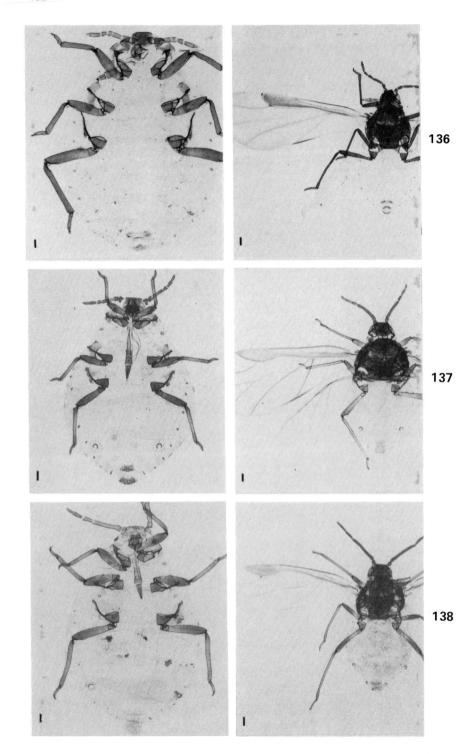

463

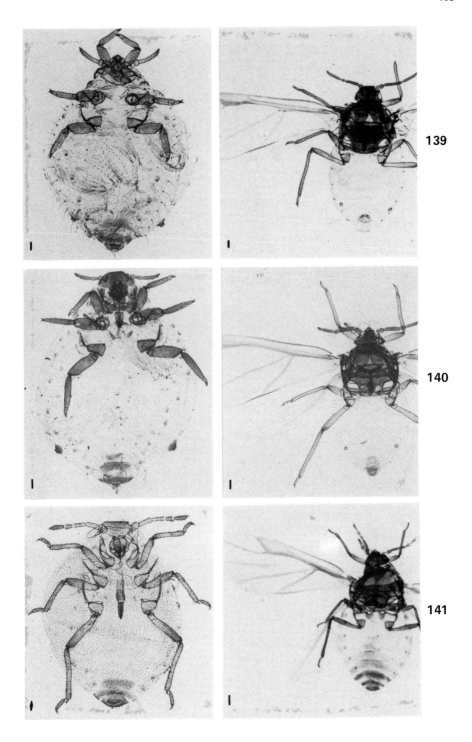

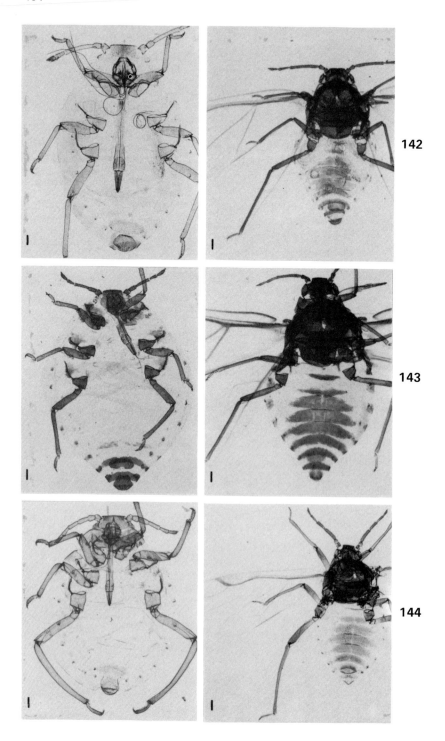

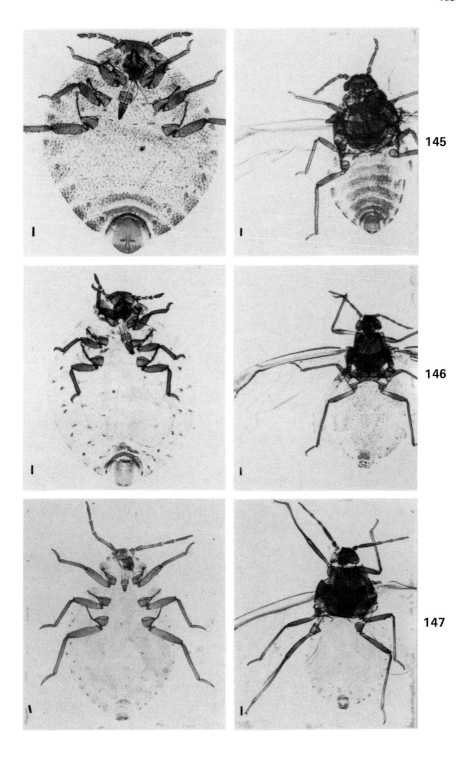

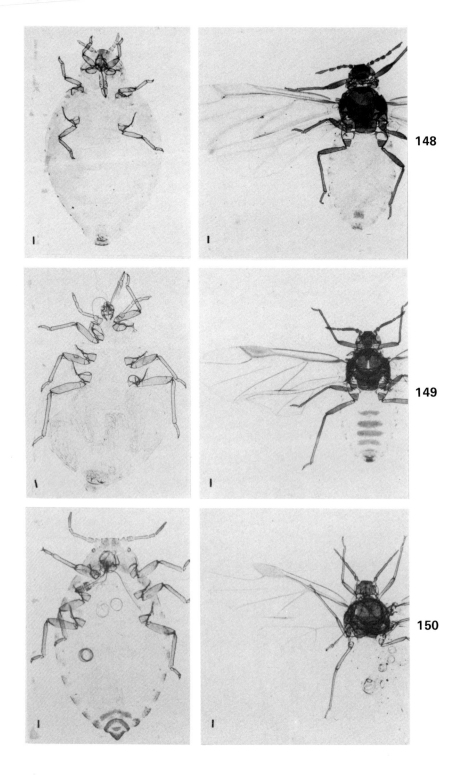